Protist Diversity and Geographical Distribution

TOPICS IN BIODIVERSITY AND CONSERVATION

Volume 8

http://www.springer.com/series/7488

Protist Diversity and Geographical Distribution

Edited by

W. Foissner

and

David L. Hawksworth

Reprinted from *Biodiversity and Conservation*, volume 17:2 (2008)

Springer

Editors

Prof. Dr. W. Foissner
Universität Salzburg
Naturwissenschaftliche
Fakultät
Inst. Zoologie
Hellbrunnerstr. 34
5020 Salzburg
Austria

Dr. David L. Hawksworth
Universidad Complutense
Fac. Farmacia
Dept. Biologia Vegetal II
Plaza de Ramony Cajal,
28040 Madrid
Ciudad Universitaria
Spain
myconova@terra.es

ISBN: 978-90-481-2800-6 e-ISBN: 978-90-481-2801-3

DOI: 10.1007/978-90-481-2801-3

Library of Congress Control Number: 2009927286

Cover photo: Various ciliates (Stentor polymorphus, Frontonia leucas etc.) attached to a mud particle.

Printed on acid-free paper

springer.com

Contents

Protist Diversity and Geographical Distribution

Foreword

Conservation and biodiversity of protists

The conservation of biodiversity is not just an issue of plants and vertebrates. It is the scarcely visible invertebrates and myriads of other microscopic organisms that are crucial to the maintenance of ecological processes on which all larger organisms and the composition of the atmosphere ultimately depend. *Biodiversity and Conservation* endeavours to take an holistic view of biodiversity, and when the opportunity arises to issue collections of papers dealing with too-often neglected groups of organisms. The protists, essentially eukaryotes that cannot be classified in the kingdoms of animals, fungi, or plants, include some of the least-known groups of organisms on earth. They are generally treated as a separate kingdom, commonly named *Protista* (or *Protoctista*) in textbooks, but in reality they are a mixture of organisms with disparate affinities.

Some authors have hypothesized that the numbers of protists are not especially large, and that many have extraordinarily wide distributions. However, the picture that unfolds from the latest studies discussed in this issue is different. There are many species with wide ranges, and proportionately more cosmopolitan species than in macroorganism groups, as a result of their long evolutionary histories, but there are also definite patterns and geographical restrictions to be found. Further, some protists are linked to host organisms as mutualists or parasites and necessarily confined to the distributions of their hosts. It is now also becoming clear that there is a great deal of cryptic speciation and clonal selection that has not been appreciated until the advent of molecular phylogenetic approaches. The traditional use of morphospecies concepts badly underestimates protistan diversity – by two or three times in the case of foraminifera and ciliates. Molecular work in most protistan groups is very much in its infancy, but it seems likely that there are more than 300 000 species of protists on Earth. This collection of papers makes clear that protistan groups are not a special case of limited diversity, but are very imperfectly known and merit more attention than is generally accorded to them in assessments of biodiversity.

In dealing with primarily microscopic groups of organisms, conservation is a difficult task and must be two-pronged and involve both the *in situ* conservation of different habitat types as single-species plans are unlikely ever to be practical, and *ex situ* maintenance in genetic resource collections. However, focusing attention on "flagship species" that can serve as surrogates has the potential both to generate more basic knowledge and to act as a bioindicator of the situation in others.

The 15 papers presented here consider a range of protistan groups and fundamental issues, while others focus on particular types of protists, notably slime moulds, foraminifera, ciliates, desmids, diatoms, dinoflagellates, chrysophytes, and testate amoebae; there is also one addressing dispersal and biogeography in bryophytes where there are parallel issues of wide distributional ranges. Protistan fungal groups other than slime moulds are not covered as the January 2007 issue of *Biodiversity and Conservation* (**16**(1): 1–111) included eight papers devoted to the estimation of fungal diversity. The contributions have been selected and edited by Wilhelm Foissner, a long-standing member of the Editorial Board of *Biodiversity and Conservation*. I am very pleased that he rose to the challenge of assembling such a thought-provoking series of papers, which should serve to highten awareness of issues in protistan diversity in biodiversity science and conservation.

DAVID L. HAWKSWORTH
Editor-in-Chief, Biodiversity and Conservation
Universidad Complutense de Madrid, Madrid; and the Natural History Museum, London
5 January 2008

Protist diversity and distribution: some basic considerations

Wilhelm Foissner

Originally published in the journal Biodiversity and Conservation, Volume 17, No 2, 235–242.
DOI: 10.1007/s10531-007-9248-5

Abstract This essay discusses protist species number and geographic distribution, both heavily influenced by undersampling and human introductions. The features of the ubiquity model and the moderate endemicity model are compared. I recognize five main flaws of the ubiquity model, viz., the ignorance of the extraordinary possibilities protists have to speciate due to their short generation time and the likelihood that many persisted over geological time scales; that all protist species have high abundances; that their small size is a main reason for global distribution; the ignorance of human introductions; and the rejection of literature evidence on the occurrence of flagship species with restricted distribution in a wide variety of protists. Thus, the data available support the moderate endemicity model which proposes about 300,000 extant, free-living protist species, of which one third might have a restricted distribution, i.e., is not cosmopolitan in spite of suitable habitats. To sum up, the distribution of protists, flowering plants, and larger animals has much in common, but protists usually have wider ranges and thus a higher proportion of cosmopolites. Future research should reconcile morphologic, genetic, and ecological species concepts because this is crucial for determining the number of protist species. Further, greatly intensified research is required on morphospecies in heterotrophic protists because their diversity has never been investigated in large areas of the earth.

Keywords Community structures of protists and multicellular organisms · Distribution models · Moderate endemicity distribution model · Protist endemism · Protist species number · Ubiquitous distribution model · Undersampling

Introduction

The collection of papers in this issue of *Biodiversity and Conservation* was stimulated by the controversy whether or not micro-organisms have biogeographies. Indeed, several

---

Special Issue: Protist diversity and geographic distribution. Guest editor: W. Foissner.

W. Foissner (✉)
FB Organismische Biologie, Universität Salzburg, Hellbrunnerstrasse 34, 5020 Salzburg, Austria
e-mail: Wilhelm.Foissner@sbg.ac.at

W. Foissner et al. (eds.), *Protist Diversity and Geographical Distribution*
DOI: 10.1007/978-90-481-2801-3_1

colleagues suggested me to compare my "moderate endemicity model" (Foissner 1999, 2006) with the "ubiquity model" of Finlay et al. (1996, 2004). Thus, I provide here a brief essay not going in literature details which can be found in several recent reviews (Dolan 2005; Foissner 2006) and in the contributions contained in this issue.

Molecular studies greatly advanced our understanding of protist distribution and species number. However, we should not ignore the excellent evidence for protist endemism collected by generations of taxonomists, for instance, by Heimans (1969) in desmids, by Bonnet (1983) in testate amoebae, and by Dragesco and Dragesco-Kernéis (1986) in ciliates.

Distribution models

The current literature discusses two possibilities of protist distribution, viz., the "ubiquity model" of Finlay and Fenchel (Finlay et al. 1996; Fenchel and Finlay 2004) and the "moderate endemicity model" of Foissner (1999, 2004, 2006). Both models have much in common, for instance, that most protists are cosmopolites.

At first glance, the distribution debate appears rather academic because the importance of protists is still widely neglected, although the "microbial loop" has now found a home in most ecological textbooks. However, when the consequences of the two models are compared, the significance of the distribution debate becomes obvious, extending from academic to applied issues (Table 1).

Undersampling and human introductions, the keys to understand protist diversity and distribution

Why is it so difficult to obtain reliable data on protist diversity and geographic distribution? There are several reasons (Foissner 1999, 2006), but a main problem is undersampling, that is, only a small proportion of the protists can be seen at any time because most species are in a dormant (cystic) stage waiting for optimal conditions to become active. Furthermore, the samples studied are usually very small, both in quantity and size, because all investigations must be done with the microscope, which is time-consuming and needs well trained taxonomists. Although various attempts have been made to correct morphologic and molecular data for this hidden diversity (Chao et al. 2006; Hong et al. 2006), we have only a vague knowledge how many of them are overlooked either because they are inactive or do not reproduce to detectable numbers. When only a single sample from a habitat is investigated, for instance, a composite sample from 100 m^2 forest soil, undersampling may miss 70% of the species actually present (Foissner 1999; Foissner et al. 2002).

Foissner (2006) emphasized the importance of biogeographic changes due to human introductions and provided several examples, for instance, the introduction of the very distinctive alga *Hydrodictyon* to New Zealand by imported fish and water plants from East Asia. Further impressive cases are contained in some contributions of the present issue, for instance, the introduction of the diatom *Asterionella formosa* to New Zealand (see contribution of Vanormelingen et al.). Thus, human-induced biogeographic changes of protist communities are as important as in plants and animals and should get much more importance in the biogeography debate.

Table 1 Comparison of distribution in macro-organisms and free-living protists

	Features	Macro-organisms	Protists (micro-organisms)	
			Ubiquity model	Moderate endemicity model
1	Absolute abundance of individuals within morphospecies	Low	High	Low in the majority (≥90%) of species, high only in some euryoecious species
2	Rates of migration	Low	High	Low for most of the rare species, high only for some euryoecious species
3	Proportion of global species pool found locally	Low	High	Moderate; usually highly over-estimated due to undersampling, see Foissner (1999) for an example
4	Rates of allopatric speciation	High	Low	Low, but see next entry
5	Rates of non-allopatric speciation	Low	?	High, e.g., parapatry, microallopatry, isolation-by-distance (Helbig 2005)
6	Cryptic persistence of species	Variable	High	High
7	Persistence of specific morphotypes over geological time scales	Low	High	Moderate
8	Large-scale distribution determined by historical contingencies, e.g., continental drift	High	Low	Moderate
9	Time for speciation	Low	?	High
10	Relative number of endemics	High	Low/none	Moderate (∼30%)
11	Rates of species extinction	High	Low	Moderate
12	Global number of morphospecies	High	Low	High due to long time to speciate and non-allopatric speciation (see above)
13	Conservation	Needed	Not needed	Needed
14	Human introductions	Low	?	Likely high; see Foissner (2006) and several contributions in this issue

Based on Finlay et al. (2004), except for features (5, 9, 13) and the "moderate endemicity model"

Main flaws of the ubiquity model

I recognize five main flaws of the ubiquity model, viz., the ignorance of the extraordinary possibilities protists have to speciate because of their short generation times and the likelihood that many persisted over geological time scales; that all protist species have high abundances; that their small size is a main reason for global distribution; the ignorance of many human introductions both in the past and present; and the ignorance of literature data on the occurrence of species with restricted distribution in a wide variety of protists.

(1) Speciation is an ongoing process in most or even all organisms. I do not want to repeat the knowledge and problems of speciation and species delimination in general and of protists in particular. The evidence from protists which left fossils argue for similar speciation mechanisms in protists, plants, and animals. For instance, speciation can be a

slow (10^6 years) or rapid (10^3 years) process (Norris 2001). In the present context, two issues are of paramount importance, both suggesting a high number of protist species.

Protists have short generation times, fostering mutations and thus speciation, although their genetic isolation is possibly less strict than in most plants and animals due to their wider distribution. Thus, there must be a large stock of relatively young species which not yet fully explored their potential area, simply for the lack of time to distribute. These then appear, inter alia, as endemics in our species lists. Considering that plants and animals can speciate in 10–20,000 years, for instance, fish in lakes (Martens 1997) and plants and animals in postglacial areas (Schluter 1998), there is no logical reason to assume that protists behave different, especially when considering their short generation time. Accordingly, local and regional endemics should be widespread. Unfortunately, these species often will be inconspicuous and thus difficult to recognize. However, some examples from ciliates and diatoms are reviewed in Foissner (2007) and Mann and Droop (1996), respectively.

Protists possibly survived the great extinctions in the earth's history better than larger organisms due to their smaller habitats. Thus, they could accumulate diversity over hundreds of millions of years. Such scenery is not only suggested by various paleontological data but also by recent studies, indicating that micro-organisms persisted in cold refuges during periods of greenhouse conditions (Stoeck et al. 2007).

(2) When compared to plants and larger animals, protists are indeed much more numerous, but only a few species, while the vast majority (>90%) have moderate, low, or very low abundances, as is the case in plant and animal communities (see ecological textbooks). This is obvious from all investigations in which species richness and the abundances of the individual species were carefully studied (Figs. 1, 2). Curiously enough, the high similarity in the structure of protist, plant, and animal communities never played a significant role in the discussion about protist diversity and distribution. Of course, rare species can become numerous and vice versa, just as pests in plant and animal communities. Usually however, rare species are rare throughout time and space, while abundant species are numerous at many times and globally; the latter are those we usually recognize and make protist samples from, e.g., Europe and Australia so similar. The rare species, many of which have been not yet discovered (Chao et al. 2006) and may be endemic, are not recognized in ordinary surveys due to undersampling discussed above.

(3) Often, the wide or cosmopolitan distribution of protists is assumed to be associated with their small size and high numbers (Fenchel 1993; Finlay et al. 1996, 2006), and Wilkinson (2001) suggested that this is especially likely for organisms having a size of below 100 μm. However, this is disproved by macrofungi, mosses and ferns, many of which have small areals in spite of appropriate habitats and minute spores (<50 μm) produced in high numbers (see explanation to Fig. 3). This was first recognized by Foissner (2006) and is supported by the detailed data of Frahm in this issue. Further, seeds of higher plants often have small size and special morphologic adaptations for air dispersal, but are not cosmopolites, although many of these "exotics" grow well in our home gardens. Interestingly, morphological adaptations for air dispersal are unknown in cysts and spores of micro-organisms, suggesting that this kind of distribution never played a major role.

Wilkinson (2001) founded the hypothesis on testate amoebae whose small resting cysts (usually <100 μm) often remain in the much larger and rather robust test. However, all testless protists have only the small resting cysts for large scale distribution because the active specimens are too fragile. There are few resting cysts with a size of >100 μm, and thus size is possibly only one of several reasons for cosmopolitan or restricted distribution.

(4) Generations of taxonomists provided convincing evidence for restricted distribution of some protists, using so-called flagship species considered as "ultimate" proof of protist

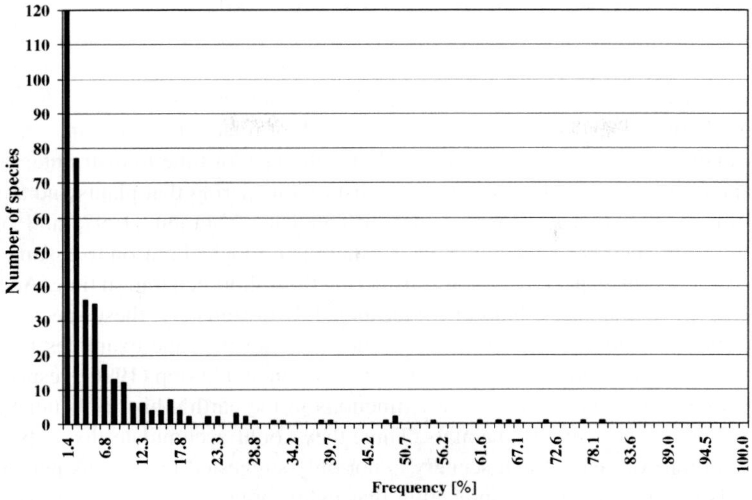

Fig. 1 Frequencies of 365 ciliate species in 73 samples from terrestrial habitats of Namibia (from Foissner et al. 2002)

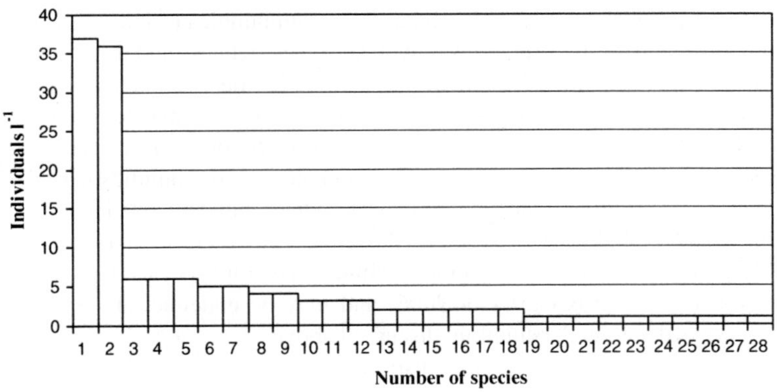

Fig. 2 Number of ciliate species and individuals in the free water of a pond in the Austrian Central Alps about 2,000 m above sea-level. The pond is ca 2,000 m^2 in size, but the average depth of the water is only 17 cm. Ciliates were rare in the free water, both qualitatively and quantitatively, possibly due to the acidic (pH 4.8–5.4) and dystrophic water. The individual numbers are averages of 14 sampling occasions between 10 July and 19 August. The dominant species were two common cosmopolites, viz., *Cyrtolophosis mucicola* and a species of the *Vorticella aquadulcis* complex

endemism by Foissner (2006). Tyler (1996) has summarized the reasons why flagship species have the greatest probability of real endemism: "Because they are so showy, or so novel, it is unlikely that such species would be overlooked if indeed they were widely distributed. If the Australian endemics occurred in Europe or North America they would have been seen there, long ago". Foissner (2006) put together flagships from various protist groups, and some are shown in the contributions to this issue.

The ubiquity model has ignored all these evidence or rejected the data as caused by undersampling and misidentification (Mitchell and Meisterfeld 2005). This stimulated more detailed research showing, e.g., endemism of some testate amoebae beyond reasonable doubt (Smith and Wilkinson 2007).

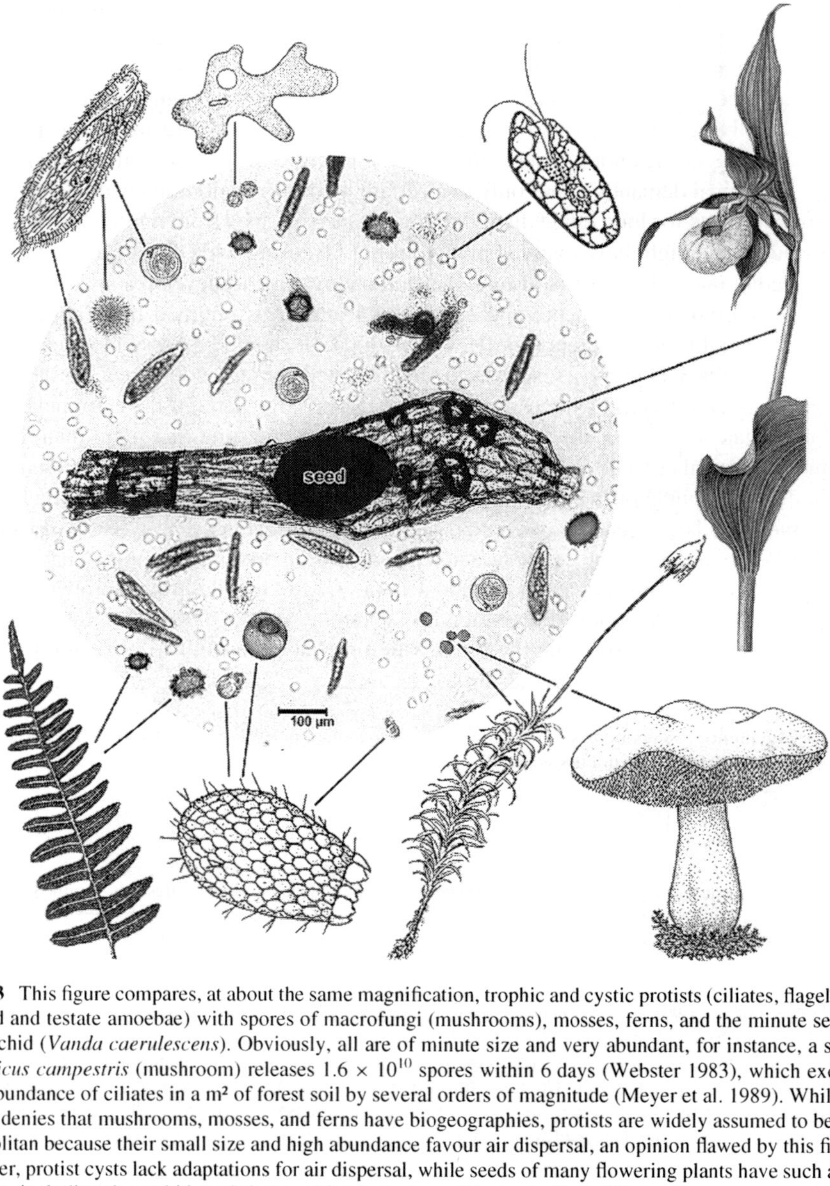

Fig. 3 This figure compares, at about the same magnification, trophic and cystic protists (ciliates, flagellates, naked and testate amoebae) with spores of macrofungi (mushrooms), mosses, ferns, and the minute seed of an orchid (*Vanda caerulescens*). Obviously, all are of minute size and very abundant, for instance, a single *Agaricus campestris* (mushroom) releases 1.6×10^{10} spores within 6 days (Webster 1983), which exceeds the abundance of ciliates in a m² of forest soil by several orders of magnitude (Meyer et al. 1989). While nobody denies that mushrooms, mosses, and ferns have biogeographies, protists are widely assumed to be cosmopolitan because their small size and high abundance favour air dispersal, an opinion flawed by this figure. Further, protist cysts lack adaptations for air dispersal, while seeds of many flowering plants have such adaptations, including the orchid seed shown which has wings of large-sized, air-filled cells

Protist species number and distribution: main challenges

The papers in this issue of *Biodiversity and Conservation* leave no doubt on the occurrence of endemic protists, i.e., of species with restricted distribution in spite of suitable habitats in other regions of the earth. We lack solid information on the number of protist species with restricted distribution, and the percentage highly depends on the species concept applied; my estimation of 30% (Table 1, Foissner 2006) of morphologic and/or genetic and/or molecular endemics is a very crude figure based on some better known groups, such as testate amoebae, ciliates, and diatoms; further, it includes 15% undescribed rare and very rare species which bear the greatest probability for restricted distribution.

The existence of endemic protists evokes the first main question: why did they not spread globally, as the majority of species? Likely, the reasons are manifold: perhaps, many are young species not having sufficient time to disperse globally; others might have specific ecological demands found only in a certain habitat or region; many do not produce stable resting cysts for long range dispersal, for instance, protists from rainforests (Foissner 2006); and others might have evolved in regions not favouring wide dispersal.

The second main challenge is of more general nature, viz., to develop a species concept reconciling morphologic, genetic, and ecological features, as outlined by Weisse in this issue. Although this is a different task (Hey et al. 2003), it should be possible to reach some agreement for practical purposes, such as biodiversity and conservation issues. Further, morphological research has to be intensified greatly because large parts of the earth never have been investigated for, especially, heterotrophic protists, suggesting that more than 50% of their morphological diversity is still undescribed (Foissner 2006; Cotterill et al. in this issue). Likely, this will double or treble the number of species in many groups, such as ciliates and naked amoebae. Thus, genetic, molecular, and ecological features will possibly double or treble this figure again (for an example, see the contribution on ciliates in this issue).

Corliss (2000) estimated about 90,000 extant, free-living described protist species. Applying the figures mentioned above and a synonymy rate of 20%, we might arrive at about 300,000 species, excluding the fungi which probably constitute over a million species (Hawksworth 2001; Taylor et al. 2006).

Acknowledgements Supported by the Austrian Science Foundation (FWF, project P19699-B17). The technical assistance of Mag. Gudrun Fuss is greatly acknowledged.

References

Bonnet L (1983) Interet biogeographique et paleogeographique des thecamoebiens des sols. Annls Stn limnol Besse 54:777–784

Chao A, Li PC, Agatha S et al (2006) A statistical approach to estimate soil ciliate diversity and distribution based on data from five continents. Oikos 114:479–493

Corliss JO (2000) Biodiversity, classification, and numbers of species of protists. In: Raven PH, Williams T (eds) Nature and human society. The quest for a sustainable world. National Academy Press, Washington, pp 130–155

Dolan JR (2005) An introduction to the biogeography of aquatic microbes. Aquat Microb Ecol 41:39–48

Dragesco J, Dragesco-Kernéis A (1986) Ciliés libres de l'Afrique intertropicale. Faune Trop 26:1–559

Fenchel T (1993) There are more small than large species? Oikos 68:375–378

Fenchel T, Finlay BJ (2004) The ubiquity of small species: patterns of local and global diversity. BioScience 54:777–784

Finlay BJ, Corliss JO, Esteban G et al (1996) Biodiversity at the microbial level: the number of free-living ciliates in the biosphere. Q Rev Biol 71:221–237

Finlay BJ, Esteban GF, Fenchel T (2004) Protist diversity is different? Protist 155:15–22

Finlay BJ, Esteban GF, Brown S et al (2006) Multiple cosmopolitan ectotypes within a microbial eukaryote morphospecies. Protist 157:377–390

Foissner W (1999) Protist diversity: estimates of the near-imponderable. Protist 150:363–368

Foissner W (2004) Ubiquity and cosmopolitanism of protists questioned. SIL News 43:6–7

Foissner W (2006) Biogeography and dispersal of micro-organisms: a review emphasizing protists. Acta Protozool 45:111–136

Foissner W (2007) Dispersal and biogeography of protists: recent advances. Jpn J Protozool 40:1–16

Foissner W, Agatha S, Berger H (2002) Soil ciliates (Protozoa, Ciliophora) from Namibia (Southwest Africa), with emphasis on two contrasting environments, the Etosha Region and the Namib Desert. Denisia 5:1–1459

Hawksworth DL (2001) The magnitude of fungal diversity: the 1.5 million species estimate revisited. Mycol Res 105:1422–1432

Heimans J (1969) Ecological, phytogeographical and taxonomic problems with desmids. Vegetatio 17:50–82

Helbig AJ (2005) A ring of species. Heredity 95:113–114

Hey J, Waples RS, Arnold ML et al (2003) Understanding and confronting species uncertainty in biology and conservation. Trends Ecol Evol 18:597–603

Hong SH, Bunge J, Jeon SO et al (2006) Predicting microbial species richness. Proc Natl Acad Sci USA 103:117–122

Mann DG, Droop SJM (1996) Biodiversity, biogeography and conservation of diatoms. Hydrobiologia 336:19–32

Martens K (1997) Speciation in ancient lakes. TREE 12:177–182

Meyer E, Foissner W, Aescht E (1989) Vielfalt und Leistung der Tiere im Waldboden. Öst Forstz 3:15–18

Mitchell EAD, Meisterfeld R (2005) Taxonomic confusion blurs the debate on cosmopolitanism versus local endemism of free-living protists. Protist 156:263–267

Norris RD (2001) Pelagic species diversity and evolution. Paleobiol 26(Suppl):236–258

Schluter D (1998) Ecological causes of speciation. In: Howard DJ, Berlocher SH (eds) Endless forms. Univ Press, New York, pp 114–129

Smith HG, Wilkinson DM (2007) Not all free-living microorganisms have cosmopolitan distributions—the case of *Nebela (Apodera) vas* Certes (Protozoa: Amoebozoa: Arcellinida). J Biogeogr 34:1822–1831

Stoeck T, Kasper J, Bunge J et al (2007) Protistan diversity in the artic: a case of paleoclimate shaping modern biodiversity? PLoS ONE 8:1–16

Taylor JW, Turner E, Townsend JP et al (2006) Eukaryotic microbes, species recognition and the geographic limits of species: examples from the kingdom fungi. Phil Trans R Soc 361:1947–1963

Tyler PA (1996) Endemism in freshwater algae with special reference to the Australian region. Hydrobiologia 336:1–9

Webster J (1983) Pilze. Eine Einführung. Springer, Berlin

Wilkinson DM (2001) What is the upper size limit for cosmopolitan distribution in free-living microorganisms? J Biogeogr 28:285–291

Distribution and diversity of aquatic protists: an evolutionary and ecological perspective

Thomas Weisse

Originally published in the journal Biodiversity and Conservation, Volume 17, No 2, 243–259.
DOI: 10.1007/s10531-007-9249-4

Abstract Assessment of the distribution and diversity of free-living protists is currently hampered by a limited taxonomic resolution of major phyla and by neglecting the significance of spatial and temporal scaling for speciation. There is a tremendous physiological and ecological diversity that is hidden at the morphological level and not apparent at the level of conserved genes. A conceptual framework linking the various levels of diversity is lacking. Neutral genetic markers are useful indicators of population structure and gene flow between populations, but do not explain adaptation to local habitat conditions. The correspondence between protein-coding genes, ecophysiological performance, and fitness needs to be explored under natural conditions. The area and the associated typical temporal dimension of active cells (their 'home range') are much smaller, respectively shorter, than the area and time period potentially covered during passive dispersal of protist resting stages. The assumptions that dispersal rates are generally high in free-living protists and that extinction of local populations is, therefore, infinitesimally small wait rigorous testing. Gene flow may be uncoupled largely from dispersal, because local adaptation and numerical effects of residents may strongly reduce or even prevent successful invasion (immigration). The significance of clonal selection depends on the as yet unknown frequency and timing of sexual reproduction, and on the stability of the environment. The extent of local adaptation and the fitness-related ecophysiological divergence are critical for the speciation process and, hence, for defining protist species.

Keywords Clones · Dispersal · Distribution · Diversity · Free-living protists · Home range concept · Sexual reproduction · Speciation

---

Special Issue: Protist diversity and geographic distribution. Guest editor: W. Foissner.

T. Weisse (✉)
Institute for Limnology of the Austrian Academy of Sciences, Mondseestr. 9, 5310 Mondsee, Austria
e-mail: thomas.weisse@oeaw.ac.at

Introduction

Protist diversity and their geographic distribution, which are ultimately linked to each other, were controversially debated in recent years (Finlay and Fenchel 1999; Foissner 1999; Lachance 2004) despite, or perhaps because of, a limited understanding of the determining processes. The current debate centres on the old hypothesis by the Delft school of microbiology that free-living microorganisms are globally distributed and ubiquitous, i.e., that they can be found wherever a suitable habitat exists (see Quispel 1998 for a short review of the historic background, and Whitfield 2005 and Foissner 2004, 2006 for brief summaries of the "everything is everywhere" view). The controversial discussion and the advent of novel, mainly molecular tools stimulated research on microbial diversity and their potential biogeography over the past decade (reviewed by Foissner 2006, 2007; Hahn 2006; Weisse 2006a). In spite of the increased efforts, the range and extent of protist dispersal, the genotypic and phenotypic diversity, the frequency of sexual reproduction and, accordingly, the extent of reproductive isolation are at present unknown for the vast majority of extant, free-living protist species.

The special issue on "Biogeography of Freshwater Algae" (Kristiansen 1996a), the theme section on "Biogeography of aquatic microbes" (Dolan 2005), the recent reviews by Foissner (2006, 2007) and several articles in this issue of BIODIVERSITY and CONSERVATION present numerous examples on limited dispersal and biogeographic differences among major phyla of terrestrial and aquatic protists. The purpose of the present paper is to discuss the distribution and diversity of protists from an ecological and evolutionary perspective. In line with Horner-Devine et al. (2004) and Hughes Martiny et al. (2006), I follow the hypothesis that protist biogeography and diversity is, in principal, similar to biogeography and diversity of macroorganisms. In particular, I will highlight the population genetic literature on parthenogenetic zooplankton and suggest that many of the processes shaping populations in protists will be common to these organisms. This conceptual article will focus on heterotrophic aquatic protists but also include some findings from autotrophs and from terrestrial environments where appropriate.

Distribution of free-living protists

Significance of the taxonomic resolution

Any census of biological diversity across phylogenetically diverse taxa relies on a meaningful unit that renders the diversity of the different taxa comparable. Traditionally, most estimates of biodiversity have used the species as the 'common currency', although it is well known that species-based approaches ignore other levels of biodiversity (e.g., Maddock and Du Plessis 1999; Giller et al. 2004). As outlined by Kristiansen (1996b) and Schlegel and Meisterfeld (2003), the species problem is still unresolved for major protist phyla. It is important, in this context of protist diversity, to differentiate between the meaning of species as an evolutionary unit ('real' species, see Claridge et al. 1997; Bachmann 1998; Hey 2001) and the meaning of species as a category. Recent work with bacteria demonstrated that the level of taxonomic resolution is crucial for detecting biogeographic differences among microbes (Horner-Devine et al. 2004; Schauer et al. 2006; Ramette and Tiedje 2007). Accordingly, a pragmatic definition of the species as a category has to be commonly accepted before any firm conclusion can be drawn on the species-specific distribution of protists and their biogeography (Mitchell and Meisterfeld 2005).

The rapidly growing molecular data on diatoms (Amato et al. 2007), dinoflagellates (Kim et al. 2004), cryptomonads (Hoef-Emden and Melkonian 2003), bodonid (von der Heyden et al. 2004; von der Heyden and Cavalier-Smith 2005) and chrysomonad flagellates (Boenigk et al. 2005, 2006), amoebae (Wylezich et al. 2002; Smith and Wilkinson 2007), foraminifera (Darling et al. 1999, 2000), and ciliates (Katz et al. 2005; Barth et al. 2006; Finlay et al. 2006; Gächter and Weisse 2006; Weisse 2006b; Weisse and Rammer 2006) implies that the morphospecies concept is, in many cases, too conservative when comparing protist diversity to that of most macroorganisms. Application of the morphospecies concept is also problematic in highly polymorphic species such as *Pfiesteria*-like dinoflagellates (Marshall et al. 2006).

Similarly to the morphological approach, evolutionary conserved genes such as the small subunit ribosomal DNA gene (SSU rDNA), may underestimate protist diversity. In several major protist phyla, a 1–2% sequence divergence of the SSU rDNA, which has been equated with morphospecies designations of protists (Jerome et al. 1996; Walochnik et al. 1999; Snoeyenbos-West et al. 2002; Katz et al. 2005), may not reflect a species resolution comparable to that of higher taxa (von der Heyden and Cavalier-Smith 2005). For instance, significant differences in the growth rate response to salinity and a huge genetic variation in the SSU rRNA gene were found for *Bodo designis* from seawater, freshwater and soil (Koch and Ekelund 2005). Strains of *B. designis* clustered in several groups, which were not obvious from morphological characters. Another comprehensive genetic and ecological study with the chrysophyte morphospecies *Spumella* from many different environments arrived at the same conclusion (Boenigk et al. 2007). Similarly, distinct genetic differences, up to 66% divergence of their 5.8 S rDNA and ITS sequences, and highly significant differences in growth rate response to salinity, with maximum growth rates ranging from 0.3 to 1.2 d^{-1}, have been reported for the marine flagellate *Oxyrrhis marina* (Lowe et al. 2005). Some more examples of similar recent investigations with ciliates are listed in Table 1. It appears that the variability of the ribosomal genes is different in the various protist phyla (Table 1). Alternatives to the taxonomy based upon SSU rDNA analyses and the pros and cons of the molecular approach to protist phylogeny have recently been reviewed (Schlegel 2003; Schlegel and Meisterfeld 2003).

In conclusion, there is currently no molecular 'gold standard' for protist taxonomy to solve the species problem, and the units of interest, i.e., species, cannot be recognized with standard tools across the protist realm (Amato et al. 2007). The current challenge is to decipher the correspondence between DNA sequence clusters, ecotypes (Turesson 1922) and (morphologically defined) biological species (Dini and Nyberg 1999; Foissner et al. 2001; Finlay 2004; Boenigk et al. 2005). In other words, the priority is to understand the interplay of molecular mechanisms with organismic and ecosystem biology (Jackson et al. 2002).

Terminology and the significance of scaling

Similar to the unresolved species problem, the use of ambiguous terminology contributed to the often emotionally led debate on protist diversity (Finlay et al. 2004). Terms such as 'ubiquitous', 'ubiquitous dispersal' and 'cosmopolitan' are virtually meaningless without clear spatial and temporal dimensions. The *potential* for regular, world-wide distribution (= unlimited distribution) may be high in frequent, widespread taxa that are dispersed passively; the *actual* dispersal rates, i.e., dispersal on relatively short time scales (i.e., related to the organisms' generation time), will be much lower in most taxa (Bohonak and Jenkins 2003).

Table 1 Examples for sequence divergence and ecophysiological differences in morphologically indistinguishable (at the light microscopical level) free-living protist taxa

Taxon investigated	Origin of strains	Genes investigated	Sequence divergence (%)	Physiological variation in	Reference
Karenia brevis (Dinophyceae), 5 strains	M, from Texas and Flv	18S rRNA; ITS-1, 5.8S, and ITS-2	0 0	Max. growth rate, toxin content, salinity tolerance	Loret et al. (2002)
Oxyrrhis marina (incertae sedis), 11 strains	M, TP; from 3 cont	18S rRNA; ITS-1, 5.8S, and ITS-2	0.6–10.5 0–65.7	Growth rate, salinity tolerance	Lowe et al. (2005)
Bodo designis (Kinetoplastida), 9 strains	FW, P, M, S; from 3 cont	18S rRNA	Huge, seq. fall into 4 different phylogentic clusters	Growth rate, salinity tolerance	Koch and Ekelund (2005)
Paramecium spp., 3 species, 19 clones	P, from 3 cont	ITS-1, 5.8S, and ITS-2; COI*	0–2.4 6.0–9.5	n.d.	Barth et al. (2006)
Meseres corlissi, (Ciliophora, Oligotrichea), 5–10 clones	FW, from 4 cont	18S rRNA, ITS-1, ITS-2	<1 0–2	Max. growth rate, temperature and pH tolerance, cell volume	Gächter and Weisse (2006), Weisse et al. (2007)
'*Spumella*-like' flagellates, 3 strains	FW, from 3 cont	18S rRNA	0	Bacterial grazing rate, growth rate	Boenigk et al. (2004)
'*Spumella*-like' flagellates from FW and soil, >50 strains	FW, P, M, S; from 5 cont	18S rRNA	0–10	Growth rate, temperature tolerance	Boenigk et al. (2005, 2006, 2007)

FW, Freshwater; M, marine; P, pond; S, soil; TP, tide pool; cont, continents; FL, Florida; seq., gene sequences

*For *Paramecium* spp., sequence variation denotes to variation within one species

Note that even actual dispersal rate is not identical with distribution, because actual dispersal becomes *effective* dispersal only after successful establishment of immigrants. Although the meaning of effective dispersal and distribution of a species may be largely identical, there is a conceptual difference, because only the former refers to a process. Similarly, there are important conceptual differences between colonization, invasion and immigration (Table 2).

It is the interplay of passive dispersal, which may be a primarily physically mediated (i.e., dispersal by air, water currents) or primarily biologically mediated process (e.g., dispersal by water fowl), with various physico-chemical and biological processes in the new habitat that determine the distribution of species with their temporal (frequency and persistence of occurrence) and spatial (extention of occurrence) dimension. Temperature, salinity and pH are major physico-chemical variables controlling the occurrence of protist species (reviewed by Weisse 2006b). Biological processes such as food limitation, predation pressure and infection by parasites contribute to temporarily or permanently low abundances in some otherwise suitable habitats.

Species-to-area curves plot the (log) number of species of a given taxon versus the (log) area sampled; typically, there is an increase in species numbers when progressing from local to global spatial dimensions. With some exceptions (Hillebrand et al. 2001), species-to-area curves of protists are virtually unknown. An increase of species number at continent-wide and global levels has recently been reported for the autotrophic synurophycean genus *Mallomonas* (Řezáčová and Neustupa 2007). If phylogenetically diverse organisms of widely differing size such as protists and mammals are to be compared (Finlay 2002), species-to-area curves must relate habitat size to the body size of the organisms to yield statistically sound results (Hillebrand et al. 2001; Hillebrand and Blenckner 2002). The area, where the organisms usually range in the course of a day or season, may be termed their home range. The home range concept was coined by Burt (1943) for territorial mammals. The size (area) of the home range is of utmost biological importance, because it sets the upper limits of biological interactions. This concept is similar to Cornell and Lawton's (1992) definition of local scales of species richness, where ecological factors such as competition and grazing determine species richness. Free-living protists have typical home ranges varying between 10^{-3} and 10^{2} m^2, corresponding to linear dimensions ranging from a few cm to several m. The typical linear home range dimensions of macroorganisms with known biogeographies range from several m to several km. The home range can be roughly equated with the 'local' dimension typical for active organisms of different sizes. It does not include the potentially much larger area covered by passive dispersal of inactive resting stages. In those taxa that disperse primarily with asexual propagules, the equivalent of the home range for a given genotype (clone or 'genet', i.e., all individuals or 'ramets' belonging to this clone) may be orders of magnitude larger.

The significance of spatial scaling is obvious for the speciation process. A swimming flagellate, with a typical generation time of one to a few days in temperate lakes (Weisse 1991, 1997), has virtually no chance to meet its congener dwelling at the other shore of a small lake of ~ 100 m length (Weisse 2006a). Accordingly, two individuals of a fish population living in a small lake are likely to be sympatric; however, two individuals of a given flagellate species living in the same lake at the same time may be sympatric, parapatric or even allopatric. It is, therefore, misleading to relate the term 'sympatric speciation' to a well defined geographic area such as, e.g., Lake Victoria, without specifically referring to the organisms under study (Gavrilets 2003).

A corollary of the foregoing considerations is that the 'local' dimensions, at which α-diversity (species richness at a single site) is assessed, may be worlds apart for passively

Table 2 Operational definitions of key terms used in this article

Terms and definition	Source
Potential dispersal: The random, unlimited movement of individuals as active cells or resting stages (cysts) across population boundaries at all spatial scales. This definition is similar to 'ubiquitous dispersal' as defined by Finlay et al. (2004).	This study
Actual dispersal: The movement of individuals as active cells or resting stages (cysts) across population boundaries per unit time (e.g., d^{-1}, $season^{-1}$, y^{-1}, generation $time^{-1}$).	This study
Effective dispersal: The movement and successful establishment of individuals as active cells or resting stages (cysts) across population boundaries per unit time.	This study
Distribution: The area, over which a species has been recorded at any time. The temporal dimension of distribution is referred to as occurrence.	This study
Specific distribution: The variability of habitats, over which a species has been recorded at any time. Species with a narrow specific distribution occur in only a few, peculiar habitats.	This study
Cosmopolitan species: A species that thrives wherever its required habitat is realized.	Finlay et al. (2004)
Gene flow: The exchange of genes between populations, or 'gene pools.'	Bohonak and Jenkins (2003)
Metapopulation: A set of populations that (1) are spatially discrete, (2) may differ in size, demography and carrying capacity, (3) may be subject to extinction and recolonization, and (4) interact via dispersal and gene flow.	Bohonak and Jenkins (2003)
Colonization: The initial step of establishing a new population by one or more dispersers.	Modified after Bohonak and Jenkins (2003)
Invasion: Colonization that impacts other species already inhabiting an area.	Bohonak and Jenkins (2003)
Immigration: Result of successful invasion, i.e., the propagules of an invader must have reproduced and survived over many generations, thus establishing a new (sub)population within the new habitat. In contrast to colonization, invasion will be successful only if, among the flow of invaders, there are some cells better adapted (i.e., with a higher fitness) to the specific local environment than those already present. Different from effective dispersal, immigration is not necessarily a rate, because the temporal dimension of immigration remains, in many cases, unknown.	This study
Local adaptation: The ability of individuals within a population to survive and reproduce better than immigrants from other populations. If local adaptation exists, immigrants can affect some, but not all ecological and evolutionary processes.	Bohonak and Jenkins (2003)

dispersed, asexual microorganisms and for sexually reproducing microorganisms with much lower passive dispersal capacity. For the latter, the processes resulting in biogeographic patterns may be more comparable to those of macroorganisms, although at much reduced spatial and temporal scales.

Unknown rates of dispersal and survival

Local diversity in any habitat depends on the size of the reservoir community (metapopulation) from which new immigrants originate; if this is large, even physically identical habitats will harbour different communities (Curtis and Sloan 2004). The characteristic features of metapopulations (reviewed by Hanski 1999) have been summarized in Table 2. Metapopulation structure has been demonstrated for protist species with patchy distributions, which may be caused by physical factors or pronounced predator–prey cycles (Holyoak and Lawler 1996; Holyoak 2000; Montagnes et al. 2002). It is a fact that there are rare soil and aquatic protist species which will not be encountered permanently in each suitable habitat (Foissner et al. 2002; Foissner 2006, 2007). To some extent, this may reflect inadequate sampling techniques. With the sampling gear and counting methods typically applied in taxonomical and ecological studies, it is difficult to detect species with abundances that are 3 or more orders of magnitude lower than those of frequent species. Furthermore, the fixation techniques and enrichment cultures used for estimating the abundance and species composition of protists are all more or less selective (e.g., Bloem et al. 1986; Modigh and Castaldo 2005; Foissner 2005). However, in spite of these caveats it remains that some taxa occur in presumed suitable habitats, if at all, only at very low levels.

There is no doubt that many wide-spread protist species comprise vast population sizes. A medium sized lake (e.g., 20 km in length, 12.5 km in width, thermocline in 20 m) with an epilimnetic volume of 5 km^3 may, for instance, harbour 5×10^{15} ciliates of a common species (assuming a mean abundance of 1 ciliate ml^{-1}) and 5×10^{17} flagellates of a common chysomonad species. Similarly, 1 g of soil may contain some 10,000–100,000 active ciliates and similar abundances of naked amoebae and flagellates (Foissner 2005). On longer time scales (decades to centuries), dispersal of most free-living protists may result in their world-wide distribution, due to passive drifting by water currents, turbulence and mixing, and by transportation via wind, waterfowl and other migrating animals (reviewed by Kristiansen 1996c), and even more exotic vectors such as ballast water of ships (Hallegraeff and Bolch 1992; Hülsmann and Galil 2002). It is important to note that a flagellate or ciliate has no chance to travel over a long distance as an active cell, simply because the lifespan of most protists is in the order of hours to days and active cells are vulnerable to desiccation during overland transport. It must therefore be assumed that long distance dispersal is mediated via resting stages such as cysts known from many soil protozoa, oligotrich freshwater ciliates or chrysophyte flagellates (reviewed by Corliss and Esser 1974; Cronberg and Sandgren 1986; Foissner 1987; Ekelund and Rønn 1994; Müller 2000). Cysts of soil protozoa from temperate and arid environments are generally assumed to be able to withstand harsh environmental conditions for decades (Ekelund and Rønn 1994; Foissner 2005), but resting stages of soil protozoa produced in more or less permanently moist rainforests may survive droughts for only a few weeks (Foissner 2005, 2006). Foissner (2007) concluded that cyst viability is likely much more restricted than widely assumed. Cysts of the oligotrich freshwater ciliate species *Meseres corlissi* were highly resistant to adverse conditions including inorganic and organic solvents (Foissner et al. 2005). Another recent study with the same species revealed that encystment and excystment of two temperate strains from Austria were triggered by the presence of soil extract (Müller et al. 2006), while temperature was the primary factor inducing cyst formation in a tropical strain of the same species (Weisse 2004). Cysts of another oligotrich freshwater ciliate, *Pelagostrombidium* sp., lost their viability rapidly when stored under cold (1–6°C) and dark conditions (Müller 2002).

The above examples illustrate that the factors controlling encystment, excystment and viability of cysts may differ among closely related protist species and probably even between clones of the same species. Further, the few experimental studies currently available do not allow to generalize; the significance of resting stages as part of the protist life strategy and the viability of cysts remain at present unknown for >99% of all contemporary protist species.

Investigations of airborne dispersal date back to Antony van Leeuwenhoek and C. G. Ehrenberg; the latter was probably the first who demonstrated the occurrences of unicellular algae in wind and dust in the mid 19th century (reviewed by Gregory et al. 1955; Corliss 1996). Quantification of viable microbes composing the 'aeroplankton' is difficult. Rogerson and Detwiler (1999) used a filtration sampler to collect particles in the 2.0–20.0 μm size range in near-surface urban air and determined the number of viable protist cysts after enrichment cultivation in soil extract medium. Abundances of 25 different morphotypes ranged from below detection (<0.05 m^{-3}) to 1.08 cysts m^{-3}, with an overall mean of 0.25 m^{-3}. Cysts of flagellates and naked amoebae were the commonest found; ciliates were rare. A similar recent study confirmed the generally low cyst numbers in near-surface air (Feldmann 2007); only amoebal cysts were found, with a mean abundance of 0.05 m^{-3}. It is important to note that these seemingly low numbers suggest *a tremendous potential significance of airborne dispersal* in at least some protist taxa.

Terrestrial studies on immigration and succession of soil organisms in newly colonized habitats demonstrated that immigration by air is characteristic for protists such as testate amoebae, flagellates, and small ciliates (e.g., Wanner et al. 1998; Wanner and Dunger 2002; Wanner and Xylander 2005). Similarly, the few studies that used artificial microcosms to investigate the between-habitat dispersal rate of aquatic protists (Warren 1996a, b; Östman et al. 2006) assumed that airborne dispersal of protists may play a major role in natural environments. However, the rates of dispersal of common protist taxa across disconnected habitats are virtually unknown.

In theory, rates of dispersal can be inferred from estimates of gene flow between (meta)populations. However, common measures of gene flow estimate dispersal followed by successful establishment of immigrants. Allozyme and molecular DNA markers have been particularly useful to decipher rates of dispersal of freshwater cladocerans (primarily *Daphnia*), rotifers, and insects, but all gene flow estimates depend on models with underlying assumptions (e.g., equilibrium conditions) that may not be met in the field (Boileau et al. 1992; Bohonak and Jenkins 2003). The latter authors concluded from their literature review that for freshwater invertebrates (including many small crustaceans and rotifers <1 mm in size) the currently available evidence contradicts the notion that dispersal in most freshwater taxa is frequent and widespread on relatively short time scales. Freeland and co-workers (Freeland et al. 2000, 2001) reported negligible to low levels of spatial gene flow between (meta)populations of asexually propagating bryozoans, but low to high levels of within-population genetic diversity; these authors concluded that temporal gene flow via resting stages (statoblasts) at a single site is more important than spatial gene flow across different habitats. In contrast, De Meester et al. (2002) reported that high dispersal capacity of both freshwater animals (cladocerans, rotifers and bryozoans) and macrophytes contrasts with the abundant evidence of pronounced genetic differentiation among neighbouring populations in many pond-dwelling organisms. Since the size, mode of dispersal and life strategy, with intermittent production of sexual, asexual and resting stages, of larger protists are similar to that of rotifers and small cladocerans, I conclude that many populations of freshwater protists in neighbouring habitats may also be genetically different. High genetic differences, in the SSU rDNA and ITS regions, have been reported recently for neighbouring populations of the dinoflagellate *Peridinium limbatum* from Northern Wisconsin freshwater bodies (Kim et al. 2004). However, compared to cladocerans and rotifers, the study of population genetics of protists is still in its infancy (Weisse 2006b). In conclusion, population genetic studies with both common and rare free-living protist species are urgently needed to measure their effective dispersal rates; this is a key issue for the current debate about protist biogeography.

Dispersal and immigration are neither random nor neutral processes

Physical processes such as wind and water currents are primarily responsible for the uneven distribution of soil and aquatic microbes. Similar to migrating birds, the physical processes are non-random. Like water currents, airborne dispersal is uneven in areas with a prevailing wind direction. Even in areas that are permanently subjected to high wind stress, there will be some sheltered places of refuge with lowered chances of passive invasion.

Immigration and community assembly are non-neutral processes; the more similar an invader is to the resident population, the more it is inhibited and, in most cases, prevented from establishing a new population in the same habitat (Fargione et al. 2003). The strength of biological interactions such as competition, predation and parasitism are all density-dependent and tend to peak in climax communities. Accordingly, colonization favours generalists while immigration into already colonized habitats favours specialists within a seemingly uniform guild (i.e., a population, a species, a functional group such as, e.g., bacterivores). Each event of immigration may lead to a shift in the gene pool of the population (microevolution), thus promoting directional selection. Directional selection may lead to adaptive speciation sensu Gavrilets (2005), i.e., to "a speciation process in which genetic changes underlying divergence and reproductive isolation are driven by selection (as opposed to changes driven by mutation and random genetic drift)". Under which circumstances natural selection is strong enough to enhance the intraspecific genetic differences between different populations by selective sweeps remains at present unknown.

This important aspect, that every population that is everywhere is at some phenotypic and, most likely, also at some (functional) genotypic level different from each other has received little attention in the biogeography debate. In fact, phenotypic plasticity may be even a prerequisite of global distribution of protist species. Although random dispersal may be supported by the 'unified neutral theory of biodiversity and biogeography' (Bell 2001; Hubbel 2001), we cannot ignore the significance of biological interactions for everything that is smaller than 1 mm. In the neutral theory, there is no feedback from a resident population to invaders. This is a reasonable assumption in newly colonized habitats such as inland dunes or reclaimed opencast mining areas, which are characterized by a lack of biotic interactions among invading soil protists and where typical 'pioneer populations' dominate (Wanner et al. 1998; Wanner and Xylander 2005). If the assumption holds that probabilities of per capita reproduction, mortality, and migration to a neighbouring site are all identical, there is no difference in fitness among the individuals of a population and even among a species. Then, and only then indeed, are dispersal and, in consequence, distribution of species reduced to a size-related, strictly stochastic process.

Diversity of free-living protists

Protist speciation—is parapatric speciation the rule?

There is increasing evidence that allopatric speciation is not the sole speciation process in free-living protists. Within-habitat variation, a prerequisite of sympatric speciation, has already been demonstrated for freshwater bacteria and was attributed to the isolation of water by lake shoreline features such as bays or narrow constrictions (Yannarell and Triplett 2004). Peripatric speciation has been suggested as "a subset of allopatric speciation in which a peripherally isolated population diverges to become a new species" (Losos and Glor 2003). Parapatric speciation, the intermediate case between allopatric and sympatric speciation, with limited migration or dispersal between divergent (sub-)populations, has recently been assumed as the most general mode of speciation (Gavrilets et al. 2000; Gavrilets 2003). In contrast to peripatric speciation, parapatric speciation results from divergent evolution of adjacent populations (Losos and Glor 2003). Parapatric, as well as sympatric speciation require that there is some, although limited gene flow between the subpopulations. Recently, Amato et al. (2007) reported reproductive isolation among sympatric cryptic species in marine diatoms.

The mode of speciation is ultimately linked to the occurrence and extent of sexual reproduction in a population. For some taxonomic groups, such as diatoms, regular sexual reproduction will be the rule. This is, because due to their peculiar mode of cell division, average cell size of diatoms decreases in the course of continued asexual reproduction; for most diatoms, sexual reproduction, with auxospore formation, is "the only way out of this miniaturization trap" (Amato et al. 2007). For other taxa, such as ciliates, sexual reproduction has been studied in detail in the laboratory (reviewed by Dini and Nyberg 1993), but data on the frequency with which sexual reproduction occurs in situ are rare and controversial. While Doerder et al. (1995) reported that pond dwelling *Tetrahymena thermophila* mate often, Lucchesi and Santangelo (2004) found that mating among ciliates in marine temperate sandy shores is quite low. These authors concluded that conjugation is an erratic sexual phenomenon.

Clonal selection, local adaptation and unknown significance of sexual reproduction

Local adaptation results in improved fitness of individuals in a resident population relative to their invading congeners. Adaptation can be measured as increased survival and reproduction rates of the residents in response to the ambient environmental conditions. To increase the fitness of individuals, the genetic shift must affect quantitative traits; neutral (non-functional) changes in allele frequencies do not lead to improved performance. Local adaptation is a dynamic process, mainly resulting from mutation, natural selection and (limited) gene flow. If large population sizes are typical for most protist species, the absolute number of beneficial mutations in a population is presumably higher than in most macroorganisms. The opposite holds true for genetic drift, which is of minor importance in large populations. However, if the population structure is largely clonal, effective population size may be low (for quantification of the degree of clonality consult De Meester et al. 2006). Natural selection can only lead to a shift in beneficial allele frequencies in a population if gene flow is limited. If new immigrants are permanently swamping the gene pool of a resident population, local adaptation to a specific habitat is hindered.

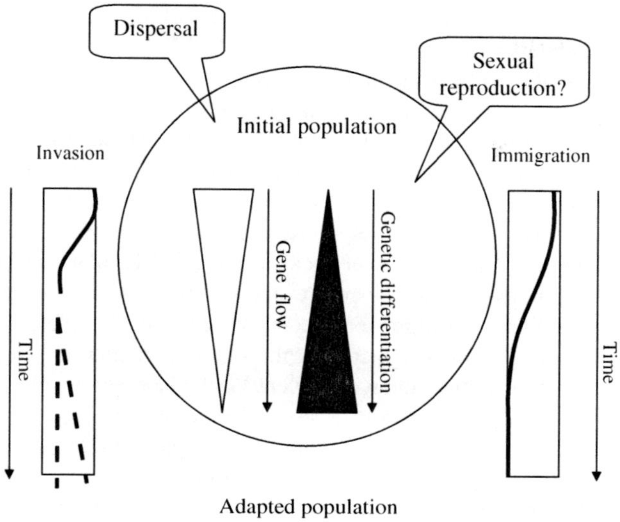

Fig. 1 A conceptual model showing the transition from an initial, unadapted population towards a population adapted to the local habitat conditions. Invasion (left, solid line) may be high at the beginning, but must then decline to reduce immigration (right) and gene flow (central, left). Invasion may remain low or increase again during the later stages of the adaptation process (left, dashed lines). Reduced gene flow and directional selection will lead to increased genetic differentiation and increased local adaptation (central, right). The speed and extent of local adaptation depend on the rate of dispersal (top, left) and the frequency of sexual reproduction

The conditions, under which local adaptation is likely to develop, and the major uncertainties in our current understanding of the protist population structure can be summarized in a conceptual model (Fig. 1). The initial population in Fig. 1 can be a founder population or any population in a given habitat at an arbitrary time zero. The gene pool of this population will be affected by a dynamic interplay of mutations, gene flow and natural selection. Marked shifts in the allele frequencies, leading to genetic differentiation towards a better adapted state of the population (with increased fitness), require that gene flow is reduced (Fig. 1, central left). Accordingly, there should be a minimum in the rate of invasion during the initial steps of the adaptation (Fig. 1, left). Invasion may stay low or increase again during the later stages of the adaptive process; irrespective of the rate of actual dispersal, the rate of immigration may continuously decline (Fig. 1, right) in response to the increased genetic differentiation (Fig. 1, central right). This is a consequence of continuously reduced competitive ability (and, therefore, lowered fitness) of the new invaders, relative to the residents, which become better and better adapted with time (De Meester et al. 2002). Note that genetic differentiation in this model does not denote genetic variation of the gene pool (which will be reduced in the course of adaptation), but the relative deviation from the initial, unadapted state. If the initial population of Fig. 1 were clonal, it should represent a 'general-purpose genotype' with broad ecological tolerances (see Vrijenhoek 1998 for review) that would slowly evolve towards a more specialized genotype, provided that the environment remains relatively stable. This conceptual model is similar to the 'Monopolization Hypothesis' (MH) proposed for pond-dwelling organisms (De Meester et al. 2002), according to which early colonists develop large, rapidly adapting populations which prevent immigration by new genotypes. The MH emphasizes the importance of numerical effects of effective dispersal, which may be reinforced in some taxa such as *Daphnia* or bryozoans (Freeland et al. 2000, 2001) by dormant stages hatching in this habitat.

There are two major unknowns in the conceptual model outlined in Fig. 1. First, invasion, directly, and immigration, indirectly, depend on the overall rate of actual dispersal, which includes release of new individuals into the habitat from excysting resting stages that were deposited earlier. Secondly, the frequency of sexual reproduction is also unknown for >99% of all free-living, extant protist species. If occurring, recombination and non-random mating in the course of sexual reproduction may largely affect the gene pool of a population and the process of local adaptation. Regular sexual reproduction, which can be inferred from relatively close agreement of allele frequencies with the Hardy–Weinberg equilibrium, will prevent a constant reduction in genetic variation in each generation. Accordingly, while sex increases the efficacy of natural selection (Goddard et al. 2005), continued recombination will prevent local adaptation by clonal lineages with improved fitness. It is the combination of sexual and clonal reproduction (cyclical parthenogenesis) that is very powerful to generate local adaptation (De Meester et al. 2002). Frequent sexual reproduction at the beginning of the growing season or during the initial stages of colonization will generate a high genetic diversity. If sexual reproduction is then replaced by clonal reproduction and diversifying clonal selection, this will efficiently promote local adaptation (De Meester et al. 2002). In obligate asexually propagating organisms the increasing average fitness of the residents with time may result from strong clonal selection, favouring the clones of the initial clone pool or new invaders that were already best (pre-)adapted to the conditions in the new habitat, rather than from adaptive mutational changes within one or a few clones (De Meester et al. 2002).

To my knowledge, no evidence is available that would allow an estimate of the relative significance of the processes enhancing or impeding local adaptation among free-living protist species. Among aquatic organisms, small-scale changes in (neutral) allele frequencies (microevolution) have been reported mainly for freshwater zooplankton, in particular for cladocerans and rotifers (reviewed by De Meester 1996; De Meester et al. 2002). It was demonstrated primarily with *Daphnia* that changes in an adaptive, quantitative trait may occur over relatively few generations (Hairston et al. 1999; Cousyn et al. 2001). The invasion resistance of resident populations has been measured experimentally with crustacean and rotifer species (Korpelainen 1986; Shurin 2000). In Shurin's (2000) study, a surprisingly high proportion, >91% of the species introduced, immediately became extinct. If the knowledge that has been gained primarily with cyclical and obligate parthenogens (*Daphnia*, rotifers) is valid for obligate or facultative asexual protist species, I infer that (some degree of) measurable local adaptation may be the rule, rather than the exception in most protist populations.

In conclusion, it appears at present impossible to unify all different aspects of genotypic variation and phenotypic plasticity for a given protist species. Pertinent questions for future research on protist diversity are, (1) how much variability beyond phenotypic plasticity and flexibility is necessary to establish a new species, and (2) how do we compare and rank the various levels (molecular, morphological, physiological, ecological) of variation for species designation. In order to understand natural protist diversity, it is more important to analyse the phenotypic properties of individuals, clones, populations and species than to characterize their genetic potential. From an ecological point of view, I suggest that a species is undergoing (parapatric or peripatric) speciation if (i) local adaptation has progressed to an extent that immigration into a particular habitat by its congeners is no longer possible, and if (ii) ecophysiological criteria (thresholds) can be defined that do not overlap between the different populations.

Acknowledgements I thank Manfred Wanner for providing literature on dispersal and colonization of protist species. Jens Boenigk and Ulrike Scheffel commented on a draft version of this manuscript. Numerous constructive comments by Luc De Meester and an anonymous reviewer are gratefully acknowledged.

References

Amato A, Kooistraa WHCF, Ghiron JHL et al (2007) Reproductive isolation among sympatric cryptic species in marine diatoms. Protist 158:193–207

Bachmann K (1998) Species as units of diversity: an outdated concept. Theory Biosci 117:213–230

Barth D, Krenek S, Fokin SI et al (2006) Intraspecific genetic variation in Paramecium, revealed by mitochondrial *COI* sequences. J Eukaryot Microbiol 53:20–25

Bell G (2001) Neutral macroecology. Science 293:2413–2418

Bloem J, Bär-Gilissen M-JB, Cappenberg TE (1986) Fixation, counting, and manipulation of heterotrophic nanoflagellates. Appl Environ Microbiol 52:1266–1272

Boenigk J, Stadler P, Wiedlroither A et al (2004) Strain-specific differences in the grazing sensitivities of closely related ultramicrobacteria affiliated with the *Polynucleobacter* cluster. Appl Environ Microbiol 70:5787–5793

Boenigk J, Pfandl K, Stadler P et al (2005) High diversity of the 'Spumella-like' flagellates: an investigation based on the SSU rRNA gene sequences of isolates from habitats located in six different geographic regions. Environ Microbiol 7:685–697

Boenigk J, Pfandl K, Garstecki T et al (2006) Evidence for geographic isolation and signs of endemism within a protistan morphospecies. Appl Environ Microbiol 72:5159–5164

Boenigk J, Jost S, Stoeck T et al (2007) Differential thermal adaptation of clonal strains of a protist morphospecies originating from different climatic zones. Environ Microbiol 9:593–602

Bohonak AJ, Jenkins DG (2003) Ecological and evolutionary significance of dispersal by freshwater invertebrates. Ecol Lett 6:783–796

Boileau MG, Hebert PND, Schwartz SS (1992) Non-equilibrium gene frequency divergence: persistent founder effects in natural populations. J Evol Biol 5:25–39

Burt WH (1943) Territoriality and home range concepts as applied to mammals. J Mammal 24:346–352

Claridge MF, Dawah HA, Wilson MR (1997) Practical approaches to species concepts for living organisms. In: Claridge MF, Dawah HA, Wilson MR (eds) Species – the units of biodiversity. Chapman & Hall, London, pp 1–15

Corliss JO (1996) Christian Gottfried Ehrenberg (1795–1876): glimpses into the personal life of this most exemplary early protistologist. In: Schlegel M, Hausmann K (eds) Christian Gottfried Ehrenberg-Festschrift. Leipziger Universitätsverlag, Leipzig, pp 31–46

Corliss JO, Esser SC (1974) Comments on the role of the cyst in the life cycle and survival of free-living protozoa. Trans Am Micros Soc 93:578–593

Cornell HV, Lawton JH (1992) Species interactions, local and regional processes, and limits to the richness of ecological communities: a theoretical perspective. J Anim Ecol 61:1–12

Cousyn C, De Meester L, Colbourne JK et al (2001) Rapid, local adaptation of zooplankton behavior to changes in predation pressure in the absence of neutral genetic changes. Proc Natl Acad Sci USA 98:6256–6260

Cronberg G, Sandgren CD (1986) A proposal for the development of standardized nomenclature and terminology for chrysophycean statospores. In: Kristiansen J, Andersen RA (eds) Chrysophytes: aspects and problems. Cambridge University Press, Cambridge, pp 317–328

Curtis TP, Sloan WT (2004) Prokaryotic diversity and its limits: microbial community structure in nature and implications for microbial ecology. Curr Opin Microbiol 7:221–226

Darling KF, Wade CM, Kroon D et al (1999) The diversity and distribution of modern planktonic foraminiferal small subunit ribosomal RNA genotypes and their potential as tracers of present and past ocean circulation. Paleoceanography 14:3–12

Darling KF, Wade CM, Stewart IA et al (2000) Molecular evidence for genetic mixing of Arctic and Antarctic subpolar populations of planktonic foraminifers. Nature 405:43–47

De Meester L (1996) Local genetic differentiation and adaptation in freshwater zooplankton populations: patterns and processes. Ecoscience 3:385–399

De Meester L, Gómez A, Okamurac B et al (2002) The Monopolization Hypothesis and the dispersal – gene flow paradox in aquatic organisms. Acta Oecol 23:121–135

De Meester L, Vanoverbeke J, De Gelas K et al (2006) Genetic structure of cyclic parthenogenetic zooplankton populations – a conceptual framework. Arch Hydrobiol 167:217–244

Dini F, Nyberg D (1993) Sex in ciliates. In: Jones JG (ed) Advances in microbial ecology. Plenum Press, New York, London, pp 129–144

Dini F, Nyberg D (1999) Growth rates of marine ciliates on diverse organisms reveal ecological specializations within morphospecies. Microb Ecol 37:13–22

Doerder FP, Gates MA, Eberhardt FP et al (1995) High frequency of sex and equal frequencies of mating types in natural populations of the ciliate *Tetrahymena thermophila*. Proc Natl Acad Sci USA 92:8715–8718

Dolan JR (2005) Biogeography of aquatic microbes. Aquat Microb Ecol 41:39–48

Ekelund F, Rønn R (1994) Notes on protozoa in agricultural soil, with emphasis on heterotrophic flagellates and naked amoebae and their ecology. FEMS Microbiol Rev 15:321–353

Fargione J, Brown CS, Tilman D (2003) Community assembly and invasion: an experimental test of neutral versus niche processes. Proc Natl Acad Sci USA 100:8916–8920

Feldmann S (2007) Untersuchungen zur Verbreitung von Protozoen über die Luft. Diploma Thesis, University of Cologne

Finlay BJ (2002) Global dispersal of free-living microbial eukaryote species. Science 296:1061–1063

Finlay BJ (2004) Protist taxonomy: an ecological perspective. Phil Trans R Soc Lond B 359:599–610

Finlay BJ, Fenchel T (1999) Divergent perspectives on protist species richness. Protist 150:229–233

Finlay BJ, Esteban GF, Fenchel T (2004) Protist diversity is different? Protist 155:15–22

Finlay BJ, Esteban GF, Brown S et al (2006) Multiple cosmopolitan ecotypes within a microbial eukaryote morphospecies. Protist 157:377–390

Foissner W (1987) Soil protozoa: fundamental problems, ecological significance, adaptations in ciliates and testaceans, bioindicators, and guide to the literature. Progr Protistol 2:69–212

Foissner W (1999) Protist diversity: estimates of the near-imponderable. Protist 150:363–368

Foissner W (2004) Ubiquity and cosmopolitanism of protists questioned. SILnews 43:6–7

Foissner W (2005) Protozoa. In: Hillel D, Hatfield JL, Rosenzweig C et al (eds) Encyclopedia of soils in the environment, vol 3. Elsevier, pp 336–347

Foissner W (2006) Biogeography and dispersal of microorganisms: a review emphasizing protists. Acta Protozool 45:111–136

Foissner W (2007) Dispersal and biogeography of protists: recent advances. Jpn J Protozool 40:1–16

Foissner W, Stoeck T, Schmidt H et al (2001) Biogeographical differences in a common soil ciliate, *Gonostomum affine* (Stein), as revealed by morphological and RAPD-fingerprint analysis. Acta Protozool 40:83–97

Foissner W, Agatha S, Berger H (2002) Soil ciliates (Protozoa, Ciliophora) from Namibia (Southwest Africa), with emphasis on two contrasting environments, the Etosha region and the Namib Desert. Denisia 5:1–1459

Foissner W, Müller H, Weisse T (2005) The unusual, lepidosome-coated resting cyst of *Meseres corlissi* (Ciliophora: Oligotrichea): light and scanning electron microscopy, cytochemistry. Acta Protozool 44:201–215

Freeland JR, Rimmer VK, Okamura B (2001) Genetic changes within freshwater bryozoan populations suggest temporal gene flow from statoblast banks. Limnol Oceanogr 46:1121–1129

Freeland JR, Noble LR, Okamura B (2000) Genetic consequences of the metapopulation biology of a facultatively sexual freshwater invertebrate. J Evol Biol 13:383–395

Gächter E, Weisse T (2006) Local adaptation among geographically distant clones of the cosmopolitan freshwater ciliate *Meseres corlissi*. I. Temperature response. Aquat Microb Ecol 45:291–300

Gavrilets S (2003) Perspective: models of speciation: what have we learned in 40 years? Evolution 57:2197–2215

Gavrilets S (2005) Adaptive speciation – it is not that easy: a reply to Doebeli et al. Evolution 59:696–699

Gavrilets S, Li H, Vose MD (2000) Patterns of parapatric speciation. Evolution 54:1126–1134

Giller P, Hillebrand H, Berninger U-G et al (2004) Biodiversity effects on ecosystem function: emerging issues and their experimental test in aquatic communities. Oikos 104:423–436

Goddard MR, Godfray HCJ, Burt A (2005) Sex increases the efficacy of natural selection in experimental yeast populations. Nature 434:636–640

Gregory PH, Hamilton ED, Sreeramulu T (1955) Occurrence of the green alga *Gleocapsa* in the air. Nature 176:1270

Hahn MW (2006) The microbial diversity of inland waters. Curr Opin Biotechnol 17:256–261

Hairston NGj, Lampert W, Caceres CE et al (1999) Rapid evolution revealed by dormant eggs. Nature 401:446

Hallegraeff G, Bolch C (1992) Transport of dinoflagellate cysts in ship's ballast water: implications for plankton biogeography and aquaculture. J Plankton Res 14:1067–1084
Hanski I (1999) Metapopulation ecology. Oxford University Press, Oxford
Hey J (2001) Genes, categories and species. Oxford University Press, Oxford
Hillebrand H, Blenckner T (2002) Regional and local impact on species diversity – from pattern to processes. Oecologia 132:479–491
Hillebrand H, Watermann F, Karez R et al (2001) Differences in species richness patterns between unicellular and multicellular organisms. Oecologia 126:114–124
Hoef-Emden K, Melkonian M (2003) Revision of the genus *Cryptomonas* (Cryptophyceae): a combination of molecular phylogeny and morphology provides insights into a long-hidden dimorphism. Protist 154:371–409
Holyoak M (2000) Effects of nutrient enrichment on predator–prey metapopulation dynamics. J Anim Ecol 69:985–997
Holyoak M, Lawler SP (1996) The role of dispersal in predator–prey metapopulation dynamics. J Anim Ecol 65:640–652
Horner-Devine MC, Lage M, Hughes JB et al (2004) A taxa–area relationship for bacteria. Nature 432:750–753
Hubbel SP (2001) The unified neutral theory of biodiversity and biogeography. Princeton University Press, Princeton, NJ
Hughes Martiny JB, Bohannan BJM, Brown JH et al (2006) Microbial biogeography: putting microorganisms on the map. Nat Rev 4:102–112
Hülsmann N, Galil BS (2002) Protists – a dominant component of the ballast-transported biota. In: Leppäkoski E, Gollasch S, Olenin S (eds) Invasive aquatic species of Europe. Distribution, impacts and management. Kluwer Academic Publishers, Dordrecht, pp 20–26
Jackson RB, Linder CR, Lynch M et al (2002) Linking molecular insight and ecological research. Trends Ecol Evol 17:409–414
Jerome CA, Simon EM, Lynn DH (1996) Description of *Tetrahymena empidokyrea* n. sp., a new species in the *Tetrahymena pyriformis* sibling species complex (Ciliophora, Oligohymenophorea), and an assessment of its phylogenetic position using small-subunit rRNA sequences. Can J Zool (Rev Can Zool) 74:1989–1906
Katz LA, McManus GB, Snoeyenbos-West OLO et al (2005) Reframing the 'Everything is everywhere' debate: evidence for high gene flow and diversity in ciliate morphospecies. Aquat Microb Ecol 41:55–65
Kim E, Wilcox L, Graham L et al (2004) Genetically distinct populations of the dinoflagellate *Peridinium limbatum* in neighboring Northern Wisconsin Lakes. Microb Ecol 48:521–527
Koch TA, Ekelund F (2005) Strains of the heterotrophic flagellate *Bodo designis* from different environments vary considerably with respect to salinity preference and SSU rRNA gene composition. Protist 156:97–112
Korpelainen H (1986) Competition between clones: an experimental study in a natural population of *Daphnia magna*. Hereditas 105:29–35
Kristiansen J (ed) (1996a) Biogeography of freshwater algae. Hydrobiologia 336:1–161
Kristiansen J (1996b) Biogeography of freshwater algae – conclusions and perspectives. Hydrobiologia 336:159–161
Kristiansen J (1996c) Dispersal of freshwater algae – a review. Hydrobiologia 336:151–157
Lachance M-A (2004) Here and there or everywhere. BioScience 54:884
Loret P, Tengs T, Villareal TA et al (2002) No differences found in ribosomal DNA sequences from physiologically diverse clones of *Karenia brevis* (Dinophyceae) from the Gulf of Mexico. J Plankton Res 24:735–739
Losos JB, Glor RE (2003) Phylogenetic comparative methods and the geography of speciation. Trends Ecol Evol 18:220–227
Lowe CD, Day A, Kemp SJ et al (2005) There are high levels of functional and genetic diversity in *Oxyrrhis marina*. J Eukaryot Microbiol 52:250–257
Lucchesi P, Santangelo G (2004) How often does conjugation in ciliates occur? Clues from a seven-year study on marine sandy shores. Aquat Microb Ecol 36:195–200
Maddock A, Du Plessis MA (1999) Can species data only be appropriately used to conserve biodiversity? Biodivers Conserv 8:603–615
Marshall HG, Hargraves PE, Burkholder JM et al (2006) Taxonomy of Pfiesteria (Dinophyceae). Harmful Algae 5:481–496
Mitchell EAD, Meisterfeld R (2005) Taxonomic confusion blurs the debate on cosmopolitanism versus local endemism of free-living protists. Protist 156:263–267

Modigh M, Castaldo S (2005) Effects of fixatives on ciliates as related to cell size. J Plankton Res 27:845–849

Montagnes DJS, Wilson D, Brooks SJ et al (2002) Cyclical behaviour of the tide-pool ciliate *Strombidium oculatum*. Aquat Microb Ecol 28:55–68

Müller H (2000) Evidence of dormancy in planktonic oligotrich ciliates. Verh Internat Verein Limnol 27:3206–3209

Müller H (2002) Laboratory study of the life cycle of a freshwater strombidiid ciliate. Aquat Microb Ecol 29:189–197

Müller H, Foissner W, Weisse T (2006) The role of soil in the life cycle of *Meseres corlissi* (Ciliophora: Oligotrichea): experiments with two clonal strains from the type locality, an astatic meadow pond. Aquat Microb Ecol 42:199–208

Östman Ö, Kneitel JM, Chase JM (2006) Disturbance alters habitat isolation's effect on biodiversity in aquatic microcosms. Oikos 114:360–366

Quispel A (1998) Lourens G. M. Baas Becking (1895–1963), inspirator for many (micro)biologists. Int Microbiol 1:69–72

Ramette A, Tiedje JM (2007) Biogeography: an emerging cornerstone for understanding prokaryotic diversity, ecology, and evolution. Microb Ecol 53:197–206

Řezáčová M, Neustupa J (2007) Distribution of the genus *Mallomonas* (Synurophyceae) – ubiquitous dispersal in microorganisms evaluated. Protist 158:29–37

Rogerson A, Detwiler A (1999) Abundance of airborne heterotrophic protists in ground level air of South Dakota. Atmos Res 51:35–44

Schauer M, Jiang J, Hahn MW (2006) Recurrent seasonal variations in abundance and composition of filamentous SOL cluster bacteria (*Saprospiraceae, Bacteroidetes*) in oligomesotrophic Lake Mondsee (Austria). Appl Environ Microbiol 72:4704–4712

Schlegel M (2003) Phylogeny of Eukaryotes recovered with molecular data: highlights and pitfalls. Eur J Protistol 39:113–122

Schlegel M, Meisterfeld R (2003) The species problem in protozoa revisited. Eur J Protistol 39:349–355

Shurin JB (2000) Dispersal limitation, invasion resistance, and the structure of pond zooplankton communities. Ecology 81:3074–3086

Smith HG, Wilkinson DM (2007) Not all free-living microorganisms have cosmopolitan distributions—the case of *Nebela* (*Apodera*) *vas* Certes (Protozoa: Amoebozoa: Arcellinida). J Biogeogr 34:1822–1831

Snoeyenbos-West OLO, Salcedo T, McManus GB et al (2002) Insights into the diversity of choreotrich and oligotrich ciliates (Class: Spirotrichea) based on genealogical analyses of multiple loci. Int J Syst Evol Microbiol 52:1901–1913

Turesson G (1922) The genotypic response of the plant species to the habitat. Hereditas 3:211–350

von der Heyden S, Chao EE, Vickerman K et al (2004) Ribosomal RNA phylogeny of bodonid and diplonemid flagellates and the evolution of Euglenozoa. J Eukaryot Microbiol 51:402–416

von der Heyden S, Cavalier-Smith T (2005) Culturing and environmental DNA sequencing uncover hidden kinetoplastid biodiversity and a major marine clade within ancestrally freshwater *Neobodo designis*. Int J Syst Evol Microbiol 55:2605–2621

Vrijenhoek RC (1998) Animal clones and diversity. BioScience 48:617–628

Walochnik J, Hassle A, Simon K et al (1999) Isolation and identification by partial sequencing of the 18S ribosomal gene of free-living amoebae from necrotic tissue of *Basiliscus plumifrons* (Sauria: Iguanidae). Parasitol Res 85:601–603

Wanner M, Dunger W (2002) Primary immigration and succession of soil organisms on reclaimed opencast coal mining areas in Eastern Germany. Eur J Soil Biol 38:137–143

Wanner M, Xylander WER (2005) Biodiversity development of terrestrial testate amoebae – is there any succession at all? Biol Fertil Soils 41:428–438

Wanner M, Dunger W, Schulz H-J et al (1998) Primary immigration of soil organisms on coal mined areas in Eastern Germany. In: Pižl V, Tajovský K (eds) 4th Central European workshop on soil zoology. České Budějovice, pp 267–275

Warren PH (1996a) Dispersal and destruction in a multihabitat system: an experimental approach using protist communities. Oikos 77:317–325

Warren PH (1996b) The effects of between-habitat dispersal rate on protist communities and metacommunities in microcosms at two spatial scales. Oecologia 105:132–140

Weisse T (1991) The annual cycle of heterotrophic freshwater nanoflagellates: role of bottom-up versus top-down control. J Plankton Res 13:167–185

Weisse T (1997) Growth and production of heterotrophic nanoflagellates in a meso-eutrophic lake. J Plankton Res 19:703–722

Weisse T (2004) *Meseres corlissi*: a rare oligotrich ciliate adapted to warm water and temporary habitats. Aquat Microb Ecol 37:75–83

Weisse T (2006a) Biodiversity of freshwater microorganisms – achievements, problems, and perspectives. Pol J Ecol 54:633–652

Weisse T (2006b) Freshwater ciliates as ecophysiological model organisms – lessons from *Daphnia*, major achievements, and future perspectives. Arch Hydrobiol 167:371–402

Weisse T, Rammer S (2006) Pronounced ecophysiological clonal differences of two common freshwater ciliates, *Coleps spetai* (Prostomatida) and *Rimostrombidium lacustris* (Oligotrichida), challenge the morphospecies concept. J Plankton Res 27:55–63

Weisse T, Scheffel U, Stadler P et al (2007) Local adaptation among geographically distant clones of the cosmopolitan freshwater ciliate *Meseres corlissi*. II. Response to pH. Aquat Microb Ecol 47:289–297

Whitfield J (2005) Biogeography: is everything everywhere? Science 310:960–961

Wylezich C, Meisterfeld R, Meisterfeld S et al (2002) Phylogenetic analysis of small subunit ribosomal RNA coding regions reveal a monophyletic lineage of testate filopod amoebae (Testaceafilosea). J Euk Microbiol 49:108–118

Yannarell AC, Triplett EW (2004) Within- and between-lake variability in the composition of bacterioplankton communities: investigations using multiple spatial scales. Appl Environ Microbiol 70:214–223

“Missing” protists: a molecular prospective

Slava Epstein · Purificación López-García

Originally published in the journal Biodiversity and Conservation, Volume 17, No 2, 261–276.
DOI: 10.1007/s10531-007-9250-y

Abstract Molecular ecology methods based on 18S rRNA amplification and sequencing have revealed an astounding diversity of microbial eukaryotes in every environment sampled so far. This is certainly true of new species and genera, as essentially every new survey discovers a wealth of novel diversity at this level. This is almost certain for taxa that are higher in taxonomic hierarchy, as many molecular surveys reported novel clades within established protistan phyla, with some of these clades repeatedly confirmed by subsequent studies. It may also be that the molecular approaches discovered several lineages of the highest taxonomic order, but this claim has not been vigorously verified as yet. Overall, the field of protistan diversity remains in its infancy. The true scale of this diversity is unknown, and so are the distribution of this diversity, its patterns, spatial and temporal dynamics, and ecological role. The sampled diversity appears to be just the tip of the iceberg, and this offers outstanding opportunities for microbial discovery for the purposes of both basic and applied research.

Keywords 18S rRNA · Cryptic species · Eukaryotic phylogeny · Molecular ecology · Species richness

Introduction

Molecular analyses based on the rRNA approach have become an affordable and practical tool that is used routinely to study microbial diversity in environmental samples. This type

---

Special Issue: Protist diversity and geographic distribution. Guest editor: W. Foissner.

S. Epstein
Marine Science Center, Northeastern University, Nahant, MA 01908, USA
e-mail: slava.epstein@gmail.com

P. López-García (✉)
Unité d’Ecologie, Systématique et Evolution, CNRS UMR8079, Université Paris-Sud, Orsay Cedex 91405, France
e-mail: puri.lopez@u-psud.fr

of analyses was designed to bypass inability to cultivate the majority of prokaryotic organisms in vitro, and provided an excellent tool to classify the organisms by placing the sequences of this marker gene in a phylogenetic tree of life (Olsen et al. 1986; Woese 1987). Results accumulated over the past two decades are impressive. They are more than "just" 300,000 SSU rDNA prokaryotic sequences now available from GenBank,—they have changed our view of the biosphere (Pace 1997).

Compared to prokaryotes, the application of the SSU rRNA gene-based approach to study protistan diversity is very recent (Díez et al. 2001; López-García et al. 2001; Moon-van der Staay et al. 2001; Moreira and López-García 2002). However, even the limited information obtained so far shows certain similarities as well as dissimilarities with what is known about prokaryotic molecular diversity. Contrary to prokaryotic microbiologists, some protozoologists maintain that the principal protistan taxa have been identified during the era of α-taxonomy, and they question the very possibility of discovering novel high ranking candidate protistan taxa (e.g. Finlay 2002; Cavalier-Smith 2004). This is not necessarily so for novel subgroups, at perhaps the class level down to species, where the newly discovered diversity appears impressive (Moreira and López-García 2002; Berney et al. 2004; Richards and Bass 2005). Furthermore, although still quite unstudied, the genetic diversity within protist morphospecies might be surprisingly high as revealed by SSU rRNA gene sequences (Boenigk et al. 2005), the combination of these with intergenic spacer sequences (ITS) (Katz et al. 2005; Rodriguez et al. 2005), and multilocus sequence analysis (Katz et al. 2006; Slapeta et al. 2006a, b). These findings call into question the validity of the phenotype only-based species concept, and open new avenues to explore protist biogeography (Foissner 2006), ecological specialization, and mechanisms underlying protist speciation. In the present review, we briefly summarize recent data on the molecular diversity of microbial eukaryotes.

The molecularly identified but undescribed diversity of protists

Power and pitfalls of molecular phylogenetic identification

The first molecular eukaryotic diversity studies in the early 2000s revealed three major categories of SSU rRNA gene sequences: sequences closely related to known species and genera genes, sequences forming divergent groups within known phyla, and sequences without a clear affiliation to any described eukaryotic phyla. The latter were more frequently identified from anoxic habitats, which have been traditionally less studied, such as various anoxic sediments (Dawson and Pace 2002; Stoeck and Epstein 2003), including hydrothermal sediments (Edgcomb et al. 2002; López-García et al. 2003), and the anoxic water column in, e.g., the Cariaco Basin of the coast of Venezuela (Stoeck et al. 2003). Some authors claimed that certain highly divergent lineages could possibly represent new eukaryotic kingdoms (Dawson and Pace 2002). This claim was subsequently contested (Berney et al. 2004; Cavalier-Smith 2004). The reanalysis of sequences deposited in databanks until 2003 suggested that there had probably been an overestimation regarding the 'novelty' of many of these sequences, due to their misplacement in eukaryotic phylogenies (Berney et al. 2004). There are 3 principal factors that account for such misplacement:

1. PCR of mixed DNA templates generates, with certain frequency, chimeric sequences. Partial treeing analyses is widely used to identify such sequences, together with visual inspection of alignments, and the use of several software programs e.g., CHIMERA_CHECK and BELLEROPHON, both available at the RDP (Maidak et al. 2001).

2. Long-branch attraction due to heterogeneous rates of evolution between lineages (Philippe et al. 2000; Brinkmann et al. 2005). This artifact can be minimized by using methods that are less prone to it (e.g., maximum likelihood or Bayesian analyses), including an appropriate selection of slowly-evolving lineages and avoiding the inclusion of other fast-evolving sequences and distant outgroups. The combined use of several phylogenetic markers, such as concatenations of multiple protein-coding genes generated from EST (expressed sequence tags) of cultivated protists (O'Brien et al. 2007), or the combination of small and large subunit rRNA genes (Moreira et al. 2007) can significantly improve problematic molecular phylogenies.
3. Insufficient taxon sampling (Philippe et al. 2000), an artifact especially important in protistan phylogeny because a significant number of known eukaryotic taxa presently lack representative rRNA gene sequences in databases. A case in point is several protistan taxa that reached a stable phylogenetic position once a significant number of sequences of other members of that group were included (see Table 1 in Berney et al. 2004). Pertinent examples include AT4-11 (Apusozoa) (López-García et al. 2003), BOLA187 and BOLA366 (*'Mastigamoeba invertens'* group) (Dawson and Pace 2002), CS_E022 (jakobid) and C1_E027 *'Retortamonas/Carpediemonas'* group) (Edgcomb et al. 2002). There are also several examples of orphan SSU rRNA sequences that eventually found a "home" taxon in what originally simply lacked representative sequences (Fig. 1). For instance, DH145-KW16, a clone that branched at the base of the Acantharea initially (López-García et al. 2002), and CS_E043 (Edgcomb et al. 2002), were shown to be members of the Taxopodida once the sequence of the heliozoan-like *Sticholonche zanclea* was determined (Nikolaev et al. 2004). Further, Fig. 1 shows that the deep-sea radiolarian-like sequence DH45-HA2 (López-García et al. 2002) turned out to cluster with the Nassellaria and various other radiolarian-like sequences from deep-sea plankton (DH148-EKD30), anoxic sediments (E32, E16, E215, E191, E196) and hydrothermal sediment (CS_E004, C1_E045) with a group of single-celled spumellarians, both of which lacked representative sequences until recently (Kunitomo et al. 2006). Another example relates to the kathablepharids, an important flagellate component of marine and freshwater systems distantly related to the cryptophytes, and for which SSU rRNA gene sequences were not available until recently (Okamoto and Inouye 2005). Upon their determination, a number of environmental sequences retrieved in various previous studies could be affiliated to this flagellate group (Slapeta et al. 2006a, b). A more extreme case is the phylotypes DH145-EKD11 from deep-sea plankton (López-García et al. 2001) and CCW75 from anoxic marine sediment (Stoeck and Epstein 2003). These were initially placed as basal unresolved lineages, but once the first representative sequences (*Myrionecta* and *Mesodinium* spp.) became available, they were shown to cluster with the Mesodinidae, a group of fast-evolving ciliates (Johnson et al. 2004).

An additional problematic aspect of rRNA surveys is that they do not necessarily identify live cells, and are likely to detect allochthonous species of little importance for the community under study. An alternative is to target rRNA via cDNA libraries, an approach that has only recently been employed in the study of microbial eukaryotes (Stoeck et al. 2007b).

Despite the above problems, molecular environmental analyses of protist diversity are revealing a large variety of protists within high-level taxa, and are helping to define their genetic diversity in nature. However, we can also predict that large protistan diversity

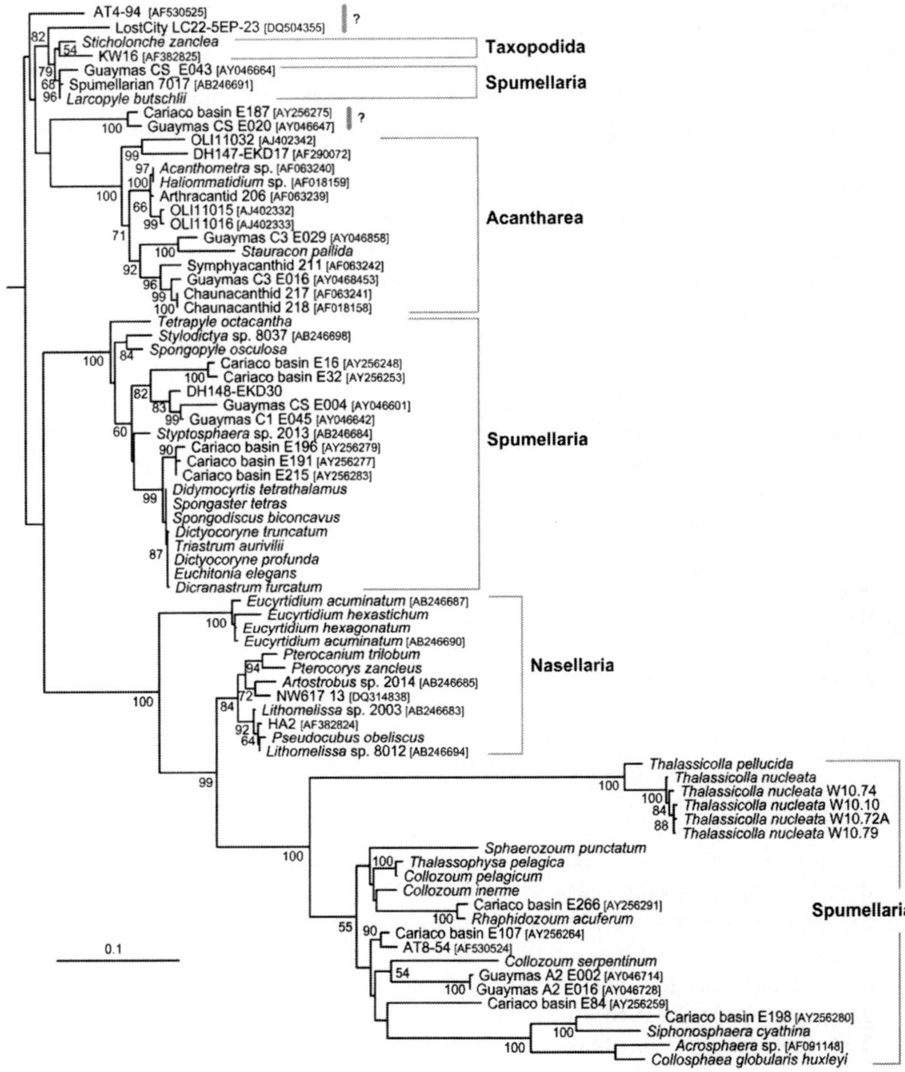

Fig. 1 Phylogenetic tree showing the position of environmental radiozoan 18S rRNA gene sequences retrieved from several oceanic locations by molecular diversity analyses. The tree was constructed by maximum likelihood (PhyML) using 901 unambiguously aligned positions, applying a GTR + Γ + I model of sequence evolution. Bootstrap values (500 replicates) higher than 50% are shown at nodes. The tree was rooted with ten cercozoan sequences

escapes molecular identification due to primer biases and other PCR-related biases (Stoeck et al. 2006; Stoeck et al. 2007a). It is quite possible that some groups are rarely, or never, captured by environmental libraries that employ general eukaryotic primers (or even combinations thereof), e.g. many fast evolving parasitic lines (gregarines, excavate parasites), many amoeboid lineages, or rhodophytes. The use of taxon-specific primers can reveal a much richer diversity, as was recently shown for the Cercozoa (Bass and Cavalier-Smith 2004).

Novel protistan lineages

Undeniably, molecular environmental surveys of eukaryotes have uncovered a wide range of SSU rRNA phylotypes that belong to previously described phyla. Some of them are represented by single sequences unrelated to others within the phylum, and since the position of singletons is often unstable, their true phylogeny remains to be validated. Others form novel clusters composed of sequences reported by independent studies, which is strongly indicative of their reality. One such example is the cluster formed by DH148-5-EKD18, CS_R003, BOLA048, Sey010 and Sey017 from deep-sea plankton (López-García et al. 2001), hydrothermal sediment (Edgcomb et al. 2002), anoxic tidal sediment (Dawson and Pace 2002) and river sediment (Berney et al. 2004, see their Fig. 2). However, the branch leading to this clade is very long suggesting that the whole group may not represent a truly novel taxon but be misplaced from its real position by the long-branch attraction artifact. Some of the clusters remain good candidates for new phylum-level groups (Berney et al. 2004), although validation by complementary approaches, including in situ fluorescence hybridization methods (FISH), and ultimately, morphological and structural characterization, will be required.

In addition, there are several cases of larger clusters of sequences, also generated by independent studies, that branch together and form defined monophyletic groups with no described representatives. The first molecular diversity studies carried out on marine picoplankton revealed the presence of two large clusters of alveolate sequences, initially termed marine alveolates Group I and II (Díez et al. 2001; López-García et al. 2001; Moon-van der Staay et al. 2001; Moreira and López-García 2002). Sequences from both groups have so far been identified exclusively in marine environments. The genetic diversity within both groups is substantial and comparable to that of, for example, the ciliates. The sequence of *Amoebophrya* sp., a parasite of the dinoflagellate *Gymnodinium sanguineum* (Gunderson et al. 1999) from the order Syndiniales, clusters with those of Group II, suggesting that at least some Group II alveolates may be parasites, and their registered diversity may be linked to that of their host organisms. Indeed, it had been previously suggested that, based on SSU rRNA gene sequences, *Amoebophrya ceratii* was a complex of species (Janson et al. 2000). However, while Group II can be identified with at least some known species, marine alveolate Group I cannot. Only recently, the first hint came about their biology, i.e., the possible symbiotic nature of its members, as Group I alveolate sequences were amplified from single radiolarian cells (Dolven et al. 2007).

Within the stramenopiles (heterokonts), several clusters of SSU rRNA sequences were identified with no known representatives; these originated mostly from the upper (photic) ocean environment. Massana and co-workers described up to 12 clusters of likely heterotrophic stramenopiles, with four MAST clades comprising approximately 75% of the total number of stramenopile sequences deposited in public databases (Massana et al. 2004a, b). In a recent study, Richards and Bass (2005) registered up to 16 stramenopile clusters. We note, however, that the entire clade of stramenopiles is unstable, casting doubt in the reality of so many independent stramenopile groups. Some of these newly detected clades are supposed to be heterotrophic because they branched at the base of the stramenopile clade, far from typical photosynthesizing lineages. The heterotrophic lifestyle was later confirmed by FISH, which suggested that at least some of these stramenopiles are bacterivorous (Massana et al. 2002, 2006). In addition to heterotrophic lines, a novel clade of photosynthetic stramenopiles was detected in Artic waters (Lovejoy et al. 2006). The MAST clades remain an exciting target for microbial discovery, and first steps have been

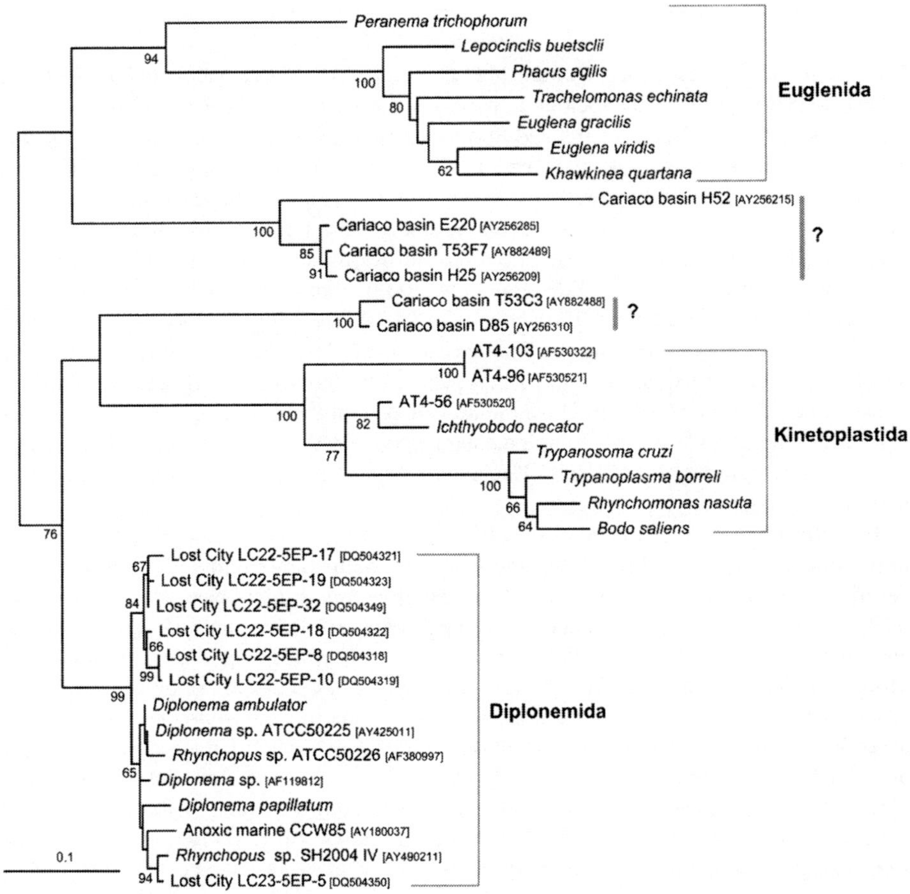

Fig. 2 Phylogenetic tree showing the position of environmental euglenozoan 18S rRNA gene sequences retrieved by molecular diversity analyses from deep-sea plankton, deep-sea hydrothermal vent areas (Lost City and Rainbow, Mid-Atlantic Ridge) and the anoxic Cariaco basin. The tree was constructed by maximum likelihood (PhyML) using 500 unambiguously aligned positions (the limited number of positions used is imposed by the presence of a few partial sequences), applying a GTR + Γ + I model of sequence evolution. Bootstrap values (500 replicates) higher than 50% are shown at nodes

taken towards recovering and visualizing cells of this elusive group (Kolodziej and Stoeck 2007).

While most of the newly discovered protistan diversity falls within known clades, such as alveolates and stramenopiles, there are several lineages possibly representing novel protistan groups of higher taxonomic rank. Typically, these lineages consist of few sequences obtained by a limited number of studies. For instance, several sequences retrieved from deep-sea plankton and fluid-seawater interfaces in the Lost City hydrothermal site formed a robust clade sister to the diplonemids within the Euglenozoa (López-García et al. 2007). This clade was clearly distinct from other euglenozoan sequences identified by Stoeck et al. (2006) in anoxic sediments (Fig. 2). The sequences RT5iin14 and RT5iin16 could represent a group at the base of the opisthokonts (Amaral-Zettler et al. 2002). Finally, various divergent phylotypes might form clades within Amoebozoa or even outside the six recognized eukaryotic supergroups (e.g. in Dawson and Pace 2002; Edgcomb et al. 2002;

López-García et al. 2003; Stoeck and Epstein 2003). However, the true extent of their novelty is not clear, primarily because of the long branch attraction phenomenon.

Oceanic planktonic protists

The ocean was among the first ecosystems to be explored by the rRNA approach, and remains one of the better studied environments. Most environmental surveys have focused on the photic zone picoplankton (<2 μm) and the smallest fractions of nanoplankton (2–20 μm), usually separated by prefiltration steps. The existence of picoeukaryotic algae had been recognized for some time (Courties et al. 1994; Guillou et al. 1999; Moon-van der Staay et al. 2000), but the real extent and ecological importance of picoeukaryotes was not revealed until the first molecular studies of the smallest planktonic fractions (Worden et al. 2004). In general, the same taxa are identified in various oceanic locations, though the relative proportions of lineages fluctuate between studies. Members of the Prymnesiophyta (haptophytes), Pelagophyta (stramenopiles) and Prasinophyta (green algae) appear among the more abundant picoeukaryotic photosynthesizes in SSU rRNA gene libraries from various photic oceanic locations, including the equatorial Pacific, Mediterranean, Arctic, and the Baltic Sea (Díez et al. 2001; Moon-van der Staay et al. 2001; Guillou et al. 2004; Lovejoy et al. 2006; Medlin et al. 2006; Worden 2006). Prasinophytes, especially the genera *Ostreococcus* or *Micromonas*, appear particularly frequently in the libraries, and although common PCR-based methods are not quantitative, their relative abundance in clone libraries is suggestive of their importance in the marine ecosystem (Guillou et al. 2004; Worden 2006). A few studies have analyzed the SSU rRNA gene diversity in planktonic fractions larger than 5 μm. In this case, the photosynthetic component in clone libraries is dominated by diatoms or photosynthetic dinoflagellates (Savin et al. 2004), although members of picoplanktonic size were also be detected (Yuan et al. 2004).

The likely heterotrophic fraction of surface picoplankton encompasses sequences belonging to early-branching stramenopiles, alveolates Group I and II, ciliates and radiolarians. Groups present less frequently include cercozoans, cryptophytes, and fungi. Dinoflagellate phylotypes are detected regularly but it is not clear whether they are photosynthetic, heterotrophic, or mixotrophic. Stramenopiles appear to dominate heterotrophic picoplankton, accounting for up to 35% of clone libraries targeting the total heterotrophic flagellates (Díez et al. 2001; Massana et al. 2006). In several other cases, ciliates and Group II (Syndiniales) alveolates were the most abundant in libraries (Medlin et al. 2006). These differences can be explained by environmental variables, as well as variation in spatial and temporal parameters.

Deep-sea environment

Molecular diversity of deep-sea picoplankton has been addressed in only a handful of studies. López-García et al. (2001) focused on the aphotic water column from 250 to 3,000 m depth in the Antarctic Polar Front, as well as picoplankton from the fluid-seawater interface around deep-sea vents in the Atlantic (López-García et al. 2003; López-García et al. 2007). Compared to surface waters, deep-sea samples are characterized by a lack of typical photosynthetic lineages, whose occasional presence may be attributed to sinking organisms. Alveolates, particularly Groups I and II, appear to dominate deep-sea planktonic samples, together with radiolarians, a group also well represented in clone libraries constructed from deep-sea samples.

Protists from hydrothermal vent communities have attracted considerable attention. Deep-sea vents are generally associated with mid oceanic ridges typically found between 1,500 and 3,000 m below the surface. They have fascinated the scientific community because they might be analogous to the earliest ecosystems (Reysenbach and Cady 2001). The study of protists from vent areas started only several years ago (Atkins et al. 2000), and has been limited to hydrothermal sediments in the Guaymas basin in the Pacific (2,000 m depth) (Edgcomb et al. 2002), hydrothermal sediments, fluid-seawater mixtures, and experimental substrates exposed in the Mid-Atlantic Ridge in 1,600 and 2,200 m depth (López-García et al. 2003), the anoxic surroundings of shallow fumaroles at the Kagoshima bay (200 m depth) (Takishita et al. 2005), and the carbonates and fluid-seawater interface in the alkaline Lost City vent field (750 m depth) located off-axis at the Mid Atlantic Ridge (López-García et al. 2007). Of these, only hydrothermal anoxic sediments appear to contain potentially new lineages of the highest taxonomic level, some of which have been detected more that once by independent researchers (Berney et al. 2004). The rRNA signatures from other hydrothermal habitats vary by site, and are often dominated by alveolates, especially of apparently parasitic life style, such as Syndiniales, Perkinsozoa and Gregarina (Moreira and López-García 2003), and ciliates. The latter may be bacterivorous as they are typical in surfaces colonized by bacteria. These are followed by the Euglenozoa, both kinetoplastids and diplonemids, fungi, radiolarian and cercozoan sequences (López-García et al. 2007).

Sediments and other oxygen-depleted environments

Anoxic environments hold a special promise for the discovery of microbial eukaryotes (Sogin et al. 1989) because eukaryotic cells might have evolved before oxygen in the atmosphere reached its present level oxygen concentration (Fenchel and Finlay 1995; Brocks et al. 1999). The exploration of 18S rRNA gene diversity of anoxic habitats started only 5 years ago, and is represented by 10 published surveys of anoxic marine water column (Stoeck et al. 2003; Behnke et al. 2006; Stoeck et al. 2006; Zuendorf et al. 2006), shallow marine sediments (Dawson and Pace 2002; Stoeck and Epstein 2003; Takishita et al. 2005), deep-sea hydrothermal vent systems (Edgcomb et al. 2002; López-García et al. 2003), and freshwater sulfidic spring (Luo et al. 2005). While in no way comprehensive, these surveys indicate a degree of uniqueness of protistan diversity uncovered (Richards and Bass 2005), and their collective information allows some preliminary analyses of this diversity.

The most obvious feature of microbial eukaryotes from anoxic environments is a very substantial richness of their communities. The published clone libraries tend to be relatively large, often in excess of 1,000 clones sequenced, with some constructed using multiple PCR primers (Behnke et al. 2006; Stoeck et al. 2006; Zuendorf et al. 2006). Nonetheless, none of surveys came even close to phylotype saturation, indicating a substantial diversity missed (Fig. 3; see more in last section). Equally suggestive is an extremely low overlap between species lists obtained using different PCR primer sets (Behnke et al. 2006; Stoeck et al. 2006; Zuendorf et al. 2006), further indicating that more comprehensive surveys would have uncovered richness several to many times that actually reported.

It is of particular interest to analyze which specific groups of protists were discovered in the anoxic environments. From the beginning of such research (Dawson and Pace 2002; Edgcomb et al. 2002), the surveys of anoxic environments typically reported representatives of essentially all known eukaryotic clades, with an especially high diversity of alveolate, stramenopile, and fungal 18S rRNA gene sequences.

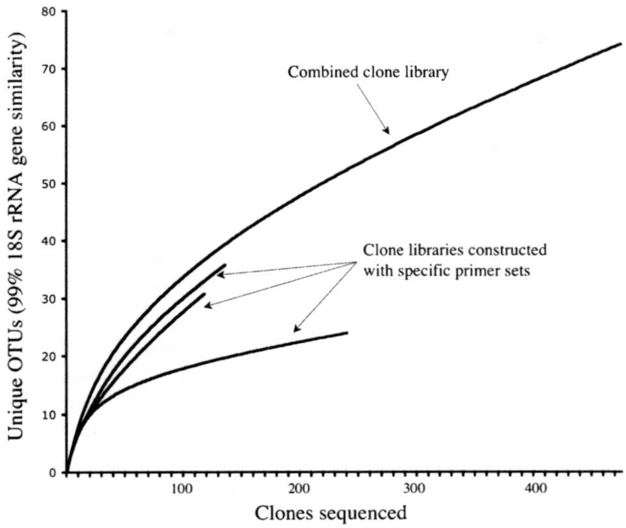

Fig. 3 Accumulation of unique phylotypes as a function of the number of clones sequenced from 18S rRNA gene clone libraries (Bunge et al. unpublished). The sequence data are from Stoeck et al. (2006). Three clone libraries were constructed using different eukaryotic primer sets. The DNA source was a single 3-l sample from anoxic water column in Cariaco Basin, off the coast of Venezuela in the Caribbean

Several specific groups appear particularly well represented in these surveys: Alveolate Group I and ciliates (López-García et al. 2003; Stoeck and Epstein 2003; Stoeck et al. 2003, 2006; Luo et al. 2005; Behnke et al. 2006; Zuendorf et al. 2006); Alveolate Group 3 (Stoeck et al. 2003); Euglenozoa, including Diplonemida (López-García et al. 2003; Stoeck et al. 2003, 2006), Cercozoa (Luo et al. 2005; Stoeck and Epstein 2003; Takishita et al. 2005), and Jakobida (Behnke et al. 2006). Perhaps more interestingly, several studies reported 18S rRNA gene sequences that appeared at the base of several important clades, such as Cercozoa and/or cercomonads (Dawson and Pace 2002; Stoeck and Epstein 2003), Labyrinthulida (Stoeck et al. 2003; Behnke et al. 2006), possibly even stramenopiles in general (Edgcomb et al. 2002; Stoeck and Epstein 2003); as well as several ciliate classes (Stoeck et al. 2003; Behnke et al. 2006).

Very exciting, and equally controversial, are the claims of discovery of rRNA gene signatures unrelated to known extant eukaryotes, as well of novel clades that appear emerge at the basal region of the 18S rRNA tree. Dawson and Pace (2002) proposed "… seven lineages … at the kingdom-level", and pointed out that several rRNA sequences obtained not only branched deeply on the eukaryotic tree, but also appeared rather slowly evolving. Over a dozen of deeply branching lineages unrelated to known eukaryotes were reported from hydrothermal vent and shallow water volcanic systems (Edgcomb et al. 2002; López-García et al. 2003; Takishita et al. 2005). The rRNA signatures were claimed to be found representing sister groups to Euglenozoa, Entamoeba, and diplonemids (Stoeck et al. 2003, 2006) as well as euglenids in general (López-García et al. 2003; Behnke et al. 2006; Stoeck et al. 2006), along with reports on novel clades of an apparently more recent origin (Stoeck et al. 2003; Zuendorf et al. 2006). There is little doubt that at least some of these findings will prove artifactual, and better taxon sampling will uncover the true position on the evolutionary tree of the sequences in question (see section *Power and pitfalls of molecular phylogenetic identification* above). Today, it remains plausible, however, that anoxic environments do indeed harbor an undescribed diversity of microbial eukaryotes of the highest taxonomic order, including perhaps "true" early branching lineages.

Freshwater systems

The rRNA surveys tend to focus on extreme environments, and the first freshwater system studied was the highly acidic (pH ~ 2.5) Rio Tinto (Amaral-Zettler et al. 2002). A surprisingly high diversity of all major eukaryotic groups was found, with green and red algae, diatoms and *Euglena* spp. as photosynthesizes, and fungi, nucleariid-related members, cercozoans, heterotrophic stramenopiles and Amoebozoa as heterotrophic constituents. Berney et al. (2004) studied the eukaryotic diversity in the river Seymaz (Switzerland), finding sequences in their libraries mostly related to the fungi, ciliates, stramenopiles and Cercozoa, and to a lesser extent, apicomplexans, metazoa, green algae, Amoebozoa, and a lineage of unclear affiliation previously identified in deep-sea plankton and anoxic settings. Freshwater anoxic or suboxic running waters, such as those of a sulfur spring (Zodletone) revealed sequences of stramenopiles, fungi, alveolates, and, at lower frequency, Cercozoa, jakobids and diplonemids (Luo et al. 2005). Several studies surveyed standing waters of ponds and lakes. Slapeta et al. (2005) compared the microbial eukaryotic diversity in sediment and plankton of two different size range (>5 μm and <5 μm) from two—oxic and suboxic—ponds from the same geographic area. Ciliates dominated libraries from the suboxic system, and included members of the ciliate classes Oligohymenophorea, Prostomatea, Plagiopylea, Phyllopharyngea, Litostomatea and Spirotrichea, whereas planktonic cryptophytes dominated those from the oxic system. In the oligotrophic lake George (USA), rRNA surveys detected an abundance of stramenopiles, cryptomonads and alveolates. Interestingly, several rRNA gene sequence clusters appeared specific to freshwater environments, while others contained signatures of marine species indicating a wider distribution (Richards et al. 2005). A more recent study of the picoplankton of the meromictic lake Pavin (France) also suggested the presence of freshwater-specific clades, with alveolates, fungi and cercozoans being the most represented taxa in clone libraries (Lefèvre et al. 2007). Takishita et al. (2006) analyzed the anoxic sediment from a coastal meromictic lake in Japan, detecting phylotypes known from other anoxic environments, i.e. a variety of cercozoans and several sequences related to parasitic lineages. These data are still fragmentary, preventing a thorough comparison and ecological analyses. In this regard, noteworthy is the first comparative study of eukaryotic diversity in the picoplankton of three lakes (oligotrophic, oligomesotrophic and eutrophic) and showing differences between them as a function of their trophic status (Lefranc et al. 2005).

Soils

Despite the recognized richness of microbial soil communities, and in contrast to aquatic systems and their associated sediments, the diversity of microbial eukaryotes in soils has been rarely approached by molecular methods. This may be partly explained by the technical difficulty imposed by the overwhelming presence of fungal hyphae, and abundance of animals masking unicellular eukaryotes in clone libraries. In fact, molecular surveys targeting specifically fungal communities show that fungi are extremely diverse in soils, and particularly forest soils (Anderson and Cairney 2004; O'Brien et al. 2005). Nonetheless, several recent molecular analyses show the feasibility of group-specific primer approach also in studying the rest of the eukaryotic microbial community. For example, Moon-van der Staay and colleagues amplified the 18S rRNA genes from a wide variety of eukaryotic phyla from agricultural soils, including fungi, cercozoans, green algae, ciliates, and a large variety of amoeboid protists including heteroloboseans and amoebozoans (Moon-van der Staay et al. 2006). In spite of this progress, soils remain among the most poorly studies habitat of unicellular eukaryotes.

The unidentified diversity of protists

If there is anything that the molecular approaches to microbial diversity showed with certainty is that the earlier notion of protozoologists having access to "... a full deck, or at least as full as we could" (Baldauf 2003) might have been rather naive. However, attaching specific numbers to this qualitative observation has proved challenging.

Estimates of the total number of protistan species are not based on statistical theory and are emphatically speculative. This is one reason why they vary so widely: while some argue that the global richness of free-living protozoa is less than 20,000 species (Finlay and Fenchel 1999), others maintain that a single phylum of ciliates comprises more than 30,000 species (Foissner 1999). There are multiple reasons behind this disagreement. One of them is the well known problem with species definition in microorganisms. This problem goes outside of the scope of this paper, but we note that it can be pragmatically resolved by considering rRNA gene sequence-based operational taxonomic units (OTUs). Of relevance here is the universal observation that the observed OTU frequency distribution is characterized by a large number of OTUs registered only once. Accumulation curves in molecular surveys are far from saturation (Fig. 3). This indicates that the complete inventory is a daunting task, and that the protistan richness even in relatively simple communities can be only estimated.

There are two principal groups of statistical approaches for estimating how many OTUs have been missed by a survey. Both have been used in macro- and prokaryotic biology, but it was not until 2004 that protozoologists began their adaptation to the needs of protistan biodiversity research. Nonparametric methods typically employ nonparametric coverage based estimators developed by Anne Chao, such as S_{Chao1}, ACE and ACE1 (Chao 1984, 2005). Massana et al. (2004a, b) employed S_{Chao1} and estimated the total eukaryotic (including metazoan) richness in four 10-l samples of surface waters in NW Mediterranean as seasonally varying between 72 and 171 OTUs per sample (OTUs defined as RFLP patterns with the 18SrRNA gene sequence similarity within phylotype ranging from 96.3% to 99.9%). Countway et al. (2005) used a similar approach and applied it to a collection of eight 4-l surface seawater samples from NW Atlantic and predicted the total eukaryotic (including metazoan) richness as 162–282 OTUs (defined as groups of sequences sharing over 95% 18S rRNA gene similarity). It is not clear, however, if the use of nonparametric estimators was optimal in these cases. These estimators are known to perform well under conditions of high coverage of the total richness by the empirical data (Chao and Bunge 2002). This condition is almost never met in microbial diversity studies (e.g., the number of OTUs registered only once is very high). Under these circumstances, the coverage-based estimators are likely to underestimate the total richness, and do so by an unknown factor. A notable exception may be a recent study of soil ciliate diversity by Chao et al. (2006). The painstakingly collected empirical dataset reported there provided an unusually high coverage of the total ciliate diversity, possibly justifying the application of nonparametric estimators.

The second group of methods employs parametric distribution to describe the frequency distribution of detected OTUs, and project the distribution so as to estimate how many OTUs must have been missed (Bunge and Fitzpatrick 1993). The application of these methods in biodiversity (including microbial diversity) research typically does not include estimation of model parameters using maximum likelihood, goodness-of-fit assessment, or correct maximum likelihood standard errors (Hong et al. 2006). It is also common to assume that the frequency distribution of the detected OTUs follows the log-normal model, which has little justification in microbial diversity research (i.e., even if frequency distribution of species in nature is lognormal, the same is not necessarily true of that of PCR products, or clones in clone libraries). In light of that, we developed an empirical approach (Hong et al. 2006), informed by modern statistical theory, which makes no a priori assumptions on the nature of OTU frequency distribution. The approach is based on systematic application of several parametric models to the datasets, choosing the one that fits the data the best and gives a biologically meaningful standard error. Jeon et al. (2006) used this approach to estimate protistan richness in a single 3-l sample from anoxic waters of the Cariaco Basin off the coast of Venezuela, and predicted the total richness of the sample to be 398 ± 156 OTUs (defined as clusters sharing over 99% of 18S rRNA gene sequence similarity). Behnke et al. (2006) applied the same approach to the 18S rRNA gene survey data on stratified water column of a Norvegian fiord, and estimated the total protistan richness in habitats with different oxygen and sulfide regimes: 64 ± 15 OTUs in sulfide-free layer, 147 ± 46 OTUs at the upper sulfide boundary, and 27 ± 8 OTUs in the highly sulfidic layer.

The amount of information on the pool of species "missed" by standard clone libraries is thus very limited, and this prevents a thorough analysis of protistan richness at any spatial scale. As a result, we do not really know the total number of protistan species per sample, habitat, environment, etc. The few available statistical estimates, and qualitative picture of protistan diversity that transpires from the collection curves published, provide only a suggestive view of this diversity. A recent study (Stoeck et al. 2007a) compiled and re-analyzed via a collection of parametric and nonparametric approaches the data for marine sediments and marine anoxic environments (Dawson and Pace 2002; Edgcomb et al. 2002; López-García et al. 2003; Stoeck and Epstein 2003, Stoeck et al. 2003, 2006). An emerging pattern is that samples from more extreme environments (High Arctic and hydrothermal vents) contain more to many more phylotypes at the approximately the species level (OTUs formed at 97–99% rRNA gene sequence identity) than temperate zone tidal flats (hundreds vs dozens in one to several grams of sediment). Most of these phylotypes seem unique and rarely exhibit an exact match to organisms reported from elsewhere. If even small volume individual samples show degrees of uniqueness, the grand total of global protistan richness must be very large indeed. Presently, it would be imprudent to attach a number to this richness, but it does seem likely that genetic diversity exceeds projections based on morphological criteria—compare for example (Fenchel et al. 1990, 1995 vis a vis Zuendorf et al. 2006).

Additionally, recent observations show that even organisms that are essentially identical morphologically may be phylogenetically very different (Boenigk et al. 2005; Katz et al. 2005, 2006; Rodriguez et al. 2005; Slapeta et al. 2006a, b) and/or ecologically quite distinct (see for example Nanney et al. 1998; Foissner 1999; Coleman 2002; Lowe et al. 2005). The number of these so-called 'cryptic species' is completely unknown. Collectively, this makes it reasonable to hypothesize that pre-molecular inventories significantly underestimated global protistan richness. This might have produced a skewed picture of protists as a group of cosmopolitan forms of limited diversity ("everything is every where", Beijerinck 1913). The recent data lend a tentative support to the opposing view on protists as a remarkable world of diverse species with biogeographies (Baldauf 2003; Foissner 1999, 2006).

References

Amaral Zettler LA, Gomez F, Zettler E et al (2002) Microbiology: eukaryotic diversity in Spain's River of Fire. Nature 417:137

Anderson IC, Cairney JW (2004) Diversity and ecology of soil fungal communities: increased understanding through the application of molecular techniques. Environ Microbiol 6:769–779

Atkins MS, Teske AP, Anderson OR (2000) A survey of flagellate diversity at four deep-sea hydrothermal vents in the Eastern Pacific Ocean using structural and molecular approaches. J Eukaryot Microbiol 47:400–411

Baldauf SL (2003) The deep roots of eukaryotes. Science 300:1703–1706

Bass D, Cavalier-Smith T (2004) Phylum-specific environmental DNA analysis reveals remarkably high global biodiversity of Cercozoa (Protozoa). Int J Syst Evol Microbiol 54:2393–2404

Behnke A, Bunge J, Barger K et al (2006) Microeukaryote community patterns along an O2/H2S gradient in a supersulfidic anoxic fjord (Framvaren, Norway). Appl Environ Microbiol 72:3626–3636

Beijerinck MW (1913) De infusies en de ontdekking der backteriën. Müller, Amsterdam, the Netherlands

Berney C, Fahrni J, Pawlowski J (2004) How many novel eukaryotic 'kingdoms'? Pitfalls and limitations of environmental DNA surveys. BMC Biol 2:13

Boenigk J, Pfandl K, Stadler P et al (2005) High diversity of the 'Spumella-like' flagellates: an investigation based on the SSU rRNA gene sequences of isolates from habitats located in six different geographic regions. Environ Microbiol 7:685–697

Brinkmann H, van der Giezen M, Zhou Y et al (2005) An empirical assessment of long-branch attraction artefacts in deep eukaryotic phylogenomics. Syst Biol 54:743–757

Brocks JJ, Logan GA, Buick R et al (1999) Archaean molecular fossils and the early rise of eukaryotes. Science 285:1025–1027

Bunge J, Fitzpatrick M (1993) Estimating the number of species—a review. J Am Stat Assoc 88:364–373

Cavalier-Smith T (2004) Only six kingdoms of life. Proc R Soc Lond B Biol Sci 271:1251–1262

Chao A (1984) Non-parametric estimation of the number of classes in a population. Scand J Statist 11:265–270

Chao A (2005) Species richness estimation. In: Balakrishnan C, Read B, Vidakovic B (eds) Encyclopedia of statistical sciences. Wiley, New York, pp 7907–7916

Chao A, Bunge J (2002) Estimating the number of species in a stochastic abundance model. Biometrics 58:531–539

Chao A, Li PC, Agatha S et al (2006) A statistical approach to estimate soil ciliate diversity and distribution based on data from five continents. Oikos 114:479–493

Coleman AW (2002) Microbial eukaryote species. Science 297:337; author reply 337

Countway PD, Gast RJ, Savai P et al (2005) Protistan diversity estimates based on 18S rDNA from seawater incubations in the Western North Atlantic. J Eukaryot Microbiol 52:95–106

Courties C, Vaquer A, Troussellier M et al (1994) Smallest eukaryotic organisms. Nature 370:255

Dawson SC, Pace NR (2002) Novel kingdom-level eukaryotic diversity in anoxic environments. Proc Natl Acad Sci USA 99:8324–8329

Díez B, Pedrós-Alió C, Massana R (2001) Study of genetic diversity of eukaryotic picoplankton in different oceanic regions by small-subunit rRNA gene cloning and sequencing. Appl Environ Microbiol 67:2932–2941

Dolven JK, Lindqvist C, Albert VA et al (2007) Molecular diversity of alveolates associated with Neritic North Atlantic radiolarians. Protist 158:65–76

Edgcomb VP, Kysela DT, Teske A et al (2002) Benthic eukaryotic diversity in the Guaymas Basin hydrothermal vent environment. Proc Natl Acad Sci USA 99:7658–7662

Fenchel T, Finlay BJ (1995) Ecology and evolution in anoxic worlds. Oxford Univ Press, Oxford

Fenchel T, Kristensen LD, Rasmussen L (1990) Water column anoxia: vertical zonation of planctonic protozoa. Mar Ecol Prog Ser 62:1–10

Fenchel T, Bernard C, Esteban G et al (1995) Microbial diversity and activity in a Danish fjord with anoxic deep water. Ophelia 43:45–100

Finlay BJ (2002) Global dispersal of free-living microbial eukaryote species. Science 296:1061–1063

Finlay BJ, Fenchel T (1999) Divergent perspectives on protist species richness. Protist 150:229–233

Foissner W (1999) Protist diversity: estimates of the near-imponderable. Protist 150:363–368

Foissner W (2006) Biogeography and dispersal of micro-organisms: a review emphasizing protists. Acta Protozool 45:111–136

Guillou L, Chretiennot-Dinet MJ, Medlin LK et al (1999) *Bolidomonas*: a new genus with two species belonging to a new algal class, the Bolidophyceae (Heterokonta). J Phycol 35:368–381

Guillou L, Eikrem W, Chretiennot-Dinet MJ et al (2004) Diversity of picoplanktonic prasinophytes assessed by direct nuclear SSU rDNA sequencing of environmental samples and novel isolates retrieved from oceanic and coastal marine ecosystems. Protist 155:193–214

Gunderson JH, Goss SH, Coats DW (1999) The phylogenetic position of *Amoebophrya* sp. infecting *Gymnodinium sanguineum*. J Eukaryot Microbiol 46:194–197

Hong SH, Bunge J, Jeon SO, Epstein SS (2006) Predicting microbial species richness. Proc Natl Acad Sci USA 103:117–122

Janson S, Gisselson LA, Salomon PS et al (2000) Evidence for multiple species within the endoparasitic dinoflagellate *Amoebophrya ceratii* as based on 18S rRNA gene-sequence analysis. Parasitol Res 86:929–933

Jeon SO, Bunge J, Stoeck T et al (2006) Synthetic statistical approach reveals a high degree of richness of microbial eukaryotes in an anoxic water column. Appl Environ Microbiol 72:6578–6583

Johnson MD, Tengs T, Oldach DW et al (2004) Highly divergent SSU rRNA genes found in the marine ciliates *Myrionecta rubra* and *Mesodinium pulex*. Protist 155:347–359

Katz LA, McManus GB, Snoeyenbos-West OLO et al (2005) Reframing the 'Everything is everywhere' debate: evidence for high gene flow and diversity in ciliate morphospecies. Aquat Microbiol Ecol 41: 55–65

Katz LA, Snoeyenbos-West O, Doerder FP (2006) Patterns of protein evolution in *Tetrahymena thermophila*: implications for estimates of effective population size. Mol Biol Evol 23:608–614

Kolodziej K, Stoeck T (2007) Cellular identification of a novel uncultured marine stramenopile (MAST-12 Clade) small-subunit rRNA gene sequence from a norwegian estuary by use of fluorescence in situ hybridization-scanning electron microscopy. Appl Environ Microbiol 73:2718–2726

Kunitomo Y, Sarashina I, Iijima M et al (2006) Molecular phylogeny of acantharian and polycystine radiolarians based on ribosomal DNA sequences, and some comparisons with data from the fossil record. Eur J Protistol 42:143–153

Lefèvre E, Bardot C, Noël C et al (2007) Unveiling fungal zooflagellates as members of freshwater picoeukaryotes: evidence from a molecular diversity study in a deep meromictic lake. Environ Microbiol 9:61–71

Lefranc M, Thenot A, Lepere C et al (2005) Genetic diversity of small eukaryotes in lakes differing by their trophic status. Appl Environ Microbiol 71:5935–5942

López-García P, Rodríguez-Valera F, Pedrós-Alió C et al (2001) Unexpected diversity of small eukaryotes in deep-sea Antarctic plankton. Nature 409:603–607

López-García P, Rodríguez-Valera F, Moreira D (2002) Towards the monophyly of Haeckel's Radiolaria: 18S rRNA environmental data support the sisterhood of Polycystinea and Acantharea. Mol Biol Evol 19:118–121

López-García P, Philippe H, Gaill F et al (2003) Autochthonous eukaryotic diversity in hydrothermal sediment and experimental micro-colonizers at the Mid-Atlantic Ridge. Proc Natl Acad Sci USA 100:697–702

López-García P, Vereshchaka A, Moreira D (2007) Eukaryotic diversity associated with carbonates and fluid-seawater interface in Lost City hydrothermal field. Environ Microbiol 9:546–554

Lovejoy C, Massana R, Pedros-Alio C (2006) Diversity and distribution of marine microbial eukaryotes in the Arctic Ocean and adjacent seas. Appl Environ Microbiol 72:3085–3095

Lowe CD, Day A, Kemp SJ et al (2005) There are high levels of functional and genetic diversity in *Oxyrrhis marina*. J Eukaryot Microbiol 52:250–257

Luo Q, Krumholz LR, Najar FZ et al (2005) Diversity of the microeukaryotic community in sulfide-rich Zodletone Spring (Oklahoma). Appl Environ Microbiol 71:6175–6184

Maidak BL, Cole JR, Lilburn TG et al (2001) The RDP-II (Ribosomal database project). Nucleic Acids Res 29:173–174

Massana R, Guillou L, Diez B et al (2002) Unveiling the organisms behind novel eukaryotic ribosomal DNA sequences from the ocean. Appl Environ Microbiol 68:4554–4558

Massana R, Balagué V, Guillou L et al (2004a) Picoeukaryotic diversity in an oligotrophic coastal site studied by molecular and culturing approaches. FEMS Microbiol Ecol 50:231–243

Massana R, Castresana J, Balague V et al (2004b) Phylogenetic and ecological analysis of novel marine stramenopiles. Appl Environ Microbiol 70:3528–3534

Massana R, Terrado R, Forn I et al (2006) Distribution and abundance of uncultured heterotrophic flagellates in the world oceans. Environ Microbiol 8:1515–1522

Medlin LK, Metfies K, Mehl H et al (2006) Picoeukaryotic plankton diversity at the Helgoland time series site as assessed by three molecular methods. Microb Ecol 52:53–71

Moon-van der Staay A-Y, van der Staay GWM, Guillou L et al (2000) Abundance and diversity of prymnesiophytes in the picoplankton community from the equatorial Pacific Ocean inferred from 18S rDNA sequences. Limnol Oceanogr 45:98–109

Moon-van der Staay SY, De Wachter R, Vaulot D (2001) Oceanic 18S rDNA sequences from picoplankton reveal unsuspected eukaryotic diversity. Nature 409:607–610

Moon-van der Staay SY, Tzeneva VA, van der Staay GW et al (2006) Eukaryotic diversity in historical soil samples. FEMS Microbiol Ecol 57:420–428

Moreira D, López-García P (2002) Molecular ecology of microbial eukaryotes unveils a hidden world. Trends Microbiol 10:31–38

Moreira D, López-García P (2003) Are hydrothermal vents oases for parasitic protists? Trends Parasitol 19:556–558

Moreira D, von der Heyden S, Bass D et al (2007) Global eukaryote phylogeny: Combined small- and large-subunit ribosomal DNA trees support monophyly of Rhizaria, Retaria and Excavata. Mol Phylogenet Evol: doi:10.1016/j.ympev.2006.11.001

Nanney DL, Park C, Preparata R et al (1998) Comparison of sequence differences in a variable 23S rRNA domain among sets of cryptic species of ciliated protozoa. J Eukaryot Microbiol 45:91–100

Nikolaev SI, Berney C, Fahrni JF et al (2004) The twilight of Heliozoa and rise of Rhizaria, an emerging supergroup of amoeboid eukaryotes. Proc Natl Acad Sci USA 101:8066–8071

O'Brien HE, Parrent JL, Jackson JA et al (2005) Fungal community analysis by large-scale sequencing of environmental samples. Appl Environ Microbiol 71:5544–5550

O'Brien EA, Koski LB, Zhang Y et al (2007) TBestDB: a taxonomically broad database of expressed sequence tags (ESTs). Nucleic Acids Res 35:D445–D451

Okamoto N, Inouye I (2005) The katablepharids are a distant sister group of the Cryptophyta: a proposal for Katablepharidophyta divisio nova/Kathablepharida phylum novum based on SSU rDNA and beta-tubulin phylogeny. Protist 156:163–179

Olsen GJ, Lane DJ, Giovannoni SJ et al (1986) Microbial ecology and evolution: a ribosomal RNA approach. Annu Revue Microbiol 40:337–365

Pace NR (1997) A molecular view of microbial diversity and the biosphere. Science 276:734–740

Philippe H, Lopez P, Brinkmann H et al (2000) Early-branching or fast-evolving eukaryotes? An answer based on slowly evolving positions. Proc R Soc Lond B Biol Sci 267:1213–1221

Reysenbach AL, Cady SL (2001) Microbiology of ancient and modern hydrothermal systems. Trends Microbiol 9:79–86

Richards TA, Bass D (2005) Molecular screening of free-living microbial eukaryotes: diversity and distribution using a meta-analysis. Curr Opin Microbiol 8:240–252

Richards TA, Vepritskiy AA, Gouliamova DE, Nierzwicki-Bauer SA (2005) The molecular diversity of freshwater picoeukaryotes from an oligotrophic lake reveals diverse, distinctive and globally dispersed lineages. Environ Microbiol 7:1413–1425

Rodriguez F, Derelle E, Guillou L et al (2005) Ecotype diversity in the marine picoeukaryote *Ostreococcus* (Chlorophyta, Prasinophyceae). Environ Microbiol 7:853–859

Savin MC, Martin JL, LeGresley M et al (2004) Plankton diversity in the Bay of Fundy as measured by morphological and molecular methods. Microb Ecol 48:51–65

Slapeta J, López-García P, Moreira D (2006a) Global dispersal and ancient cryptic species in the smallest marine eukaryotes. Mol Biol Evol 23:23–29

Slapeta J, López-García P, Moreira D (2006b) Present status of the molecular ecology of kathablepharids. Protist 157:7–11

Slapeta J, Moreira D, López-García P (2005) The extent of protist diversity: insights from molecular ecology of freshwater eukaryotes. Proc Biol Sci 272:2073–2081

Sogin ML, Gunderson JH, Elwood HJ et al (1989) Phylogenetic meaning of the kingdom concept: an unusual ribosomal RNA from Giardia lamblia. Science 243:75–77

Stoeck T, Epstein S (2003) Novel eukaryotic lineages inferred from small-subunit rRNA analyses of oxygen-depleted marine environments. Appl Environ Microbiol 69:2657–2663

Stoeck T, Taylor GT, Epstein SS (2003) Novel eukaryotes from the permanently anoxic Cariaco Basin (Caribbean Sea). Appl Environ Microbiol 69:5656–5663

Stoeck T, Hayward B, Taylor GT et al (2006) A multiple PCR-primer approach to access the microeukaryotic diversity in environmental samples. Protist 157:31–43

Stoeck T, Kasper J, Bunge J, Leslin C, Ilyin V, Epstein S (2007a) Protistan diversity in the arctic: a case of paleoclimate shaping modern biodiversity? PLoS ONE 2:e728

Stoeck T, Zuendorf A, Breiner HW, Behnke A (2007b) A molecular approach to identify active microbes in environmental eukaryote clone libraries. Microb Ecol 53:328–339

Takishita K, Miyake H, Kawato M et al (2005) Genetic diversity of microbial eukaryotes in anoxic sediment around fumaroles on a submarine caldera floor based on the small-subunit rDNA phylogeny. Extremophiles 9:185–196

Takishita K, Tsuchiya M, Kawato M et al (2006) Genetic diversity of microbial eukaryotes in anoxic sediment of the saline meromictic lake Namako-Ike (Japan): on the detection of anaerobic or anoxic-tolerant lineages of eukaryotes. Protist 2006 Sep 1; [Epub ahead of print]

Woese CR (1987) Bacterial evolution. Microbiol Rev 51:221–271

Worden AZ (2006) Picoeukaryote diversity in coastal waters of the Pacific Ocean. Aquat Microbiol Ecol 43:165–175

Worden AZ, Nolan JK, Palenik B (2004) Assessing the dynamics and ecology of marine picophytoplankton: The importance of the eukaryotic component. Limnol Oceanogr 49:168–179

Yuan J, Chen MY, Shao P et al (2004) Genetic diversity of small eukaryotes from the coastal waters of Nansha Islands in China. FEMS Microbiol Lett 240:163–170

Zuendorf A, Bunge J, Behnke A et al (2006) Diversity estimates of microeukaryotes below the chemocline of the anoxic Mariager Fjord, Denmark. FEMS Microbiol Ecol 58:476–491

Diversity, dispersal and biogeography of bryophytes (mosses)

Jan-Peter Frahm

Originally published in the journal Biodiversity and Conservation, Volume 17, No 2, 277–284.
DOI: 10.1007/s10531-007-9251-x

Abstract Bryophytes disperse by small unicellular spores between 7 μm und usually less than 100 μm. A large percentage of species is sterile and propagates vegetatively either by special brood bodies or fragments of whole plants. It is shown that there is no difference in the effectiveness between generative and vegetative propagation. Size and weight of the diaspores suggest that both must easily be dispersed and the species must therefore have wide ranges. This does, however, not result in ubiquitous occurrence. This is only true for part of the species. Many, even sterile species show wide transcontinental ranges. On the other hand, there are many examples of limited to very limited distribution in spite of a rich production of diaspores. These are explained by narrow ecological niches, age of taxa, local extinction or historical events such as ice ages. Conspicuously, species can loose the ability for dispersal for unknown (perhaps genetic) reasons, which may ultimately lead to extinction.

Keywords Bryophytes · Dispersal · Vegetative propagation · Spore dispersal · Endemism · Relics · Ubiquitism

Introduction

Bryophytes are the second largest group of green land plants and the only haploid land plants. They are presumably polyphyletic and do no exist as natural group, similarly so as "algae", but are composed by mosses (Bryophyta), liverworts (Marchantiophyta) and hornworts (Anthocerotophyta). Although very different in structure, they correspond, however, in their life cycle. Like ferns and allies, they are spore plants. In contrast to spermatophytes, they distribute a single cell with a haploid genome, but usually in masses. Seed plants, in contrast, distribute a diploid embryo in a seed, variously packed in a fruit. Spermatophytes have thus

Special Issue: Protist diversity and geographic distribution. Guest editor: W. Foissner.

J.-P. Frahm (✉)
Nees-Institut für Biodiversität der Pflanzen, Rheinische Friedrich-Wilhelms-Universität Bonn, Meckenheimer Allee 170, 53115 Bonn, Germany
e-mail: frahm@uni-bonn.de

taken an alternative evolutionary pathway, which was by far more successful than the traditional spore dispersal (Frahm 2001).

Bryophytes can therefore be used as a model organism for spore plants. Bryophyte spores are produced in sporangia ("capsules"). The spores are between 7 and 100 µm, max. 200 µm large, on average 10–20 µm. They are released in quantities between 4 and 5 millions per sporangium, usually several hundred thousands. The smaller spores (<25 µm) can easily be lifted into the atmosphere by warm air and distributed by air currents globally. There is, however, no doubt that also the larger spores can be transported over large distances by stronger winds or attached to animals. Spores are thus ubiquitous, they are also dispersed into regions in which the species cannot grow. This is shown by large sand or clay pits in central Europe, in which occasionally arctic species are found. An analysis of spores in rain water of a meteorological station in Finland revealed that there were spores of *Aloina brevirostris*, a species not known from Finland (Petterson 1940). The overall presence of spores is one of the reasons for the use of bryophytes as indicators for climate changes: spores of southern species are dispersed across the former borders of their ranges and can react immediately upon a warmer climate (Frahm and Klaus 2001). In addition, these spores can keep their vitality for more than 100 years. As an example, several rare moss species usually typical for xerothermic vegetation were found on loess soil along the edge of a forest south of Bonn. The rarity of these species, of which some have never been found in this region or not for 100 years, made it unlikely that they arrived by recent spore dispersal. It turned out that there were vineyards in this region before 1880 and that they were given up and forested as a consequence of the phylloxera disease (Frahm 2006).

The high potential for generative and vegetative propagation of bryophytes may lead to the assumption that all bryophytes have large ranges. This is not the case. In spite of the facility to spread very easily, many bryophyte species have very limited ranges.

Diversity

Bryophytes are a group of about 15,000 species. Traditionally, the species number was indicated in textbooks with 25,000 species, which seems to be a count of all species enumerated in the bryophyte volumes of Engler-Prantĺs "Natürliche Pflanzenfamilien". Recent worldwide monographs and revisions have distinctly reduced the number of species. Reason is especially that many species have large, for instance pantropical ranges, and have been described repeatedly from different parts of the tropics under different names. In the nineteenth century, the religious convictions of some authors did not allow the idea that a species occurs on different continents because they were assumed to be independently created by god. In contrast to other organisms, the number of bryophytes is decreasing (also and especially in the tropics) because more species are reduced into synonymy then described as new. For this reason, bryophytes provide a good base for calculation of biodiversity because they give approximate species numbers.

Dispersal

Vegetative dispersal

Two thirds of the mosses are dioecious. A fertilisation of the female egg is only possible if male plants are in close contact. The spermatozoids can swim only over a short distance

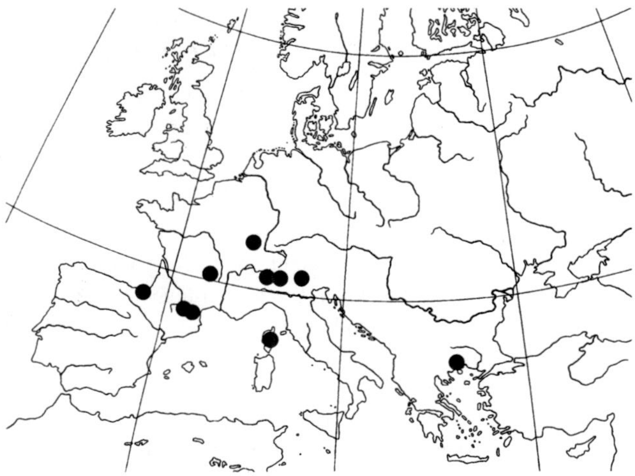

Fig. 1 Distribution of the moss *Campylopus oerstedianus* in Europe. The species is only known in sterile condition and therefore the range may indicate a relictual status of the populations. However, molecular data indicate a closer genetic relationship of the populations in the Pyrenees, the Massiv Central and the Vosges in France, suggesting a recent dispersal during the Pleistocene

and rain drops with spermatozoids can hardly reach more than one meter. If a male spore lands somewhere, it establishes a male clone, which will not have any chance for sexual propagation until it gets in contact with a female clone. To cover the time span in between sexual propagation, both sexes propagate asexually. The need for an asexual strategy of propagation has led to an enormous variety of different structures for this purpose amongst bryophytes. Basically, every part of a bryophyte plant can reproduce vegetatively (rhizoids, leaves, stems).

However, only part of the species produce spores. Many species are only known in sterile condition. This depends often on the lack of one sex. There is, however, hardly any difference in the size of ranges between sterile and fertile species. As an example, the moss *Tortula pagorum* is dioiceous. It exists only in female plants in Europe, while male plants occur in North America. It propagates by terminal clusters of brood bodies with a length of about 100 μm. Nevertheless, it has a wide range on both continents.

Bryophytes are able to propagate by a large variety of vegetative ways such as rhizoidal gemmae, axillary gemmae, brood bodies, detaching leaves or buds, leaf fragments etc. Although these organs for vegetative propagation are much larger than spores, they have similar effects for the dispersal over large distances. Recent molecular studies on disjunct populations of sterile mosses revealed that dispersal is possible even without any specialized means of vegetative propagation. The moss *Campylopus oerstedianus* is worldwide known only in sterile condition. It is distributed from Costa Rica in an arc over Jamaica, Georgia to southern Europe (Pyrenees, southern Alps, Chalkidike). It is found in addition in the Massif Central and the Vosges in France (Fig. 1). Determinations of the genetic distances between the disjunct populations in Europe show that the populations in the Massif Central and the Vosges are derived from that in the Pyrenees (Sabovljevic and Frahm in press). Therefore, dispersal over a distance of several 100 km (in the direction of the prevailing winds) is possible even for a sterile species. The effectiveness of both vegetative and generative propagation is best demonstrated by the re-colonization of Europe and North America after the glaciations in the Pleistocene from refugia in southern latitudes.

Dispersal is not only possible by wind but also by animals. The aquatic liverwort *Ricciocarpos natans* has an almost cosmopolitan range. It is presumably dispersed by water fowl.

Large ranges: relics vs. long distance dispersal

Transoceanic disjunctions of flowering plants are usually only found at genus level. In contrast, this distribution pattern is frequent in bryophytes. This includes not only Laurasian and Gondwanan ranges, but also Pangaean ranges. These distribution patterns can be the result of long distance dispersal but also of relic origin. Species equally found in North America and Europe are either dispersed (presumably from west to east with the prevailing wind system) but can also be remnants of former continuous ranges before the breakup of the Laurasian continent. The question cannot be fully clarified at present. Records of sterile bryophyte species occurring in North America and Europe in different sexes in microhabitats such as rock fissures support the relic theory. These species must have an age of 50 mio years and have not altered morphologically or anatomically. Although this might be difficult to understand by phanerogamists (since species of flowering plants are distinctly younger), this hypothesis is supported by fossil records of bryophytes in Eocene amber, which still exist and therefore must have an age of 45 mio years. On the other hand, bryophyte species common in North America but rarely found in Europe (also in different habitats) may indicate long distance dispersal. Thus the answer on this question may not be either—or but as well as.

Tropical African–South American disjunctions would be even more difficult to explain, since the break up of both continents started 100 mio years b.p. and such an age of a bryophyte species would be more difficult to believe. Even more problematic to understand are Pangaean ranges, species which are found in North America as well as in Europe, South America and Africa. In this case the species must have an age of 180 mio years and more. This hypothesis was favoured by Frey (1990). This question can be solved today by molecular studies comparing the genetic distances between the disjunct populations but have not yet been performed on such a scale.

As spore plants, bryophytes show a much smaller extent of endemism than flowering plants. Endemism on the species level is found in North America in 16% (liverworts) and 18% (mosses) of species, respectively. This is especially true for islands: in New Zealand, 86% of the flowering plants are endemic but only 28% of the mosses (van Zanten and Pócs 1981).

The dispersal strategy by small uni- or multicellular units should cause that all bryophytes have large ranges because the diaspores can easily be dispersed over long distances. In fact, however, only few species are really ubiquitous, such as *Funaria hygrometrica* and *Bryum argenteum*. They are like weeds and seem to have been distributed by man. Naturally wide ranges are found in Laurasian and Gondwanan species. A reason seems to be that the Inner Tropical Convergence (ITC) forms a barrier for spore dispersal across the equator and limits gene exchange across the equator. In the Americas, only 65 bryophyte species have a bipolar range (Ochyra 1992). They have disjunct ranges between Alaska and Chile and seem to be dispersed by migrating birds. Within the northern and southern hemisphere, many species have large transcontinental ranges: 65% of all mosses in North America are also found in Europe (Frahm and Vitt 1993). This is the reason for the lack of neophytic bryophytes from North America in Europe, as is the case in flowering plants: all species which could be dispersed by man from North America are already present in Europe. In Europe, all neophytic bryophytes come from the southern hemisphere. Species such as

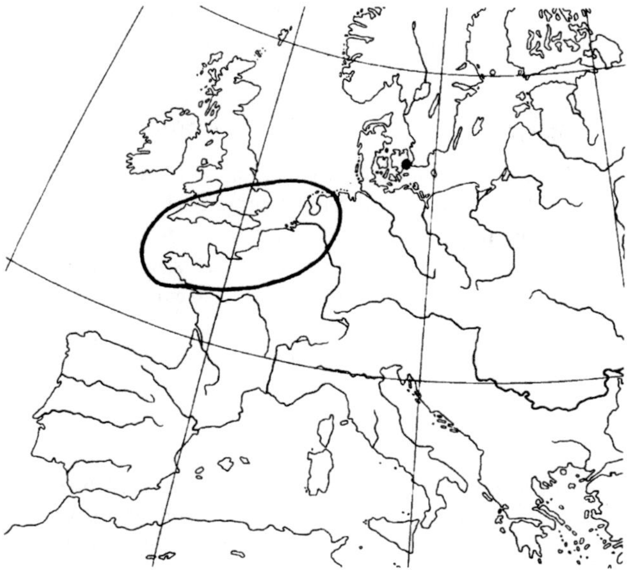

Fig. 2 Distribution of the moss *Leptodontium gemmascens*. The species produces gemmae abundantly which allows an easy dispersal (as shown by a disjunct occurrence near Copenhagen), but is confined to a (for bryophytes) comparably very small range

Orthodontium lineare or *Campylopus introflexus* have then enlarged their secondary ranges in Europe and spread over large areas in a few decades. Transcontinental ranges can be explained by either long distance dispersal (=gene exchange) or disjunctions caused by continental drift (=relics). In fact, both arguments are supported by molecular studies: for example, *Campylopus introflexus*, a moss widespread in temperate regions of the southern hemisphere, is genetically homogenous. This species is fertile and has presumably gene exchange by the circumglobal air currents (Stech and Dohrmann 2004). Other disjunct species, such as *Lopidium concinnum* in New Zealand, Tasmania and Chile seems to be relics of a former continuous Gondwanan range (Frey et al. 1999).

Biogeography

The fact that many bryophyte species (as well as macrofungi and ferns, see Foissner 2006) have limited ranges in spite of their high potential of dispersal seems paradox. For example, the moss *Leptodontium gemmascens* produces abundantly gemmae that are 80–100 μm long. In spite of this, *L. gemmascens* is confined to southern England, northern France, Belgium, Luxemburg and neighbouring parts of Germany and the Netherlands plus a disjunct occurrence in Denmark (Fig. 2). Possible reasons for restricted ranges are discussed in the following paragraphs.

Endemism

Within Laurasia, endemism amongst bryophytes is found north of the Pleistocene ice shield. Species like *Bryobrittonia longipes* are exclusively found in North America, in northern Alaska, and in the ice free corridor east of the Rocky Mountains and the Kola

Peninsula in Northeast Europe. This species has apparently lost the ability to extend after glaciation, similarly to the Nunatak effect described from the flowering plants.

Habitat size and life form

In a study of rare mosses of Alberta (Canada), Vitt and Bellard (1997) concluded that rarity depends on life form, life strategy, and habitat preference. Long-lived perennials, pleurocarpous mosses forming large mats are more common since they are more competitive. Rare species occur in small niches, for instance, rock fissures in cliffs. The smaller the habitat, the rarer the species. The authors studied, however, local rarity. Rarity is here understood as overall rarity. Therefore many reasons for rarity cannot be applied (occurrence at the borders of the range, outside the main range).

Age of taxa

We have to face that species have a limited time for existence. This time is quite high in bryophytes, as shown by fossil evidences. A high percentage of mosses known from Baltic and Saxon amber is still existing. They are 45 mio years old. Nevertheless, some species approach extinction. The most common moss in the Eocene amber forest is *Hypnodontopsis mexicana* (Frahm 2005). It is presently known only from the type locality (collected in 1928 in Mexico) and a recent collection in Uganda.

Local extinction

Especially the Pleistocene ice ages have led to a considerable extinction of taxa. This is shown by the effect that species known as fossil from the Tertiary of Europe are found today on the Macaronesian Islands (*Echinodium* spp., *Andoa bertholetiana*) or in E-Asia (*Trachycystis* spp.). Some may have survived the climatic changes in special areas such as the southern Alps, where some species (e.g. *Braunia alopecura*) can be interpreted as Tertiary relics that did not expand into the Holocene.

Local, disjunct occurrence

Disjunct occurrences can be the result of either long distance dispersal or a relic. Today, molecular studies allow to solve this question by determination of genetic distances between the populations. In such studies it turned out that both possibilities are realized, even within the same species. For instance, *Isothecium holtii* occurs in the high oceanic parts of Western Europe, but it is also found in some sites of Central Europe. These populations could be the result of dispersal or relics of a former continuous range. A molecular study revealed that the populations in two valleys of the Harz Mountains are derived from Wales and Ireland, and are thus a result of long distance dispersal, whereas the population in the Rur valley of the Eifel Mountains is a relic of a former larger range, presumably the Atlantic climate period 8,000 years bp (Sabovljevic et al. 2005).

Highly specialized habitat preferences

Cosmopolitan species are usually ubiquitous, that means they have a wide ecological range and are not much specialized. Rarity may depend on rarely available habitats. Heavy metal bryophytes can serve as an example. Habitats (mine deposits) are only found very scattered. Nevertheless obligate heavy metal bryophyte species are found on many of them. The moss *Ditrichum plumbicola* is only known from Britain and Germany. It has no special means of vegetative propagation, but is known in Germany from six places which are in 50–100 km air distances. The occurrence of the moss *Rhynchostegium rotundifolium* is often associated with the surroundings of medieval ruins. It grows below the walls of castles, where the litter has accumulated. It could be that the species has a preference for habitats rich in nitrate or phosphate. Dung mosses (Splachnaceae) grow only on substrate highly enriched with nutrients, most of them solely on dung of animals. Their occurrence is restricted by the availability of such habitats.

Lack of one sex

There are cases of rare dioiceous species, which are present on a continent in only one sex. For example, only male plants of the liverwort *Plagiochila corniculata* are found in North America, while females are restricted to Europe (Schuster 1984). Sexual propagation is thus inhibited.

Lack of ability for long distance dispersal

Long distance dispersal happens at higher altitudes in the atmosphere. Diaspores must (a) be able to get into these altitudes (difficult for species from forest floor) and (b) able to survive the harsh conditions of high altitudes with strong UV radiation and temperatures much below the freezing point. In spectacular experiments, van Zanten (1978) exposed spores of mosses from New Zealand, which occur also in Chile, to UV radiation and low temperatures. Conspicuously, the spores of species were able to germinate after this treatment. The species endemic to New Zealand failed to germinate. Endemism is in this case a matter of lacking UV and frost tolerance of spores.

Bryophytes in the context of the current debate on protist distribution

Spore plants such as bryophytes should have an overall distribution or at least very large ranges. Vegetative propagation is not a limiting factor but has the same effects. It has been developed as an alternative to sexual reproduction, for which water is necessary. It can result in similarly large ranges. As already outlined by Foissner (2006), the small size of diaspores of bryophytes does not correlate with wide ranges, which concern only part of the bryophyte species. Many species have limited to very limited ranges, which seems to be a contradiction to the easy vegetative or generative propagation. Ubiquity is thus theoretically possible amongst bryophytes (concerning the means and effectiveness of dispersal), but realized only to a small extent. There are many examples of very small ranges or local endemism. This discrepancy can be explained by exogenous effects such as narrow ecological requirements (climatic, habitat). This explanation does not work for all cases. Molecular causes such as genetic aging leading to extinction may be one possible explanation. Limitations are not the diaspores (even sterile species can have wide ranges) but climatic or ecological boundaries. Mountains and oceans are not, boundaries. If a species has a limited range in spite of existing dispersal facilities, there seems to be an endogenous barrier for dispersal. The lacking ability for dispersal seems to be especially expressed in sterile species. As shown by the wide ranges of other sterile species, sterility itself cannot be the limiting factor. A factor could, however, be the clonal reproduction of bryophyte species, which results in a loss of ability for dispersal.

References

Frahm J-P (2001) Biologie der Moose. Spektrum Akad. Verlag, Heidelberg and Berlin

Frahm J-P (2005) The genus *Hypnodontopsis* (Bryopsida, Rhachitheciaceae) in Baltic and Saxon amber. Bryologist 108:228–235

Frahm J-P (2006) Eine hundertjährige Sporenbank?—Seltsames Vorkommen thermophiler Moose. Notul bryol rhen 1:1–3

Frahm J-P, Klaus D (2001) Bryophytes as indicators for past and present climate fluctuations. Lindbergia 26:97–104

Frahm J-P, Vitt DH (1993) A comparison of the mossfloras of Europe and North America. Nova Hedwigia 56:307–333

Foissner W (2006) Biogeography and dispersal of micro-organisms: a review emphasizing prostists. Acta Protozool 45:111–136

Frey W (1990) Geno-Elemente prä-angiospermen Ursprungs bei Bryophyten. Bot Jahrb Syst 111:433–456

Frey W, Stech M, Meißner K (1999) Chloroplast DNA-relationship in palaeoaustral *Lopidium concinnum* (Hypopterygiaceae, Musci). An example of steno-evolution in mosses. Studies in austral temperate rain forest bryophytes 2. Plant Syst Evol 218:67–75

Ochyra R (1992) *Amblyodon dealbatus* (Musci, Messiaceae)—a bipolar disjunct. Frag Flor Geobot 37: 251–259

Petterson B (1940) Experimentelle Untersuchungen über die euanemochore Verbreitung der Sporenpflanzen. Acta Bot Fenn 25:1–102

Sabovljevic M, Frahm J-P, Herbiniaux U (2005) Taxonomic value, systematic position and the origin of German populations of *Isothecium holtii* Kindb, based on molecula data. Lindbergia 30:107–112

Sabovljevic M, Frahm J-P (in press) Relic or long-distance dispersal? A case study of the rare and endangered moss *Campylopus oerstedianus* (Müll. Hall.) Mitt. (Dicranaceae) in Europe. Cryptogamie Bryol

Schuster RM (1984) Phytogeography of the bryophytes. In: Schuster RM (ed), New manual of bryology, vol I. The Hattori Botanical Laboratory, Nichinan, pp 463–626

Stech M, Dohrmann J (2004) Molecular relationships and biogeography of two Gondwanan *Campylopus* species, *C. pilifer* and *C. introflexus* (Dicranaceae). Monogr Syst Bot Missouri Bot Gard 98 415–431

Vitt DH, Bellard RJ (1997) Attributes of rarity among Alberta mosses: patterns and prediction of species diversity. Bryologist 100:1–12

Zanten BO (1978) Experimental studies on trans-oceanic long-range dispersal of moss species in the Southern Hemisphere. J Hattori Bot Lab 44:455–482

Zanten BO, Pócs T (1981) Distribution and dispersal of bryophytes. Adv Bryol 1:479–562

Myxomycete diversity and distribution from the fossil record to the present

Steven L. Stephenson · Martin Schnittler · Yuri K. Novozhilov

Originally published in the journal Biodiversity and Conservation, Volume 17, No 2, 285–301.
DOI: 10.1007/s10531-007-9252-9

Abstract The myxomycetes (plasmodial slime molds or myxogastrids) are a group of eukaryotic microorganisms usually present and sometimes abundant in terrestrial ecosystems. Evidence from molecular studies suggests that the myxomycetes have a significant evolutionary history. However, due to the fragile nature of the fruiting body, fossil records of the group are exceedingly rare. Although most myxomycetes are thought to have very large distributional ranges and many species appear to be cosmopolitan or nearly so, results from recent studies have provided evidence that spatial distribution patterns of these organisms can be successfully related to (1) differences in climate and/or vegetation on a global scale and (2) the ecological differences that exist for particular habitats on a local scale. A detailed examination of the global distribution of four examples (*Barbeyella minutissima*, *Ceratiomyxa morchella*, *Leocarpus fragilis* and *Protophysarum phloiogenum*) demonstrates that these species have recognizable distribution patterns in spite of the theoretical ability of their spores to bridge continents.

Keywords Distribution patterns · Ecology · Long-distance dispersal · Microorganisms · Slime molds

---

Special Issue: Protist diversity and geographic distribution. Guest editor: W. Foissner.

S. L. Stephenson
Department of Biological Sciences, University of Arkansas, Fayetteville, AR 72701, USA

M. Schnittler (✉)
Institute of Botany and Landscape Ecology, Ernst Moritz Arndt University Greifswald, Grimmer Str. 88, 17487 Greifswald, Germany
e-mail: martin.schnittler@uni-greifswald.de

Y. K. Novozhilov
V.L. Komarov Botanical Institute of the Russian Academy of Sciences, Prof. Popov St. 2, 197376 St. Petersburg, Russia

W. Foissner et al. (eds.), *Protist Diversity and Geographical Distribution*
DOI: 10.1007/978-90-481-2801-3_5

Introduction

The myxomycetes (also called plasmodial slime molds or myxogastrids) are a group of eukaryotic microorganisms usually present and sometimes abundant in terrestrial ecosystems. Myxomycetes have been known from their fruiting bodies since at least the middle of the seventeenth century, when the first recognizable description of a member of the group (the very common species now known as *Lycogala epidendrum*) was provided by the German mycologist Thomas Panckow. Evidence from molecular studies (e.g., Baldauf and Doolittle 1997; Baldauf et al. 2000) indicates that the myxomycetes should be placed within the "crown" clade of eukaryotes, which would suggest that they have a significant evolutionary history. However, due to the fragile nature of the fruiting body, fossil records of the group are exceedingly rare. Domke (1952) described a species of *Stemonitis* and Dörfelt et al. (2003) a species of *Arcyria* from Baltic amber dating from the Eocene, whereas Waggoner and Poinar (1992) reported the fossil of a myxomycete plasmodium in amber from Eocene-Oligocene deposits in the Dominican Republic. The maximum age that could be assigned to any of these fossils would not exceed about 50 million years, which is greater than that of the few records of fossil spores that appear to be those of myxomycetes, which date only from the Oligocene and Pleistocene (Graham 1971).

Life cycle

The myxomycete life cycle (Fig. 1) encompasses two very different trophic stages, one consisting of uninucleate amoebae, with or without flagella, and the other consisting of a distinctive multinucleate structure, the plasmodium (Martin et al. 1983). Under favorable conditions, the plasmodium gives rise to one or more fruiting bodies containing spores. Identification of myxomycetes is based almost exclusively upon features of the fruiting body (Martin and Alexopoulos 1969). The fruiting bodies produced by myxomycetes are somewhat suggestive of those produced by higher fungi, although they are considerably smaller (usually no more than 1–2 mm tall). The spores of the vast majority of myxomycetes range in size from 5 to 15 μm in diameter, with most species producing spores 10 ± 2 μm in diameter. Presumably, the spores are wind-dispersed and complete the life cycle by germinating to produce the uninucleate amoeboflagellate cells. These feed and divide by binary fission to build up large populations in the various microhabitats in which these organisms occur. Ultimately, this stage in the life cycle gives rise to the plasmodium. This process can result from gametic fusion between compatible amoeboflagellates or it can be apomictic (Collins 1980, 1981). Bacteria apparently represent the main food resource for both trophic stages, but plasmodia are also known to feed upon yeasts, algae (including cyanobacteria), and fungal spores and hyphae (Stephenson and Stempen 1994). Under adverse conditions, such as drying out of the immediate environment or low temperatures, a plasmodium may convert into a hardened, resistant structure called a sclerotium, which is capable of reforming the plasmodium upon the return of favorable conditions. Moreover, amoeboflagellate cells can undergo a reversible transformation to dormant structures called microcysts. Both sclerotia and microcysts can remain viable for long periods of time and are probably very important in the continued survival of myxomycetes in some ecological situations and/or habitats, such as the bark surface of living trees and deserts.

The fruiting bodies of many species of myxomycetes can achieve macroscopic dimensions and be collected and preserved for study in much the same way as the sporocarps of

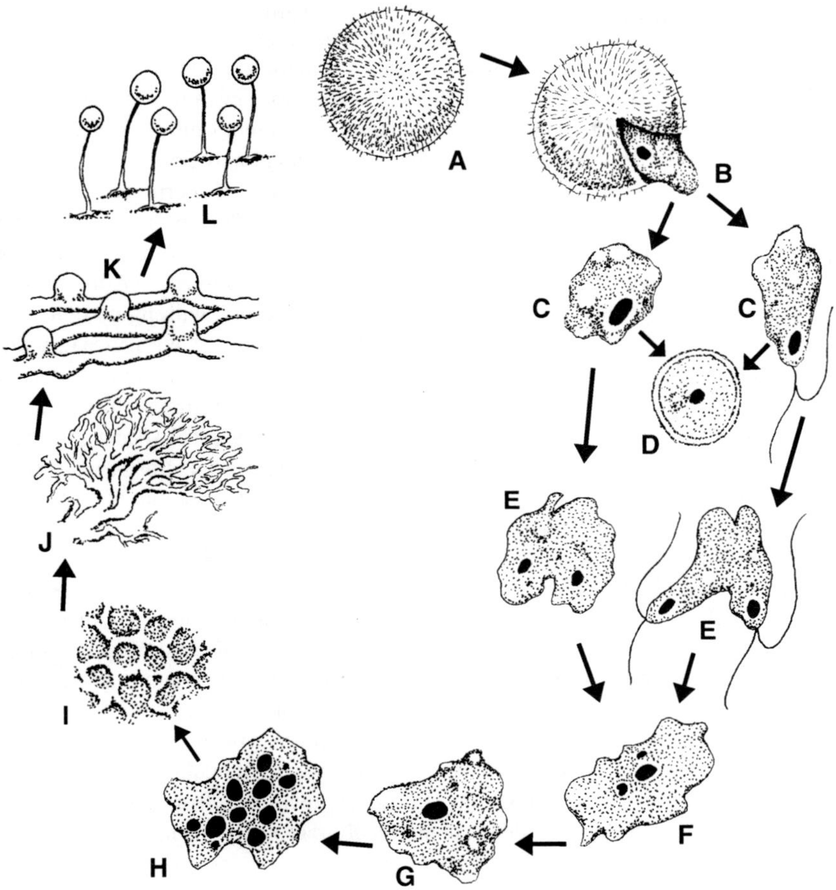

Fig. 1 Life cycle of a typical myxomycete. A, Spore. B, Germinating spore. C, Uninucleate amoeboid stage, with (right) or without (left) flagella. D, Microcyst. E–F, Fusion of two compatible amoebae to produce a single cell. G, Zygote. H, Early plasmodium. I, Sclerotium. J, Portion of a mature plasmodium. K, Beginning of sporulation. L, Mature fruiting body with spores still enclosed. (Adapted from Stephenson and Stempen 1994)

fungi or even specimens of bryophytes, lichens, and vascular plants. However, most species of myxomycetes tend to be rather inconspicuous or sporadic in their occurrence and are not always easy to detect in the field. Moreover, fruiting bodies of most species are relatively ephemeral and do not persist in nature for very long. Myxomycetes also spend a portion of their life cycle in a state where their very presence in a given habitat can be exceedingly difficult if not impossible to determine. Because of their life history strategy and inconspicuous nature, these organisms provide an immense challenge in biodiversity assessments and, consequently, often have been neglected in such studies.

Taxonomy

Approximately 875 species of myxomycetes have been described (Lado 2001), and these have been placed in six different taxonomic orders (Ceratiomyxales, Echinosteliales,

Liceales, Physarales, Stemonitales, and Trichiales). However, members of the Ceratiomyxales are distinctly different from members of the other orders, and many modern workers have removed these organisms from the myxomycetes and reassigned them to the protostelids (Olive 1970, 1975; Olive and Stoianovitch 1979). The exact evolutionary affinities of the myxomycetes are still debated, but these organisms constitute a well-defined and homogenous group. Evidence from DNA sequence analysis (Baldauf and Doolittle 1997) suggests that even what appear to be closely related taxa on the basis of morphological similarity may have diverged from each other a long time ago (Clark 2000). In the first phylogenetic study based on molecular data, Fiore-Donno et al. (2005) suggested that the Echinosteliales, which produce fruiting bodies with a simple structure, represented the most basal clade of myxomycetes, with two more advanced groups, the first with light-colored spores and consisting of the Trichiales and the (presumably not monophyletic) Liceales, and the second having dark spores and made up of the Physarales and Stemonitales. Because of their small size and the limited array of morphological characters upon which their taxonomy is based, determination of what constitutes a natural biological species, in the same sense that the concept is used for many of the more familiar groups of organisms (Mayr 1970), is sometimes rather problematic. It is now known that a number of the more common and widespread morphospecies actually consist of complexes of geographically restricted apomictic clonal lines (El Hage et al. 2000; Clark 2000; Clark and Stephenson 2000; Irawan et al. 2000). These genetically isolated lines are capable of independent evolution, which can lead to the accumulation of minor morphological differences that reflect specific adaptations to the particular set of environmental conditions in which they occur. For example, some of the forms found in special microhabitats (e.g., the inflorescences of tropical herbs) differ in some respects (e.g., color and size of the fruiting bodies) from specimens of the same species collected from more typical habitats. These almost certainly represent biotypes that are adapted to the microhabitat in question. Approximately 50% of all described species of myxomycetes are known only from the type locality or fewer than five localities worldwide. It seems likely that many of these "species" are no more than morphologically distinct biotypes present in particular habitats or confined to a certain regions of the world. If so, then the criteria that need to be applied before describing a taxon as new should be reconsidered to account for this phenomenon (Schnittler and Mitchell 2000). As shown by these authors, the annual number of species described as new to science is steadily increasing, although the morphological species concept as currently applied hardly considers reproductively isolated lines within a morphological species, as seen in culture experiments.

Limiting factors for myxomycete occurrence

Temperature and moisture are thought to be the main factors limiting the occurrence of myxomycetes in nature (Alexopoulos 1963), and species richness tends to increase with increasing diversity and biomass of the vascular plants providing the resources (various types of detritus) that support the bacteria and other microorganisms upon which the two trophic stages in the myxomycete life cycle feed (Madelin 1984; Stephenson 1989). The pH of the substrates potentially available to myxomycetes in a particular habitat also represents an important factor influencing the distribution of these organisms (Härkönen 1977; Stephenson 1989; Wrigley de Basanta 2000; Mosquera et al. 2000). Although many myxomycetes appear to have a relatively wide pH tolerance, this is not the case for all species. Härkönen (1977), who measured the pH of substrates upon which fruitings occurred in a study of the distribution patterns of myxomycetes associated with the bark of living trees in southern Finland, concluded that species of myxomycetes have different pH optima and amplitudes. In her study, some species seemed to prefer an acidic substrate, whereas others never developed under low pH conditions. Stephenson (1989) found the same to be true for both bark and forest floor litter in a study carried out in the eastern United States. In general, members of the Stemonitales developed under more acidic conditions than did members of the Physarales and the Trichiales.

Temperate forests

Much of what is known about the distribution and ecology of myxomycetes in terrestrial ecosystems has been derived from studies carried out in temperate forests of the Northern Hemisphere. In such forests, myxomycetes are associated with a number of different microhabitats. (As used herein, the term "microhabitat" simply denotes a specific portion of the total forest habitat represented by a homogenous substrate and similar microclimatic conditions (Stephenson 1989)). These include coarse woody debris on the forest floor, the bark surface of living trees, forest floor litter, the dung of herbivorous animals, and aerial portions of dead but still standing herbaceous plants. Each of these microhabitats tends to be characterized by a distinct assemblage of species (Stephenson 1988, 1989; Stephenson and Stempen 1994).

Lignicolous myxomycetes associated with coarse woody debris are the best known, since the species typically occurring in this microhabitat tend to be among those characteristically producing fruiting bodies of sufficient size to be detected in the field (Martin and Alexopoulos 1969). Many of the more common and widely known myxomycete taxa, including various species of *Arcyria*, *Lycogala*, *Stemonitis* and *Trichia*, are predominantly lignicolous. Much less is known about the myxomycetes associated with the microhabitats represented by the bark surface of living trees and forest floor litter. The primary reason for this is that many of the species involved are rather inconspicuous or sporadic in their occurrence and thus difficult to detect in the field. However, the moist chamber culture technique as it applies to myxomycetes (Gilbert and Martin 1933) provides a convenient and often very productive method of supplementing field collections when studying such microhabitats as bark and litter. Since its introduction, the technique has been used with considerable success by many researchers (e.g., Keller and Brooks 1976; Härkönen 1981; Blackwell and Gilbertson 1980; Stephenson et al. 1999). More than 100 species of "corticolous" myxomycetes have been reported from the bark microhabitat as field and/or moist chamber collections (Mitchell 1980). Many of these are also known to occur in other microhabitats, but at least some species seem restricted to bark of living trees. Prominent examples include various species of *Echinostelium*, *Licea*, and *Macbrideola* (Alexopoulos 1964; Mitchell 1980). The litter microhabitat of temperate forests is more heterogeneous than the bark microhabitat, since it usually consists of a mixture of leaves from different tree species along with other types of plant debris (e.g., pieces of bark, fragments of wood, fruits, seeds, inflorescences, and small twigs). Some of the species associated with litter also occur on bark, but others (e.g., various species of *Diderma*, *Didymium*, and *Physarum*) are found predominantly or even exclusively on litter. The assemblage of species associated with litter derived from coniferous trees tends to be distinctly different from that associated with litter from broadleaf trees (Härkönen 1981; Stephenson 1989).

Another microhabitat potentially available for myxomycetes is represented by the dung of herbivorous animals. A few species seem to occur predominantly or even exclusively on dung (Eliasson and Keller 1999). Dung is a highly complex substrate, and the exact role of myxomycetes in the communities of organisms associated with its decomposition is not yet known (Eliasson and Lundqvist 1979). However, dung has several characteristics seemly unusually favorable for myxomycetes: high moisture content, a large microbial population, and nutrient richness (Hudson 1986). Moreover, dung also has a much higher pH than most of the other substrates upon which myxomycetes are typically found (Stephenson 1989). In moist tropical forests, animal dung decomposes very rapidly and usually does not persist long enough to serve as a potential microhabitat for myxomycetes. In temperate forests, dung also tends to be of minor importance (Stephenson 1989), but in some high-latitude and desert ecosystems, it becomes one of the major substrates on which myxomycetes are encountered (Stephenson and Stempen 1994).

Myxomycetes also are known to occur in forest soils (Kalyanasundaram 1997), but their occurrence in this microhabitat has received very little study (Thom and Raper 1930; Feest 1987; Kerr 1994; Stephenson and Landolt 1996; Stephenson and Cavender 1996). Various species of *Didymium* appear to be the most widespread and abundant myxomycetes present in soil (Madelin 1990). Interestingly, myxomycetes seem to be relatively more abundant in grassland and agricultural soils than in forest soils (Feest and Madelin 1985), whereas the distribution patterns for the amoeboid cells of dictyostelids (cellular slime molds) are exactly the reverse. Just why this is the case is not yet known.

A number of instances are known of apparent ecological associations of myxomycetes and vascular plants. The ultimate basis of these associations relates to the structure of the plants in question. For example, in his studies of the myxomycetes of Hawaii, Eliasson (1991) suggested that the network of supporting tissue within decaying stems of certain cacti (*Opuntia* spp.) and other succulent vascular plants could function as a "natural moist chamber." Such decaying stems may retain moisture for several weeks after the last precipitation and the temperature beneath them may be many degrees below that of the exposed soil surface, thus providing unusually favorable conditions for myxomycetes. The enclosed space formed by the basal sheath of a palm petiole represents yet another example of such a natural moist chamber. The leaves of most palms are rather large, and the base of the petiole is expanded to form a sheath where it is attached to the stem. After abscission occurs and the leaves fall to the forest floor, the edges of the sheath become convoluted, creating a more-or-less enclosed space. Nikau palm (*Rhopalostylis sapida*) is one of the more characteristic small trees of lowland forests of northern New Zealand (Poole and Adams 1990). The microhabitat represented by the basal sheath of decaying leaves of this plant often yields collections of myxomycetes when all other substrates in the same forest and are non-productive (Stephenson 2003).

Other types of ecosystems

Our knowledge of the distribution and ecology of myxomycetes in tropical forests is still rather limited (Alexopoulos 1970, Farr 1976, Schnittler and Stephenson 2000, Stephenson et al. 2001, Stephenson et al. 2004), but available data suggest that the assemblages of species present are associated with microhabitats and exhibit ecological patterns quite different from those found in temperate forests. For example, in tropical forests, myxomycete biodiversity seems to be greatest in aerial microhabitats located above the ground, whereas in temperate (and also boreal) forests it is greatest in microhabitats associated with the forest floor. Moreover, in tropical forests, myxomycete diversity and abundance (as derived from monitoring fruiting bodies) appear to be concentrated towards the dry end of the moisture gradient and not towards the moist end as is generally the case in temperate and boreal forests. This is especially true for the species associated with the bark surface of living trees, which tend to support a much lower diversity of corticolous myxomycetes than most trees in temperate forests. Microhabitats seemingly unique to tropical forests and not known to support myxomycetes until very recently are the living inflorescences of large tropical herbs (Schnittler and Stephenson 2001), the cover of epiphyllic liverworts on living leaves of understory plants (Schnittler 2000), and the mantle of dead organic matter (literally a "canopy soil") found at the bases of vascular epiphytes growing on the trunks and larger branches of trees (Stephenson and Landolt 1998). The first of these appears to be especially rich in myxomycetes, and some species (e.g., *Physarum didermoides*) seem to show a strong preference for this microhabitat.

Boreal forests tend to be characterized by a lower biodiversity of myxomycetes than temperate forests, although certain taxa (*Fuligo septica* and several species of *Trichia*) can be exceedingly common in some situations (Eliasson 1981; Schnittler and Novozhilov 1996; Stephenson and Laursen 1998). The majority of species are lignicolous, and only a few examples (e.g., *Leocarpus fragilis*) are commonly encountered on litter. Most substrates are fairly acidic, which undoubtedly represents a limiting factor (Novozhilov et al. 1999). In general, members of the Trichiaceae are the most prominent myxomycetes in boreal forests (Novozhilov et al. 1998).

Myxomycetes are even less common in tundra than in boreal forests. Stephenson et al. (2000), who analyzed the data represented by almost 2,000 specimens of myxomycetes collected from areas of tundra and forest-tundra in Iceland, northern Russia, Alaska, and Greenland, recorded a total of 150 species, but only 33 of these were widely distributed enough to be regarded as true inhabitants of high-latitude ecosystems. The main factors for the reduction in the number of species of myxomycetes in high-latitude regions are almost certainly the unfavorable temperature conditions and the reduced range and extent of available microhabitats. In typical areas of tundra, the microhabitat represented by coarse woody debris is extremely limited but still exists in the form of small twigs and tiny branchlets of dwarf woody plants such as species of willow (*Salix*) and alder (*Alnus*). A high proportion of the species present in tundra are associated with litter, especially the litter that accumulates around the base of dwarf woody plants (Stephenson and Laursen 1993). As might be expected based on what is known for other organisms, overall biodiversity of myxomycetes appears to lowest at the very highest latitudes for which there are any data available (Stephenson et al. 2000). However, Stephenson et al. (2007) recorded 25 species from subantarctic Macquarie Island, and seven species are now known from the continent of Antarctica (Stephenson and Moreno 2006).

Because myxomycetes are almost invariably associated with relatively moist habitats, one might not expect these organisms to occur in deserts. However, the number of species reported from deserts is surprisingly high (Blackwell and Gilbertson 1980). The most productive substrates for myxomycetes in deserts include the pith skeletons of decaying cacti in areas of the world where these plants exist, the dead parts of living plants in contact with the ground, animal dung, and the bark of living shrubs and trees (Schnittler and Novozhilov 2000; Schnittler 2001). Field collections are usually limited to periods of no more than a few days or weeks immediately following a period of significant precipitation, but myxomycetes are exceedingly common in moist chamber cultures prepared with desert plant material.

One group of myxomycetes is restricted to alpine areas of mountains, where its members are found fruiting along the margins of melting snowbanks in late spring and early summer. The species that occupy this rather special and very limited habitat are usually referred to as "snowbank" or "nivicolous" myxomycetes. They constitute a distinct ecological group, since they usually produce fruiting bodies only during the relatively brief period of time when the special microenvironmental conditions associated with margins of snowbanks and apparently required for their growth and fruiting exits. During the remainder of the summer, the species of myxomycetes found in these alpine areas are very much the same as those collected at lower elevations in the same regions. Interestingly, the majority of species in some genera tend to be predominately alpine in distribution. This is true for *Dianema* (Kowalski 1967), *Lamproderma* (Kowalski 1970), and *Lepidoderma* (Kowalski 1971).

Long-distance dispersal

It has long been recognized that various small particles, including dust, spores, bacteria and other microbes, can be carried long distances by wind. For example, the British mycologist Berkeley (1857) concluded that "The trade winds, for instance, carry spores of fungi mixed with their dust, which may have traveled thousands of miles before they are deposited." In the mid-1930's, Meier (Meier and Lindbergh 1935) identified a variety of fungal spores, pollen, algae and diatoms from a series of samples collected over the North Atlantic by exposing sterile, oil-coated microscope slides directly to the air by way of a long metal arm extending from an airplane. A process that could transfer enormous numbers of microorganisms into the atmosphere was identified in the late 1990's, when satellite images revealed the astonishing magnitude by which desert soils are aerosolized into giant clouds of dust (Griffin et al. 2001, 2002; Kellogg and Griffin 2006). These clouds of dust frequently move across the Atlantic Ocean and reach the Caribbean, Central America, northern South America and the southeastern United States, where the particles they contain (including spores) are deposited. Amazingly enough, it has been demonstrated that plants living in the upper canopy of rainforests in South America actually derive a major portion of their nutrients from this dust fallout (Swap et al. 1996). Similar long-range movements of dust have been demonstrated for other parts of the world, including from Asia across the Pacific Ocean to western North America and from Australia to New Zealand. One particularly large dust cloud originating in China actually moved eastward all the way across the Pacific, North America and the Atlantic to reach Europe, thus traveling most of the way around the world (Griffin et al. 2002). Clearly, airborne spores would have the potential of being dispersed by wind over considerable distances. Muñoz et al. (2004), who evaluated the possible role that wind might play in long-distance dispersal of mosses, liverworts, lichens, and pteridophytes among land masses in the Southern Hemisphere, found that floristic similarities were more strongly correlated with global wind patterns than geographic proximity. For the most part, the land masses considered in this study were the small, rather isolated islands in the Southern Ocean, for which the groups of organisms being considered are well documented. The results reported in this study would seem to lend support for the idea that myxomycetes could have reached these same islands as a result of long-distance dispersal by wind, and the data available for one of the islands, Macquarie Island located south of Tasmania, indicate that the island is characterized by a relatively diverse myxoflora (Stephenson et al. 2007). In the same way, a diverse assemblage of myxomycetes occurs on the isolated Hawaiian Islands (Eliasson 1991).

If the spores of myxomycetes are largely wind-dispersed, as is generally considered to be the case (Alexopoulos 1963), then the global wind patterns noted above would give them considerable potential for long-distance dispersal over intercontinental distances.

However, it is possible that long-distance dispersal by wind may not be as common for myxomycetes as one might suspect. Although most myxomycetes are thought to have very large distributional ranges and many species appear to be cosmopolitan or nearly so, results from recent studies (e.g., Stephenson et al. 1993) have provided evidence that spatial distribution patterns of these organisms can be successfully related to (1) differences in climate and/or vegetation on a global scale and (2) the ecological differences that exist for particular habitats on a local scale. To demonstrate that myxomycetes have recognizable distribution patterns in spite of the theoretical ability of their spores to bridge continents, the global distribution of four species will be assessed and then discussed in the section that follows. These four species are *Barbeyella minutissima*, *Ceratiomyxa morchella*, *Leocarpus fragilis* and *Protophysarum phloiogenum*.

Examples for myxomycete distribution patterns

The four species being considered are (1) very different in their ecology, (2) taxonomically distinct and thus unlikely to be confused with any other species, and (3) display microhabitat preferences that are sufficiently well known to allow them to be detected during the course of biodiversity surveys of the myxoflora of a particular region. The records upon which our analyses are based have been filtered out of a body of more than 100,000 digitalized records of myxomycetes, encompassing the resources of the Global Biodiversity Information Facility (GBIF, http://www.secretariat.gbif.net/portal/index.jsp), the database of the Eumycetozoan Project based at the University of Arkansas (http://slimemold.uark.edu), several major herbarium collections (Beltsville/BPI, Toronto/TRTC, Ottawa/DAOM, Munich/M, St. Petersburg/LE, Madrid/MA-Fungi (see Pando et al. 2003)), and the private collections of a number of individuals, including the authors.

Barbeyella minutissima Meyl. (Echinosteliales) is a minute but very distinct myxomycete described in 1916 from the Swiss Yura Mountains. As a result of their conspicuous black color, which stands out in sharp contract to the substrate upon which they usually occur, the 0.4–0.9 mm tall sporocarps can be detected in the field; only rarely has the species been reported to occur in moist chamber culture. In one of the first studies of the distribution pattern of a particular species of myxomycete, the biogeography of *B. minutissima* was discussed in detail by Schnittler et al. (2000). Herein, we present an updated distribution map compiled from 123 records (Fig. 2). This species shows a strong preference for slimy, algae-covered and decorticated logs and seems to develop only in situations providing continuously high humidity but sheltered from direct rainfall. A good indicator for the presence of *B. minutissima* seems to be lignicolous occurrences of the liverwort *Nowellia curvifolia* (Dicks.) Mitt. (Stephenson and Studlar 1985). Virtually all records are from boreal and montane coniferous forests, especially *Picea* and/or *Abies* forests (Schnittler et al. 2000; Stephenson 2004). One exception represents a record from a moist chamber culture of *Malus* bark collected in the upper Rhine valley (Neubert et al. 1993), were spores of the species could have come from the adjacent Black Forest. The range of *B. minutissima* seems to be very fragmented, but repeated findings on the high volcanos in Mexico (the species appears to be quite common in the *Abies* forests near the summit of the Malinzi Volcano in the state of Tlaxcala (A. Estrado-Torrez, pers. comm.)) seems to provide evidence of effective dispersal by air-borne spores.

The minute myxomycete *Protophysarum phloiogenum* M. Blackw. and Alexop. is the only member of a monotypic genus, described by Blackwell and Alexopoulos (1975), that is assumed to be basal for the Physarales (Castillo et al. 1998). The tiny sporocarps, only

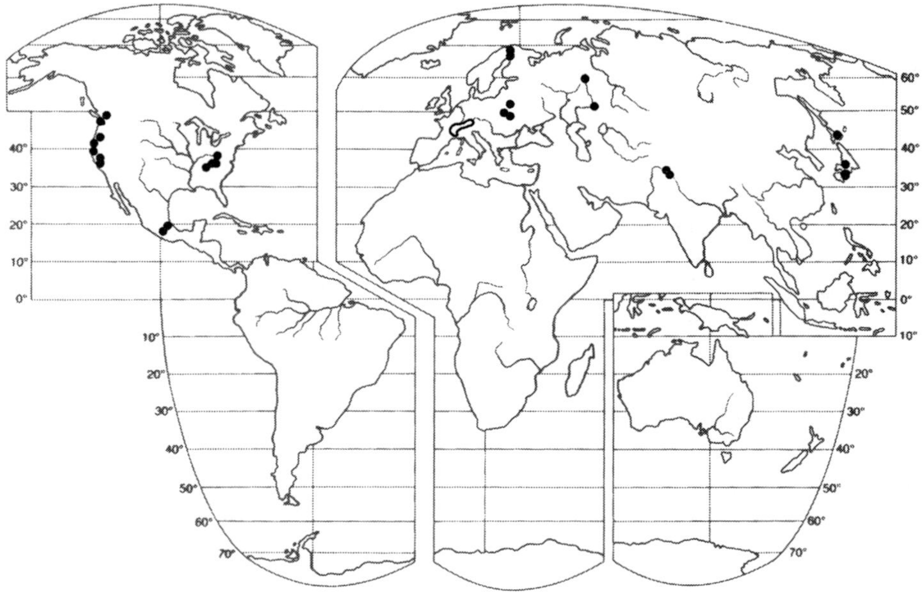

Fig. 2 World distribution map for *Barbeyella minutissima* Meyl., a myxomycete associated with montane *Picea* and/or *Abies* forests. The encircled range indicates the European Alps, with numerous records of the species

0.3–0.8 mm in total height, have been observed thus far almost exclusively in moist chamber cultures, where they are rather conspicuous by having a silvery iridescent peridium. All 72 records of which we are aware were obtained from the bark of living trees or (less frequently) shrubs. At least one distribution centre of the species seems to be represented by the arid regions of the western United States (Colorado, New Mexico, Arizona and Utah), although the species is present (albeit with a low frequency) in the less thoroughly studied deserts of Central Asia (southern Russia, Kazakhstan and Mongolia). Fragmentary records from other arid zones of the world (southern Spain, Tunesia and Oman) contribute to the picture of a species apparently restricted to arid steppes and deserts, with a (possibly questionable) record from Myanmar (formerly Burma) as the only noticeable exception (Fig. 3). Since arid habitats are generally understudied for myxomycetes, the distribution pattern presented herein is certainly fragmentary, and more intense studies in shrub deserts or gallery forests that occur along rivers in deserts and other arid areas will certainly add to the number of localities from which the species is known.

Leocarpus fragilis (Dicks.) Rost. (Physarales), which produces calcareous, orange to red sporocarps that resemble miniature grapes, is one of the most of all distinctive myxomycetes. The species is frequently recorded by non-specialists for the group. The large and robust, yellow phaneroplasmodium inhabits ground litter with an acidic pH (most often between 4 and 6). Fruitings can consist of several thousand individual sporocarps. In dry coniferous forests, exceedingly large mass fruitings can develop, and in 1993 one dry pine plantation in eastern Germany (Brandenburg, Eggsdorf near Berlin) had an estimated density of 200–400 fruitings per hectare. With a diameter of 12–14 μm, the spores are relatively large for a myxomycete, which should decrease the probability of long-distance dispersal, and local biotypes seem to exist. A form with spores in clusters of two has been described as *L. bisporus* Nann.-Bremek. and Mitch. (Nannenga-Bremekamp 1989). Some

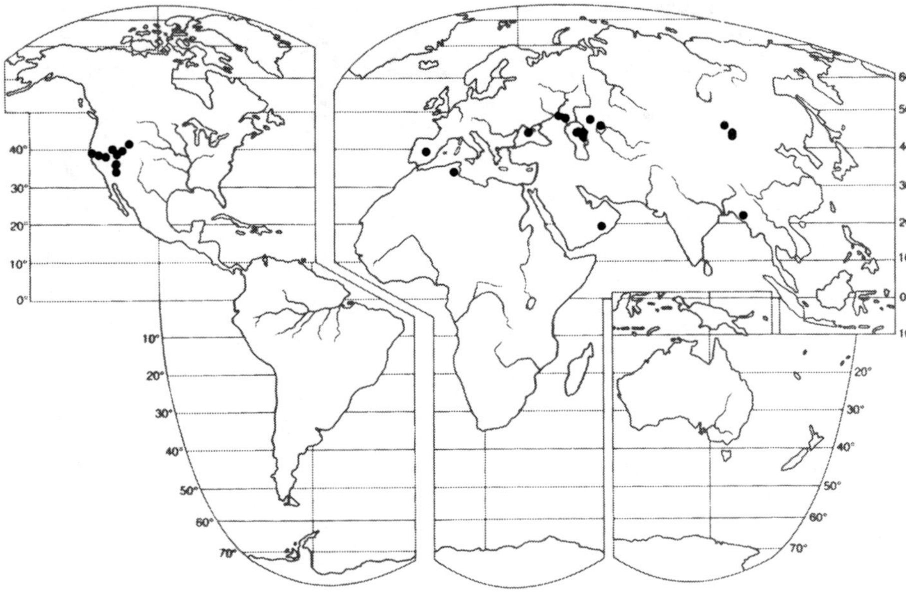

Fig. 3 Distribution of *Protophysarum phloiogenum* M. Blackw. & Alexop., a species found almost exclusively in arid environments

specimens from South Africa have almost globose sporocarps, whereas the typical form is egg-shaped (Schnittler, pers. observation of specimens from the Kew Botanical Garden, London). The distribution map (Fig. 4) was developed from 1474 records. Regions with many records were encircled with a line on the map. Two patterns seem evident when the overall distribution is considered. First, *L. fragilis* rarely occurs in arctic regions but as a litter-inhabiting species it can spread beyond the timberline. As such, its northern distribution may be limited more by climate than microhabitat availability. It is very common in the temperate zone, but seems to be less common in southern temperate zones (e.g., in the Mediterranean region or in the southeastern United States). The species appears to be absent from highly arid regions (records from Israel and Spain are from the less arid parts of these countries) and the humid tropics. However, disjunct occurrences in montane regions are found farther southward (e.g., in the Canary Islands or the African Atlas Mountains). Also, in fairly well studied tropical regions (e.g. Central America, Ecuador, Tanzania and Taiwan), *L. fragilis* has not yet been found, except for a single record from the paramo region of Costa Rica. As shown by a single record from southern Argentina (Tierra del Fuego) and a few from South Africa and southern Australia, it reappears in temperate regions of the southern hemisphere. In contrast to the statement in Martin and Alexopoulos (1969) that *L. fragilis* is a cosmopolitan species, it seems to display a clear preference for temperate zones.

Ceratiomyxa morchella Welden (Ceratiomyxales) was unknown to science until 1954. The fruiting bodies are very evanescent and start to dissipate on the day following development. However, fruitings of *C. morchella* are conspicuous, since the stalked, 2.5–4 mm high fruiting bodies can occur in large developments. The stalk is translucent, colourless and crowned by a pure white, globose head that resembles a miniature morel. The ovoid spores, 9–11 × 6–8 µm in diameter, are formed at the tip of 2–10 µm long stalks that cover the white upper portion of the fruiting body like a covering of fur. According to studies by

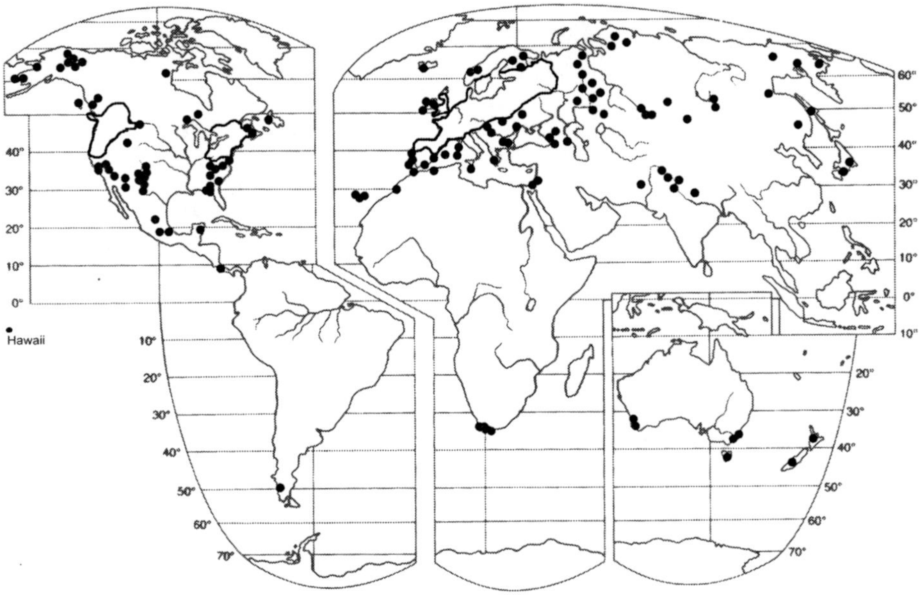

Fig. 4 World distribution map for *Leocarpus fragilis* (Dicks.) Rost., a temperate-zone myxomycete confined largely to the northern hemisphere. Encircled ranges in Europe and the United States depict regions with several hundred records each

the authors in Costa Rica and Ecuador, *C. morchella* is strongly limited to decaying wood with a very acidic pH, a rather rare microhabitat in tropical forests. Without exception, the substrate upon which fruiting occurred for the more than 100 records considered herein consisted of decorticated, moderately to strongly decayed wood of large diameter logs with measured pH values between 3.0 and 4.5. This precludes any confusion with *C. sphaerosperma* Boedijn, the second tropical species in the genus *Ceratiomyxa*, since the latter is found exclusively upon substrates with a pH above 6.5 (up to 8), most often portions of decaying of hard-shelled fruits. For *C. morchella*, fruitings with more than 10,000 individual fruiting bodies are not rare, and these typically cover the lower side of large, fallen tree trunks that are kept above the ground by their root disk or other logs. The spores are colourless and thin-walled. Lacking any UV-absorbing pigments, they may not be able to survive the long periods of exposure to light that would be associated with long-distance dispersal in the higher layers of the atmosphere. The preliminary map for *C. morchella* (Fig. 5) was compiled from only 72 records, reflecting the fact that the slimy fruiting bodies are difficult to preserve as herbarium specimens. The northernmost records are from subtropical Florida and tropical Mexico. The species is fairly common in the Caribbean region, being known from Puerto Rico, Jamaica, St. Vincent and Trinidad. Records for Costa Rica come from all parts of the country except the dry southwest and the highest mountains. It was recorded twice during a survey in western Ecuador (Schnittler et al. 2002). The real distribution centre of the species seems to be the Amazon Basin. In a study carried out in the Yasuni National Park (eastern Ecuador) it was one of the most common species (Schnittler, unpubl. data). Due to the paucity of records, it can not yet be determined if the species commonly occurs only in the Paleotropics. The species is still unknown from tropical Africa, and there is only a single record from eastern Asia (New Guinea) mentioned in the literature.

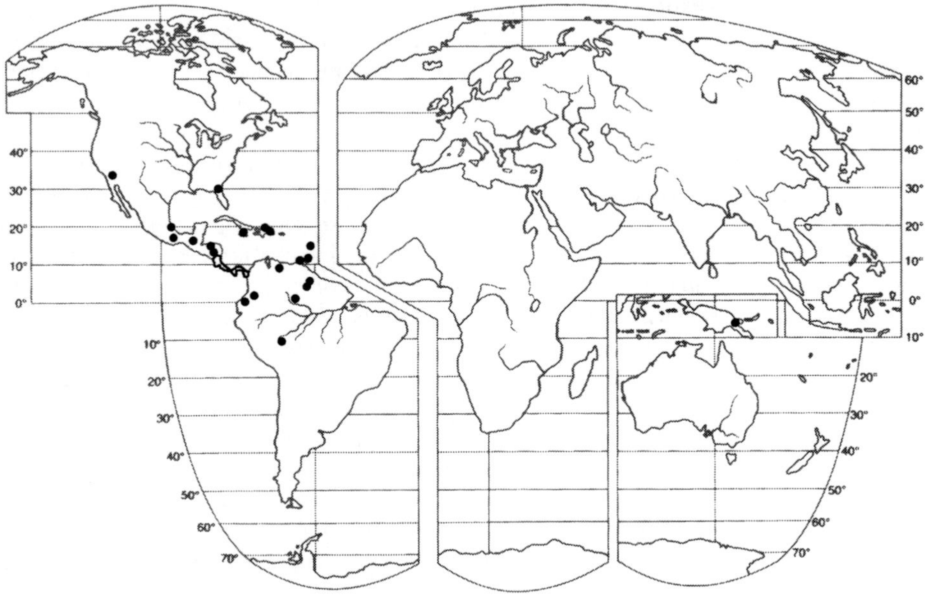

Fig. 5 Distribution map for the tropical myxomycete-like protostelid *Ceratiomyxa morchella* Welden

Myxomycetes in the context of the current debate on protist distribution

On the whole, myxomycetes would seem to be rather opportunistic or "fugitive" organisms (*sensu* Hutchinson 1951) in that they have a high reproductive potential, seem to possess effective dispersal mechanisms, and are characterized by rapid development. These properties allow them to exploit successfully habitat islands occurring both temporally and spatially in nature. Although a particular habitat within which a species of myxomycete has been established may persist for only a short period of time, the species always survives by reestablishing itself in some new habitat (which may be indeed the very same habitat if conditions once again become favorable). Although the spores of myxomycetes would appear to have considerable potential for long-distance dispersal, there is little question that some species are more common in some regions of the world than others, and the non-availability of certain microhabitats apparently imposes major constraints upon their occurrence even within a particular region. As such, it would seem that myxomycetes do not necessarily conform completely to the "ubiquity of small free-living eukaryotic species" concept as proposed by Finlay (2002) and Fenchel and Finlay (2004). The very fragmented range of *Barbeyella minutissima*, a species that appears to be confined almost exclusively to montane *Picea* and/or *Abies* forests, provides a good example. Thus, the myxomycete distribution data are consistent with the "moderate endemicity model" proposed by Foissner (2006), who suggests that about 30% of the protist species are morphological and/or genetical and/or ecological endemics.

Although at least some species of myxomycetes are very easy to recognize (for instance, *Leocarpus fragilis*) and their fruitings may well assume macroscopic conditions, there is no doubt that the distribution maps presented herein are still fragmentary. As to be expected for organisms having a limited number of active researchers, the range of a species as depicted on a distribution map will largely reflect those records from regions of the world (Europe, eastern North America and Japan) where most myxomycetologists live and work. Only recently, in the context of the Global Biodiversity of Eumycetozoans and GBIF International projects mentioned earlier in this chapter, has a systematic digitalization of records begun, and we estimate that fewer than 25% of all herbarium specimens worldwide have been databased. Only the fruiting body stage of myxomycetes was considered in the development of the distribution maps presented herein, and it is certainly possible that there are habitats where myxomycetes live as amoebal and/or plasmodial populations only and do not fruit. If this is the case, then a particular species could have a larger "true" distribution than reflected by a map compiled on the basis of records of fruiting bodies. This may well be the case for continuously moist montane rainforests, where Schnittler and Stephenson (2000), for a study carried out in Costa Rica, detected a gradient of decreasing myxomycete diversity with increasing elevation and rainfall. Direct environmental sampling with the use of molecular techniques such as DNA probes would represent a way of detecting hidden amoebal and/or plasmodial populations of myxomcyetes, which would be regarded as "sink" populations in terms of dispersal capacities. On the other hand, especially for corticolous myxomycetes such as *Protophysarum phloiogenum* that grow readily in substrate cultures (this species is characterized by a developmental time between 2 and 6 days), such "hidden" populations should be detected regularly if the appropriate survey efforts were carried out.

Acknowledgements A major part of the specimen digitalization work which formed the basis for the distribution maps presented in this paper was carried out within a grant of the seed money program of GBIF International, coordinated by the second author. Another grant funded in the context of the Global Biodiversity of Eumycetozoans project (DEB-0316284) from the US National Science Foundation added data especially for the American continent. Appreciation is extended to many of our colleagues for providing additional information on records of particular species. Among these are G. Adamonité (Lithuania), U. Eliasson (Sweden), A. Estrada-Torrez (Mexico), A. Koshelova (Russia), C. Lado (Spain), R. McHugh (Ireland), M. Meyer (France), D. Wrigley de Basanta (Spain) and I. Zemlianskaya (Russia). We also wish to thank John Shadwick for helping produce the distribution maps of the four species.

References

Alexopoulos CJ (1963) The myxomycetes II. Bot Rev 29:1–78

Alexopoulos CJ (1964) The rapid sporulation of some myxomycetes in moist chamber culture. Southwestern Nat 9:155–159

Alexopoulos CJ (1970) Rain forest myxomycetes. In: Odum HT (ed) A tropical rain forest. United States Atomic Energy Commission, Washington, DC, pp F21–F23

Baldauf SL, Roger AJ, Wenk-Siefert J et al (2000) A kingdom-level phylogeny of eukaryotes based on combined protein data. Science 290:972–977

Baldauf SL, Doolittle WF (1997) Origin and evolution of the slime molds (Mycetozoa). PNAS 94:12007–12012

Berkeley MJ (1857) Introduction to cryptogamic botany. Bailliere, London

Blackwell M, Alexopoulos CJ (1975) Taxonomic studies in the Myxomycetes IV. *Protophysarum phloiogenum*, a new genus and species of Physaraceae. Mycologia 67:32–37

Blackwell M, Gilbertson RL (1980) Sonoran desert myxomycetes. Mycotaxon 11:139–149

Castillo A, Illana C, Moreno G (1998) *Protophysarum phloiogenum* and a new family in the Physarales. Mycol Res 102:838–842

Clark J (2000) The species problem in the myxomycetes. Stapfia 73:39–53

Clark J, Stephenson SL (2000) Biosystematics of the myxomycete *Physarum melleum*. Nova Hedwigia 71:161–164

Collins OR (1980) Apomictic-heterothallic conversion in a myxomycete, *Didymium iridis*. Mycologia 72:1109–1116

Collins OR (1981) Myxomycete genetics, 1960–1981. J Elisha Mitchell Sci Soc 97:101–125

Domke W (1952) Der erste sichere Fund eines Myxomyceten im Baltischen Bernstein (*Stemonitis splendens* Rost Fa Succini fa Nov Foss). Mitteilungen aus dem Geologischen Staatsinstitut in Hamburg 21:154–161
Dörfelt H, Schmidt AR, Ullmann P et al (2003) The oldest fossil myxogastroid slime mold. Mycol Res 107:123–126
El Hage N, Little C, Clark L et al (2000) Biosystematics of *Didymium squamulosum*. Mycologia 92:54–64
Eliasson UH (1981) Patterns of occurrence of myxomycetes in a spruce forest in south Sweden. Holarctic Ecology 4:20–31
Eliasson UH (1991) The myxomycete biota of the Hawaiian Islands. Mycol Res 95:257–267
Eliasson UH, Lundqvist N (1979) Fimicolous myxomycetes. Bot Not 132:551–568
Eliasson UH, Keller HW (1999) Coprophilous myxomycetes: updated summary, key to species, and taxonomic observations on *Trichia brunnea*, *Arcyria elaterensis*, and *Arcyria stipata*. Karstenia 39:1–10
Farr ML (1976) Myxomycetes. Flora Neotropica, Monograph 16. New York Botanical Garden, New York
Feest A (1987) The quantitative ecology of soil Mycetozoa. Prog Protist 2:331–361
Feest A, Madelin MF (1985) A method for the enumeration of myxomycetes in soils and its application to a wide range of soils. FEMSW Microbiol Ecol 31:103–109
Fenchel T, Finlay BJ (2004) The ubiquity of small species: patterns of local and global diversity. Bio Science 54:777–784
Finlay BJ (2002) Global dispersal of free-living microbial eukaryotic species. Science 296:1061–1063
Fiore-Donno A-M, Berney C, Pawlowski J et al (2005) Higher-order phylogeny of plasmodial slime molds (Myxogastria) based on elongation factor 1-A and small subunit rRNA gene sequences. J Eukaryot Microbiol 52:1–10
Foissner W (2006) Biogeography and dispersal of micro-organisms: a review emphasizing protests. Acta Protozool 45:111–136
Gilbert HC, Martin GW (1933) Myxomycetes found on the bark of living trees. Univ Iowa Stud Nat Hist 15:3–8
Graham A (1971) The role of Myxomyceta spores in palynology (with a brief note on the morphology of certain algal zygospores). Review Palaeobot Palynol 11:89–99
Griffin DW, Kellogg CA, Shinn EA (2001) Dust in the wind: long range transport of dust in the atmosphere and its implications for global public and ecosystem health. Glob Change Hum Health 2:20–33
Griffin DW, Kellogg CA, Garrison VH et al (2002) The global transport of dust. Am Scientist 90:230–237
Härkönen M (1977) Corticolous myxomycetes in three different habitats in southern Finland. Karstenia 17:19–32
Härkönen M (1981) Myxomycetes developed on litter of common Finnish trees in moist chamber cultures. Nordic J Bot 1:791–794
Hudson HJ (1986) Fungal biology. Edward Arnold, Baltimore, Maryland
Hutchinson GE (1951) Copepodology for the ornithologist. Ecology 32:571–577
Irawan B, Clark J, Stephenson SL (2000) Biosystematics of the *Physarum compressum* morphospecies. Mycologia 92:884–893
Kalyanasundaram I (1997) Myxomycetes in the tropics: distribution and ecology. In: Janardhanan KK, Rajendran C, Natarajan K, Hawksworth DL (eds) Tropical mycology. Science Publishers Inc, Enfield, New Hampshire, pp 227–237
Keller HW, Brooks TE (1976) Corticolous myxomycetes V: observations on the genus *Echinostelium*. Mycologia 68:1204–1220
Kellogg CA, Griffin DW (2006) Aerobiology and the global transport of desert dust. Trends Ecol Evol 21:638–644
Kerr SJ (1994) Frequency of recovery of myxomycetes from soils of the northern United States. Canad J Bot 72:771–778
Kowalski DT (1967) Observations on the Dianemaceae. Mycologia 59:1075–1084
Kowalski DT (1970) The species of *Lamproderma*. Mycologia 62:621–672
Kowalski DT (1971) The genus *Lepidoderma*. Mycologia 63:490–516
Lado C (2001) Nomenmyx. A nomenclatural taxabase of Myxomycetes. Cuadernos de Trabajo Flora Micológica Ibérica 16:1–221
Madelin MF (1984) Myxomycetes, microorganisms and animals: a model of diversity in animal-microbial interactions. In: Anderson JN, Rayner ADA, Walton DWH (eds) Invertebrate-microbial interactions. Cambridge University Press, New York, pp 1–33
Madelin MF (1990) Methods for studying the ecology and population dynamics of soil myxomycetes. Meth Microbiol 22:405–416
Martin GW, Alexopoulos CJ (1969) The Myxomycetes. Univ of Iowa Press, Iowa City
Martin GW, Alexopoulos CJ, Farr ML (1983) The genera of myxomycetes. Univ of Iowa Press, Iowa City
Mayr E (1970) Population, species, and evolution. Belknap Press, Harvard Univ Press, Cambridge

Meier FC, Lindbergh CA (1935) Collecting micro-organisms from the Arctic atmosphere. Sci Mon 40:5–20

Mitchell DW (1980) A Key to the corticolous Myxomycetes. The British Mycological Society, Cambridge, England

Mosquera J, Lado C, Beltrán-Tejera E (2000) Morphology and ecology of *Didymium subreticulosporum*. Mycologia 92:978–983

Muñoz J, Felicisimo AM, Cabezas F et al (2004) Wind as a long-distance dispersal vehicle in the Southern Hemisphere. Science 304:1144–1147

Nannenga–Bremekamp NE (1989) Notes on myxomycetes XXIII. Seven new species of Myxomycetes. Proc Koninkl Nederl Akad Wet C 92:505–515

Neubert H, Nowotny W, Baumann K (1993) Die Myxomyceten Deutschlands und des angrenzenden Alpenraumes unter besonderer Berücksichtigung Österreichs. Band 1 *Ceratiomyxales*, *Echinosteliales*, *Liceales*, *Trichales*. Baumann, Gomaringen, Germany

Novozhilov YK, Schnittler M, Stephenson SL (1998) Analysis of myxomycete diversity of Russian subarctic and arctic areas. Mikol Fitopatol 32:27–33

Novozhilov YK, Schnittler M, Stephenson SL (1999) Myxomycetes of the Taimyr Peninsula (north–central Siberia). Karstenia 39:77–97

Olive LS (1970) The Mycetozoa: a revised classification. Bot Rev 36:59–87

Olive LS (1975) The Mycetozoans. Academic Press, New York

Olive LS, Stoianovitch C (1979) Observations of the mycetozoan genus *Ceratiomyxa*: description of a new species. Mycologia 71:546–555

Pando F et al (2003) MA Cryptogamic collections online databases. http://wwwrjbcsices/herbario/crypto/crydbhtm. Cited 01 Jan 2007

Poole AL, Adams NM (1990) Trees and shrubs of New Zealand. DSIR Publishing, Wellington, New Zealand

Schnittler M (2000) Foliicolous liverworts as a microhabitat for Neotropical myxomycetes. Nova Hedwigia 72:259–270

Schnittler M (2001) Ecology of myxomycetes of a winter-cold desert in western Kazakhstan. Mycologia 93:653–669

Schnittler M, Mitchell DW (2000) Species diversity in myxomycetes based on the morphological species concept – a critical examination. Stapfia 73:55–62

Schnittler M, Novozhilov YK (1996) The myxomycetes of boreal woodlands in Russian northern Karelia: a preliminary report. Karstenia 36:19–40

Schnittler M, Novozhilov YK (2000) Myxomycetes of the winter-cold desert in western Kazakhstan. Mycotaxon 74:267–285

Schnittler M, Stephenson SL (2000) Myxomycete biodiversity in four different forest types in Costa Rica. Mycologia 92:626–637

Schnittler M, Stephenson SL (2001) Inflorescences of Neotropical herbs as a new microhabitat for myxomycetes. Mycologia 94:6–20

Schnittler M, Lado C, Stephenson SL (2002) Rapid biodiversity assessment of a tropical myxomycete assemblage – Maquipucuna Cloud Forest Reserve, Ecuador. Fungal Divers 9:135–167

Schnittler M, Stephenson SL, Novozhilov YK (2000) Ecology and world distribution of *Barbeyella minutissima* (Myxomycetes). Mycol Res 104:1518–1523

Stephenson SL (1988) Distribution and ecology of myxomycetes in temperate forests I. Patterns of occurrence in the upland forests of southwestern Virginia. Can J Bot 66:2187–2207

Stephenson SL (1989) Distribution and ecology of myxomycetes in temperate forests II. Patterns of occurrence on bark surface of living trees, leaf litter, and dung. Mycologia 81:608–621

Stephenson SL (2003) Myxomycetes associated with decaying fronds of Nikau palm (*Rhopalostylis sapida*) in New Zealand. NZ J Bot 41:311–317

Stephenson SL (2004) Distribution and ecology of myxomycetes in southern Appalachian subalpine coniferous forests. In: Cribbs CL (ed) Fungi in forest ecosystems: diversity, ecology, and systematics. New York Botanical Garden, Bronx, pp 203–212

Stephenson SL, Cavender JC (1996) Dictyostelids and myxomycetes. In: Hall GS (ed) Methods for the examination of organismal diversity in soils and sediments. CAB International, Oxon, UK, pp 91–101

Stephenson SL, Estrada-Torres A, Schnittler M et al (2001) Distribution and ecology of myxomycetes in the forests of Yucatan. In: Gómez–Pompa A, Allen M, Fedick S et al (eds) Lowland Maya area: three millennia at the human–wildland interface. Haworth Press, New York

Stephenson SL, Kalyanasundaram I, Lakhanpal TN (1993) A comparative biogeographical study of myxomycetes in the mid-Appalachians of eastern North America and two regions of India. J Biogeogr 20:645–657

Stephenson SL, Landolt JC (1996) The vertical distribution of dictyostelids and myxomycetes in the soil/litter microhabitat. Nova Hedwigia 62:105–117

Stephenson SL, Landolt JC (1998) Dictyostelid cellular slime molds in canopy soils of tropical forests. Biotropica 30:657–661

Stephenson SL, Landolt JC, Moore DL (1999) Protostelids, dictyostelids, and myxomycetes in the litter microhabitat of the Luquillo experimental forest, Puerto Rico. Mycol Res 103:209–214

Stephenson SL, Laursen GA (1993) A preliminary report on the distribution and ecology of myxomycetes in Alaskan tundra. Biblthca mycol 150:251–257

Stephenson SL, Laursen GA (1998) Myxomycetes from Alaska. Nova Hedwigia 66:425–434

Stephenson SL, Laursen GA, Seppelt RD (2007) Myxomycetes of subantarctic Macquarie Island. Austral J Bot (in press)

Stephenson SL, Moreno G (2006) A new species of *Didymium* (Myxomycetes) from subantarctic Macquarie Island. Mycol Progr 5:255–258

Stephenson SL, Novozhilov YK, Schnittler M (2000) Distribution and ecology of myxomycetes in high-latitude regions of the northern hemisphere. J Biogeogr 27:741–754

Stephenson SL, Schnittler M, Lado C et al (2004) Studies of Neotropical mycetozoans. Syst Geogr Plants 74:87–108

Stephenson SL, Stempen H (1994) Myxomycetes: a handbook of slime molds. Timber Press, Portland, Oregon

Stephenson SL, Studlar SM (1985) Myxomycetes fruiting upon bryophytes: coincidence or preference? J Bryology 13:537–548

Swap R, Garstang M, Greco S et al (1996) Saharan dust in the Amazon Basin. Tellus 44:133–149

Thom C, Raper KD (1930) Myxamoebae in soil and decomposing crop residues. J Wash Acad Sci 20:362–370

Waggoner BM, Poinar GO Jr (1992) A fossil myxomycete plasmodium from Eocene-Oligocene amber of the Dominican Republic. J Euk Microbiol 39:639–642

Wrigley de Basanta D (2000) Acid deposition in Madrid and corticolous myxomycetes. Stapfia 73:113–120

Diversity and endemism in Rotifera: a review, and *Keratella* Bory de St Vincent

Hendrik Segers · Willem H. De Smet

Originally published in the journal Biodiversity and Conservation, Volume 17, No 2, 303–316.
DOI: 10.1007/s10531-007-9262-7

Abstract We confront patterns in the chorology and diversity of freshwater and limnoterrestrial Rotifera with predictions following from the recently revived ubiquity theorem on the distribution of microscopic organisms. Notwithstanding a strong taxonomic impediment and lack of data, both bdelloid and monogonont rotifers appear to conform to the hypothesis' predictions that local diversity is relatively high compared to global diversity and that cosmopolitism is important. To the contrary, however, a latitudinal diversity gradient is obvious, and endemicity is present, and exhibits diverse patterns. This is illustrated by the case of *Keratella* rotifers, in which we identify purported relict endemicity hotspots in the east Palaearctic (China) and in temperate and cold regions of the southern hemisphere, and a recent radiation in North America. The apparent paradox may result from an antagonism between rotifer's high population sizes and presence of potentially highly efficient propagules, versus pre-emption of habitats and local adaptation by resident populations, specific dispersal ability, and ecological and geographical factors. We conclude that distribution patterns of microscopic organisms, as represented by rotifers, most likely span the whole range of alternatives, from full cosmopolitanism to local endemism, and suggest that studying this diversity is more productive to come to an understanding of their chorology and diversity.

Keywords Micro-organisms · Ubiquity theorem · Cosmopolitanism · Freshwater · Limnoterrestrial · Marine

---

Special Issue: Protist diversity and geographic distribution. Guest editor: W. Foissner.

H. Segers (✉)
Belgian Biodiversity Platform, Freshwater Biology, Royal Belgian Institute of Natural Sciences, Vautierstraat 29, 1000 Brussels, Belgium
e-mail: Hendrik.Segers@naturalsciences.be

W. H. De Smet
Department of Biology, Section Polar Ecology, Limnology and Palaeobiology, University of Antwerp, Campus Drie Eiken, Universiteitsplein 1, 2610 Wilrijk, Belgium
e-mail: willem.desmet@ua.ac.be

Introduction

Recently, the debate on the biogeography, or lack thereof, of microscopic organisms has regained interest. Indeed, some recent molecular studies on the diversity and distribution of small-sized taxa, with body lengths of less than 1 mm, seem to corroborate the ancient adagio (e.g. Jennings 1900; Rousselet 1909) that such organisms are potentially cosmopolitan, and that it is the environment that determines the diversity and composition of local communities (Baas-Becking 1934; Finlay 2002; Fenchel and Finlay 2004). The underlying idea is that high rates of dispersal and low rates of local extinction, both related to the huge population sizes of microscopic organisms, lead to a situation in which "everything is everywhere", and that the species found in a given habitat are a function only of habitat properties and not of historical factors. Consequences of this ubiquity theorem are that a latitudinal diversity gradient would be absent or weak, that local diversity would be relatively high compared to the global diversity of the group, and that cosmopolitanism would be important (Fenchel and Finlay 2004).

Rotifera, with an adult size range of 50–2,000 μm (and far smaller propagules; for an introduction see Wallace et al. 2006), are perfectly suited to test this hypothesis, which, in fact, has already raised substantial criticism (Foissner 1999, 2006; Whitaker et al. 2003; Lachance 2004; Jenkins et al. 2007). The debate on cosmopolitanism has a long history in rotiferology (review in Segers 1996) and is still being fed by new data derived from diverse research approaches. Here, we present on overview of the knowledge on diversity and endemicity of rotifers in the light of the ubiquity theorem and illustrate the issue by a case study, namely that of the common and ubiquitous genus *Keratella* Bory de St. Vincent (Figs. 1, 2). *Keratella* rotifers are found on all continents and several species are amongst the most common pelagic rotifers. The genus as a whole is eurytopic and cosmopolitan, yet attempts to analyse its chorology have, while demonstrating some intriguing features, so far not been able to elucidate a general pattern (Pejler 1977; De Ridder 1981; Dumont 1983).

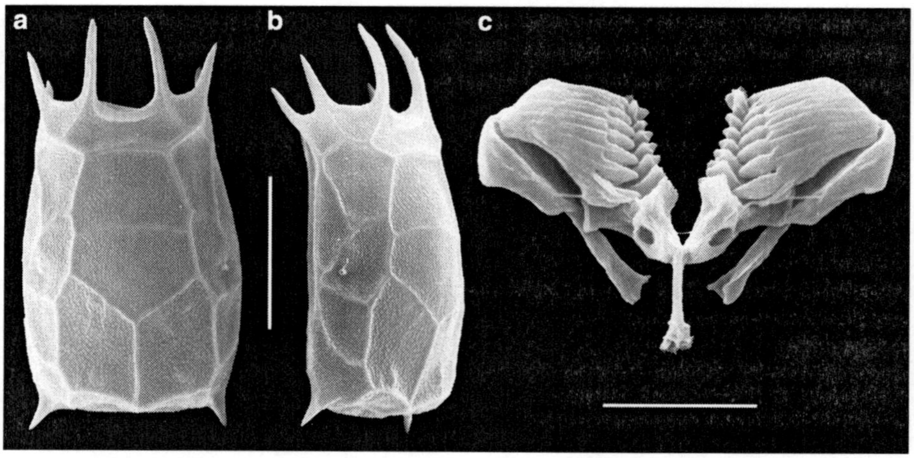

Fig. 1 *Keratella sancta*, known from the Kerguelen Islands, Macquarie I, and New Zealand. (**a**) Surface ornamentation of lorica, dorsal view; (**b**) ibidem, lateral view; (**c**) masticatory apparatus or trophi, ventral view

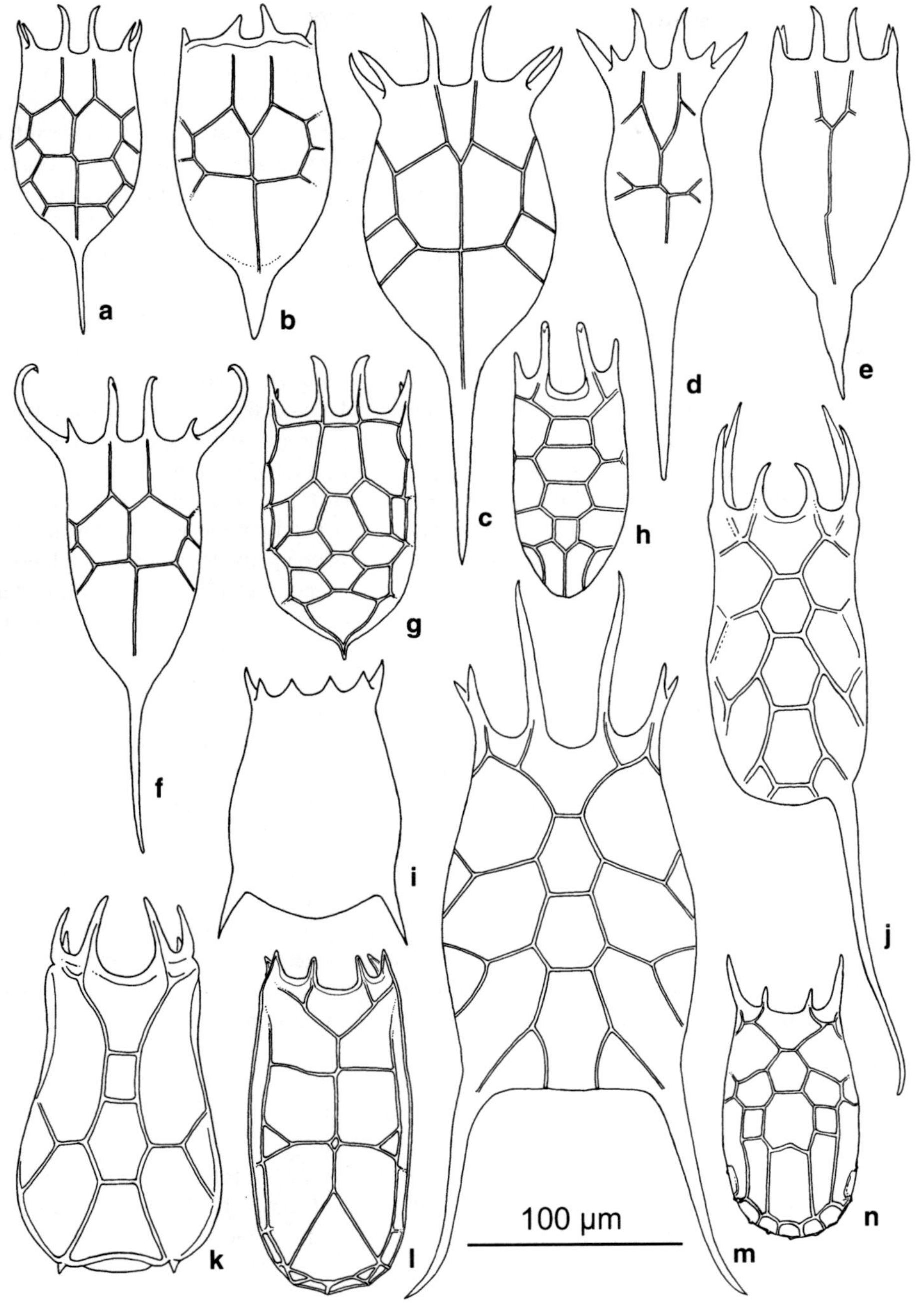

Fig. 2 *Keratella* spp. (**a**) *K. cochlearis*, cosmopolitan; (**b**) *K. cochlearis polaris*, Canadian high arctic; (**c**) *K. cochlearis* var. *faluta*, northern USA, Canada; (**d**) *K. armadura*, Michigan, USA; (**e**) *K. cochlearis pachyacantha*, tropical Africa, South America; (**f**) *K. taurocephala*, northeast North America; (**g**) *K. yamana*, Tierra del Fuego, Argentina; (**h**) *K. taksinensis*, southern Thailand; (**i**) *K. kostei*, Patagonia, Falkland Island, South Georgia; (**j**) *K. tropica* f. *taurocephala*, the Caspian; (**k**) *K. shieli*, Victoria, southern Australia; (**l**) *K. mongoliana*, inner Mongolia, PR China; (**m**) *K. quadrata* var. *adnata*, Lake Christine, Alberta, Lake Lilian, BC, Canada; (**n**) *K. reducta*, Cape region, South Africa

The taxonomic impediment

One of the issues raised in most of the relevant communications is that any analysis has to be interpreted in the context of present taxonomic knowledge, which is far from perfect in most microscopic organism groups (Fenchel and Finlay 2004; Lachance 2004). Also rotifer students have lamented the state of knowledge in taxonomy (e.g. Koste in Dumont 1980), and no treatise on rotifer chorology or diversity can circumvent the problem. The causes of confusion are well known (Koste and Shiel 1989; Ruttner-Kolisko 1989; Segers 1998) and break down to (purportedly) high morphological variability or cyclomorphosis including the rotifer's ability to phenotypic plasticity, their parthenogenetic reproduction, and, regretfully, poor quality of taxonomic work. The *Keratella cochlearis* complex is illustrative: here, caudal spine length and degree of development of lorica pustulation, amongst other features, were demonstrated as being variable relatively early (Lauterborn 1898, 1900; Ahlstrom 1943). A less desirable consequence of this discovery was that morphologically similar entities were generally lumped in encompassing species concepts, resulting in a complicated and inaccessible nomenclature including subspecies, infrasubspecific variants, and frequents shifts in rank (Koste 1978). For bdelloids, the situation is similar: their exclusive parthenogenetic reproduction, in combination with an apparently high morphological variability and a lack of students, account for the confused bdelloid taxonomy. That diversity has been underestimated is supported by recent studies that demonstrated high genetic taxonomic diversification concealed by, occasionally superficial, morphological stasis both in monogononts (e.g. *Brachionus plicatilis* group: Gomez et al. 2002; Suatoni et al. 2006) and bdelloids (Fontaneto et al. 2007). It is highly likely that such cryptic diversity is far more common than thought in rotifers.

Furthermore confounding is the obvious lack in precision of many rotifer identifications and that a multitude of records are produced during routine surveys or limnological studies. These studies are concentrated in a few regions and habitat types and focus on taxa that are sometimes mistakenly considered "easy", like *Keratella* (e.g. Bērziņš 1955). For example, De Ridder and Segers (1997) contains four full pages of records of the pelagic *K. cochlearis* from the Palaearctic region (mostly Europe), versus a quarter of a page each from the Australian and Oriental regions. The notoriously difficult genus *Encentrum* (100+ species), living in littoral or interstitial habitats, takes up seven pages in the same work. All this impedes the derivation of the area of rotifers from historical records and exhorts a critical approach to the use of literature data in the study of the distribution and diversity of rotifers.

General patterns in diversity and endemism in Rotifera

In a recent review of the global diversity and distribution of the rotifers (Segers 2007a), building on the works of Green (1972), Pejler (1977), De Ridder (1981), Dumont (1983), and Segers (1996, 2003), the following patterns could be discerned:

Diversity

- Rotifera totals 1,571 valid species of Monogononta, 461 of Bdelloidea and 3 Seisonacea, which is an underestimate of extant diversity. Only 69 monogononts, 1 bdelloid and the 3 Seisonacea are exclusively marine, all others are freshwater or brackish water and marine.

- Local species diversity of monogonont rotifers reaches about 150 species per lake in temperate, and up to 250 species in tropical regions. Consequently, 7.5–12.5% of all global species diversity can be found in a single locality. In contrast, Fontaneto et al. (2006b) found a low local species richness with strong habitat selection in both limnoterrestrial and freshwater bdelloid communities inhabiting cushions of dry mosses, and flocks of submerged mosses in running waters and lakes, but reported fairly high local diversity accounting for about 20% of all known bdelloids worldwide. Differences in scale may account for the apparent disparity.
- The relatively low global versus high local diversity is consistent with the hypothesis that small-sized organisms tend to have cosmopolitan distributions (Fenchel and Finlay 2004). On the other hand, many rotifer taxa exhibit complex patterns of distribution and endemism.
- There is a strong discrepancy between diversity and distribution patterns in pelagic habitats, typically with low species diversity and high abundances, and littoral habitats where the opposite is the case.
- Latitudinal variation in the distribution of rotifers is clear in a number of important taxa (*Lecane*, *Brachionus*), which reach their highest diversity in (sub)tropical or, at least, warm waters whereas only a few, less diverse taxa (*Synchaeta, Notholca*) are predominantly cold water; as a result, rotifer diversity in general declines from the equator to the poles. The most diverse taxa (100+ species: *Lecane, Cephalodella, Lepadella*) contain almost exclusively benthic-littoral or psammon-inhabiting species, with a majority inhabiting oligo- to mesotrophic, slightly acidic, soft waters; *Brachionus* is an exception, generally preferring alkaline and eutrophic conditions; local species diversity is especially high in littoral-benthic habitats; in contrast, pelagic diversity is much lower.
- Diversity hotspots are hard to discern, as research intensity appears to be the major factor determining available data on regional diversity and comparable data are scarce.

Endemicity

- About 22% of non-marine monogononts are regional endemics, whereas a surprising 50% of bdelloid species are purportedly endemic. Many of these, however, are only known from their type locality and have not been recorded since their description, for instance, ca. 25% of the about 300 European bdelloids (Fontaneto and Melone 2003). Endemics occur in all geographic regions and may be among the most frequently observed and abundant species, e.g. *Philodina gregaria* on the Antarctic continent and isles. Some 60 species are cosmopolitan.
- Reliable examples of endemics at higher taxonomic levels are particularly rare in rotifers; species-level endemicity occurs at diverse scales in all regions and regards all but the species-poorest genera and families.
- Endemicity hotspots appear to exist in tropical South America (especially in the genera *Anuraeopsis, Brachionus, Macrochaetus*), Australia (*Brachionus, Keratella*), northeast North America (endemic Birgeidae, *Streptognatha, Pseudoploesoma*; species of *Trichocerca, Lecane*), and Lake Baikal (*Notholca*); endemicity is relatively low in Africa (including Madagascar) and the Indian subcontinent; in contrast to Lake Baikal, records of endemics from other ancient lakes are scarce.

A case study: diversity and endemism in the genus *Keratella*

Approach

For the analysis of diversity and endemism in *Keratella*, we started off with the taxonomy as in Koste (1978), Koste and Shiel (1987) and Segers (2007b). Here, however, we include a number of taxa that are presently considered of infrasubspecific rank, but are either being characterized by morphological features that have been confirmed as taxonomically relevant in the diagnosis of species-level taxa in the genus (e.g. foundation pattern and anterior spine morphology) and/or that have reliably and repeatedly been reported from circumscribed areas. This approach is in line with preliminary results of molecular analysis (Derry et al. 2003) indicating cryptic speciation in *K. cochlearis*. As taxonomy is not the scope of this paper, we merely list them as they occur most commonly in the literature. Distributional data are based on the literature reviews of De Ridder (1986, 1991, 1993), and De Ridder and Segers (1997). Isolated regional records of species otherwise common in other regions were critically assessed and included only after verification of published illustrations or material.

Biogeography of *Keratella*

A list of the *Keratella* included in the analysis and their distribution is as in Table 1. We recognize 53 relevant taxa, rather than the 48 recorded by Segers (2007a, b). This is a significant increase since the analyses by Peljer (1977; 25 taxa) and Dumont (1983; 37 taxa). A minority only (eight or 15%) can be called cosmopolitan, and even within this group some are restricted to warmer (*K. lenzi*, *K. procurva procurva, K. tropica*) or colder (*K. quadrata*) climates. Even if some widespread taxa (e.g. *K. serrulata*) are added to this category, the proportion of true cosmopolites remains surprisingly low.

The northern hemisphere is endowed with a number of Holarctic (5: 9.5%, also *K. testudo*?), Palearctic (6: 11.3%) and Nearctic (8: 15%) endemics. In addition, all marine taxa (5: 9.5%) known to date are restricted to the northern hemisphere. However, knowledge on marine rotifers is sketchy at best, and the isolated marine records of freshwater *Keratella* from the southern hemisphere may require re-examination (De Ridder and Segers 1997). Endemism in other regions varies greatly, from two (Africa) to eight (Neotropical region). Remarkably, most endemism in the southern hemisphere is concentrated in temperate and cold regions. This holds for the single reliable African, as well as for all but one (*K. australis*) Australian/New Zealand and six out of eight Neotropical species. Many of the taxa concerned (*K. kostei, K. ona, K. yamana, K. reducta, K. sancta, K. shieli*) are phylogenetically isolated; they may represent a true Gondwanan faunal element, being represented in the temperate and cold regions of South America, South Africa, South Australia and New Zealand. This contrasts with the late Cretaceous South American-Antarctic-Australian origin purported for the tropic-centred *Brachionus* by Segers (2007a).

In addition to the, for Rotifera, unique southern hemisphere temperate fauna, two more assemblages stand out. First, endemism in the Palaearctic region is centred in the southeast (China), whereas most Nearctic endemics occur in the northern boreal and northeast temperate part. This pattern reminds of a postglacial relict distribution. However, the Chinese endemics are of diverse affiliation and may indeed be ancient relics, whereas most of the northern North American taxa are close relatives of *K. cochlearis*, by sharing this

Table 1 List of *Keratella* spp with their distribution

Cosmopolitan species

Keratella cochlearis (Gosse, 1851)

Keratella lenzi Hauer, 1953 (tropicopolitan)

Keratella procurva procurva (Thorpe, 1891) (tropicopolitan)

Keratella quadrata (Müller, 1786) (cold water)

Keratella tecta (Gosse, 1851)

Keratella testudo (Ehrenberg, 1832) (common in Nearctic and Palearctic regions; two African records (Mauretania, Congo); one Venezuelan record)

Keratella tropica (Apstein, 1907) (warm-water; several atypical cold-water records exist)

Keratella valga (Ehrenberg, 1834)

Holarctic species

Keratella hiemalis Carlin, 1943 (incl. Himalaya region)

Keratella irregularis (Lauterborn, 1898) (northern North America; Palearctic, incl. Nepal)

Keratella mixta (Oparina-Charitonova, 1924) (single illustrated Thai record)

Keratella paludosa (Lucks, 1912) (single doubtful, pre-1930 Congo record)

Keratella ticinensis (Callerio, 1921) (incl. Himalaya; doubtful pre-1940 African records)

Widespread species

Keratella cochlearis pachyacantha Thomasson, 1980 (tropical Africa, South America)

Keratella javana Hauer, 1937 (tropical Old World; New Zealand, Tasmania)

Keratella serrulata (Ehrenberg, 1838) (not in Australia)

Regional and local endemics

Afrotropical

Keratella maliensis Koste & Tobias, 1987: AFR (Mali) (doubtful taxon)

Keratella reducta (Huber-Pestalozzi, 1929) (Cape region, South Africa)

Australia

Keratella ahlstromi Russell, 1951 (South Island, New Zealand)

Keratella australis Bērziņš, 1963 (Australia incl. Tasmania)

Keratella procurva robusta Koste & Shiel, 1980 (temperate Australia incl. Tasmania)

Keratella sancta Russell, 1944 (Kerguelen, Macquarie Isl., New Zealand)

Keratella shieli Koste, 1979 (Victoria, South Australia)

Keratella slacki Bērziņš, 1963 (temperate Australia, incl. Tasmania)

Palaearctic

Keratella cochlearis nordica Kutikova, 1978 (Siberian tundra)

Keratella mongoliana Segers & Rong, 1998 (inner Mongolia, PR China)

Keratella sinensis Segers & Wang, 1997 (Lake Yaoquan, Wudalianchi; Shanghai, PR China; South Korea)

Keratella trapezoida Zhuge & Huang, 1998 (Dongting Lake, Hunan province, PR China)

Keratella wangi Zhuge & Huang, 1997 (Yangtze river near Dongting Lake, Hunan province, PR China)

Keratella zhugeae Segers & Rong, 1998 (Indian Tibet; Inner Mongolia, PR China)

Nearctic

Keratella armadura Stemberger, 1990 (Michigan, USA)

Keratella canadensis Bērziņš, 1954 (boreal region of North America)

Keratella cochlearis polaris De Smet & Bafort, 1990 (Canadian high arctic)

Table 1 continued

Keratella cochlearis var. *faluta* Ahlstrom, 1943 (northern USA, Canada)

Keratella crassa Ahlstrom, 1943

Keratella quadrata var. *adnata* Ahlstrom, 1943 (= *Keratella quadrata neali* Bērziņš, 1961; Christine Lake, Alberta; and Lilian Lake, BC, Canada)

Keratella taurocephala Myers, 1938 (northeast North America)

Keratella earlinae Ahlstrom, 1943 (northeast North America; single record from Grand Bahamas)

Neotropical

Keratella mexicana Kutikova & Silva-Briano, 1995 (Mexico, northern Brazil)

Keratella nhamundaiensis Koste, 1982 (central Amazonia, northern Argentina)

Keratella ona Boltovskoy & Urrejola, 1977 (Tierra del Fuego, Argentina)

Keratella kostei Paggi, 1981 (Patagonia, the Falkland Islands, and South Georgia Island)

Keratella thomassoni Hauer, 1958 (Tepuhuico Lake, southern Chile)

Keratella morenoi Modenutti, Diéguez & Segers, 1998 (Argentina, Patagonian plateau)

Keratella valdiviensis Thomasson, 1957 (Valdivian Lakes, Chilean Patagonia)

Keratella yamana Boltovskoy & Urrejola, 1977 (Tierra del Fuego, Argentina)

Oriental

Keratella edmondsoni Ahlstrom, 1943 (India, Thailand)

Keratella taksinensis Chittapun, Pholpunthin & Segers, 2002 (southern Thailand)

Marine taxa

Keratella cochlearis var. *recurvispina* (Jägerskiöld, 1894)

Keratella cruciformis (Thompson, 1892) (Atlantic: Europe, Canada; Laptev Sea)

Keratella eichwaldi (Levander, 1894) (Holarctic)

Keratella quadrata var. *platei* (Jägerskiold, 1894) (Black Sea, East Sea, Baltic Sea, Northern North Sea)

Keratella tropica f. *taurocephala* Koste, 1978 (southern part of Caspian Sea)

taxon's typical, and probably apomorphic, foundation pattern of lorica ornamentation. This clade, which is in dire need of revision, further contains *K. tecta*, *K. cochlearis pachyacantha*, *K. valdiviensis* and *K. cochlearis nordica* in the present analysis, and a suite of named infrasubspecific variants. The taxa in this group are undoubtedly closely affiliated and may result from a relatively recent radiation, when compared to the origin of the Chinese group.

There are few widespread tropical species, and true tropical endemics occur only in the Oriental (*K. edmondsoni*, *K. taksinensis*) and Neotropical region (*K. mexicana*, *K. nhamundaiensis*). *K. javana* can hardly be considered tropical as it occurs as far south as New Zealand and Tasmania, where it overlaps with Australia's share of southern hemisphere temperate and cold-water taxa. By this peculiar distribution and its isolated phylogenetic position, we hypothesize that it belongs to south Australia's Gondwanan *Keratella* group, but has expanded secondarily into the Australian and later Old World tropics. Another peculiar case is that of *K. americana*, which is exceedingly common in the Americas and of which isolated records exist from the eastern hemisphere; it is likely that the latter records concern introductions (Segers 2001). Peljer (1977) formulated similar comments on the presence of *K. irregularis* in North America: also this taxon is fairly common in the Palearctic, whereas records from the Nearctic are rare and recent only. The presence of *K. cochlearis pachyacantha* in Africa, where it appears to be far less common in suitable habitats than in South America, could be explained in a similar way.

When compared to other rotifer taxa for which sufficient reliable data are available (*Brachionus*: see Pejler 1977, Dumont 1983, Segers 2007a; *Lecane*: Segers 1996, *Trichocerca*: Segers 2003), *Keratella* turns out to contain the smallest fraction of cosmopolitan (sensu lato) species. We could identify two diverse, possibly ancient groups of endemics: one in the southeast Palaearctic (China) and another, probably a Gondwanan relict, in the southern hemisphere temperate and cold regions. In addition to these, there is a possibly recent radiation in the *K. cochlearis* group in North America. Whereas the two former have no counterpart in other groups, the latter is complemented, in the northeast Nearctic, by endemics in *Lecane* and *Trichocerca* (Segers 1996, 2003). Only the Nearctic group of endemics had been discerned in *Keratella* before (Bērziņš 1954; Pejler 1977).

Processes of dispersal and colonization

The main processes underpinning the extant patterns in the diversity and endemism of rotifers are, obviously, their mode of dispersal and reproduction.

Propagules

Rotifer propagules consist of dormant mictic or resting eggs, i.e. hard-shelled, encysted embryos in monogononts or equally dormant anhydrobiotic animals and parthenogenetically produced eggs in bdelloids. Both resting eggs and anhydrobiotic stages of bdelloids have a capacity for extended periods of dormancy (from days to years depending the species), are minute (most smaller than 100 μm), and are the main stages of dispersal (Gilbert 1974; Pourriot and Snell 1983; Schröder 2005).

Resting eggs are characterized by a thick shell, which often is ornamented with bristles, spines, knobs, bubbles, etc. The function of eggshell ornamentation is not well known, but it may enhance buoyancy and/or aid in dispersal. Resting eggs are resistant to drought, heat, cold, or other adverse chemical or physical conditions of the habitat for long periods, thus aiding in the dispersal of the species (Wallace et al. 2006). The thickness of the shell and its reduced porosity may also protect against digestive enzymes following ingestion, and hence resting eggs could survive gut passage of predators.

The great majority of bdelloid species are capable of anhydrobiosis (Ricci 1998; Ricci and Caprioli 2005), a form of quiescence in response to unpredictable and fast changes of water availability in unstable habitats. When subjected to drying, bdelloids withdraw their head and foot into the trunk and contract into a compact ‘barrel’. Desiccation during tun formation reduces volume and density, which results in very small and light propagules suitable for aerial transport. Survival of desiccated bdelloid eggs has been rarely studied and demonstrated in a few species only (Dobers 1915; Örstan 1995; Ricci et al. 1987). Egg desiccation survival times of 2 days to 14 months have been reported, indicating that bdelloid eggs may form airborne propagules as well.

Clearly, the nature of rotifer propagules is a strong argument supporting the validity of the ubiquity theorem in these microscopic organisms. However, viability of propagules can be reduced by environmental factors, like extended exposure or physical and chemical unsuitability of the substratum or habitat (Chittapun et al. 2005). Likewise, the implicit generalization that propagules of all rotifers species are efficient is unlikely correct, which could account for the local or regional, "ancient" endemics mentioned above. For one, the "classic" sampling methodology for bdelloids, viz. examination of rehydrated, dry moss samples may preclude discovery of possible local taxa that have inefficient anhydrobiotic capacities.

Dispersal

Rotifers achieve dispersal in space and time through passive transport by water currents, wind or animal vectors (Örstan 1998; Wallace et al. 2006). It is commonly assumed that transport by wind is most likely the major way of long-distance dispersal of the desiccation-resistant propagules, as has been demonstrated for algae and Protozoa (Gislén 1948; Schlichting 1961, 1964). Direct evidence was provided by Jenkins and Underwood (1998) who looked for rotifers in samples collected by windsocks and in rain samplers, and by Janiec (1996) who used traps filled with 4% formalin solution. Their results show that, contrary to the commonly accepted premise, rotifers may not disperse that readily by wind or rain. Indirect evidence of a potential of primary dispersal by wind and rain was provided by several dispersal-colonization experiments with artificial bodies of water (Maguire 1963; McCormick and Cairns 1990; Holland and Jenkins 1998; Cáceres and Soluk 2002).

In running waters active animals and propagules, originating from the catchment area or from the bottom of the riverbed, are transported by downstream drift (Sandlund 1982). The formation of temporary connections between stagnant waters during heavy rain-fall and flooding (Jenkins and Boulton 2003), or e.g. occasional breaching of Antarctic ponds and consequent formation of streams (Suren 1990) also aids in dispersal over 100s of metres to 1,000s of kilometres. Oceanic currents may be responsible for the cosmopolitanism of many marine species (Fontaneto et al. 2006a), and rotifers living in multi-year sea ice are transported by the 'Transpolar Drift' from the Eurasian shelf regions over the central Arctic Basin to the Greenland Sea (Friedrich and De Smet 2000).

Animal-mediated passive transport of rotifers ("zoochory"), both internal and external, has hardly been studied. Insects, amphibians and birds were found carrying propagules and/or active rotifers externally on their body (Maguire 1959, 1963; Schlichting 1960; Schlichting and Milliger 1969; Sides 1973). Internal transport in the gut of animal vectors, followed by defecation of viable stages surviving digestion, has yet not been demonstrated in rotifers; Wallace et al. (2006) erroneously state that Euliss et al. (1991) reported endozoochory in waterfowl. However, active bdelloids and developing eggs have been observed in the rectum or midgut of dipterid larvae belonging to *Culex*, *Chironomus* (Marchoux 1898) and *Simulium* (Nelder and McCreadie 2003), suggesting that internal transport, at least in insects, may occur.

Dispersal of rotifers can also be achieved through human activities, e.g. intentional stocking programs for fisheries (Pejler 1998), release from aquaria (Padilla and Williams 2004), release with bait fish (Havel and Stelzleni-Schwent 2000), release of ballast water (Bailey et al. 2003), etc. Studying potential survival after ballast exchange, Gray et al. (2005) demonstrated that the viability of freshwater rotifer resting eggs was not affected after exposure to open-ocean water (32‰), despite the low salinity tolerance of the adult stages of most species.

Colonization and adaptation

As rotifer popagules consist of resistant, dormant stages, they remain viable after dispersal, even when the target habitat is unsuitable at the moment of arrival. Moreover, due to their parthenogenetic reproduction a single specimen is able of establishing a new population. Consequently, the colonization capability of rotifers is potentially very high. Indeed, experiments studying colonization of artificial ponds (Jenkins 1995) or mesocosms (Langley et al. 2001; Cáceres and Soluk 2002) found rapid colonization by rotifers, amongst other zooplankton.

On the other hand, different factors thwart successful colonization. Competition by occupant species or populations appears to play a major role, as indicated by experiments by Shurin (2000). Similarly, the presence of a large propagule bank of locally adapted genotypes may present an effective barrier against newly arriving genotypes (De Meester et al. 2002).

Phylogenetic analysis of mitochondrial *cox1* genes by Birky et al. (2005) showed a significant correlation between geographical distance and sequence difference, but with indication of rapid dispersal. Using the rate of dispersal they found that bdelloid species could have dispersed around the world many times since their divergence, and suggested that continued independent evolution, in absence of isolation by distance, can be most easily explained by adaptation to different niches. Mills et al. (in press) found strong phylogenetic structure in the saline water monogonont *B. plicatilis* (sensu stricto), in addition to a strong correlation between genetic and geographic distance. They conclude that this is likely to have been produced through the colonization process, and that even potentially dispersive microscopic organisms can achieve substantial geographic subdivision which can lead to reproductive incompatibility. These intriguing results appear compatible with the patterns outlined above. Whereas a substantial body of data is becoming available on species-level chorology, however, hiatuses regarding cryptic taxa and phylogeography hamper our understanding of diversity and distribution of these microscopic metazoans. The debate on the ubiquity theorem will possibly continue to rage on for some time to come. However, it is clear that the diversity of distribution patterns in rotifers varies widely. Both local endemism and broad cosmopolitanism as well as intermediates occur, and either of the alternatives may be common, depending on the taxon. If chorology of rotifers is representative of that of other microscopic organisms, then it may be more productive to examine the relative importance and diversity of distribution patterns, and the factors determining these, at different taxonomic levels, in order to come to an understanding of their chorology and biodiversity. Basically, the rotifer data match the moderate endemicity model of Foissner (2006) and Chao et al. (2006) better than the ubiquity model of Finlay (2002) and Fenchel and Finlay (2004).

A special case: Seisonacea and other marine rotifers

The above treatments of dispersal and propagules hold for bdelloid and monogonont rotifers; Seisonacea are exclusively marine and composed of two genera: *Seison* with two species (*S. africanus*, *S. nebaliae*) and the monospecific *Paraseison annulatus* (Ricci et al. 1993; Sørensen et al. 2005). *Paraseison annulatus* and *S. nebaliae* both are exclusively epibiotic on species of the leptostracan crustacean *Nebalia* and show a similar distribution: Mediterranean, North Sea, European part of northern Atlantic and the Amiva Bay, Sakhalin Island, Sea of Okhotsk. *Seison africanus* was described from Gazi Bay, Indian

Ocean, Kenya; it was not attached to a host, and *Nebalia* was absent. Unidentified specimens have been found on *Nebalia* at Morro Bay, Pacific Ocean, California; McMurdo Sound, Ross Sea, Antarctica; and Magellanic Chile. *Seisonacea* are bisexual and no resting stages are present. Hardly anything is known about the life cycle and dispersal of Seisonacea, but their obligatory relation with *Nebalia*, sexual reproduction and absence of known resting stages or specialized propagules appear key to their distribution. Considering the evolutionary age and ubiquity of *Nebalia*, and the paucity of studies on Seisonacea, it is highly likely that diversity of the taxon is greatly underestimated. In general, it is presently hardly possible to derive patterns on the diversity and distribution of marine rotifers due to lack of data.

References

Ahlstrom EH (1943) A revision of the Rotatorian genus *Keratella* with descriptions of three new species and five new varieties. Bull Am Mus Nat Hist 80:411–457

Baas-Becking LGM (1934) Geobiologie of Inleiding tot de Milieukunde. Van Stockum en Zoon, The Hague, The Netherlands

Bailey SA, Duggan IC, van Overdijk et al (2003) Viability of invertebrate diapausing eggs collected from residual sediment. Limnol Oceanogr 48:1701–1710

Bērziņš B (1954) A new rotifer, *Keratella canadensis*. J Quekett Microsc Club 4:113–115

Bērziņš B (1955) Taxonomie und Verbreitung von *Keratella valga* und verwandten Formen. Ark Zool 8:549–559

Birky CW, Wolf C, Maughan H et al (2005) Speciation and selection without sex. Hydrobiologia 546:29–45

Cáceres CE, Soluk DA (2002) Blowing in the wind: a field test of overland dispersal and colonization by aquatic invertebrates. Oecologia 131:402–408

Chao A, Li P, Agatha S et al (2006) A statistical approach to estimate soil ciliate diversity and distribution based on data from five continents. Oikos 114:479–493

Chittapun S, Pholpunthin P, Segers H (2005) Restoration of tropical peat swamp rotifer communities after perturbation: an experimental study of recovery of rotifers from the resting egg bank. Hydrobiologia 546:281–289

De Meester L, Gómez A, Okamura B et al (2002) The monopolization hypothesis and the dispersal-gene flow paradox in aquatic organisms. Acta Oecol 23:121–135

De Ridder M (1981) Some considerations on the geographical distribution of rotifers. Hydrobiologia 85:209–225

De Ridder M (1986) Annotated checklist of non-marine rotifers (Rotifera) from African inland waters. Zoologische Documentatie, Koninklijk Museum voor Midden-Afrika, Tervuren, Belgium, 21, 123 pp

De Ridder M (1991) Additions to the "annotated checklist of non-marine rotifers (Rotifera) from African inland waters". Rev Hydrobiol Trop 24:25–46

De Ridder M (1993) Additions II to the "annotated checklist of non-marine rotifers (Rotifera) from African inland waters". Biol Jb Dodonaea 61:99–153

De Ridder M, Segers H (1997) Rotifera Monogononta in six zoogeographical regions after publications between 1960–1992. In: Van Goethem J (ed) Studiedocumenten van het Koninklijk Belgisch Instituut voor Natuurwetenschappen, 481 pp

Derry AM, Hebert PDN, Prepas EE (2003) Evolution of rotifers in saline and subsaline lakes: a molecular phylogenetic approach. Limnol Oceanogr 48:675–685

Dobers E (1915) Über die Biologie der Bdelloidea. Int Rev Hydrobiol Hydrograph, Suppl 7:1–128

Dumont HJ (1980) Workshop on taxonomy and biogeography. In: Dumont HJ, Green J (eds) Rotatoria. Hydrobiologia 73:205–206

Dumont HJ (1983) Biogeography of rotifers. Hydrobiologia 104:19–30

Euliss ND, Jarvis RL, Gilmer DS (1991) Feeding ecology of waterfowl wintering on evaporation ponds in California. Condor 93:582–590

Fenchel T, Finlay JB (2004) The ubiquity of small species: patterns of local and global diversity. Bioscience 54:777–784

Finlay JB (2002) Global dispersal of free-living microbial eukaryote species. Science 96:1061–1063

Foissner W (1999) Protist diversity: estimates of the near-imponderable. Protist 150:363–368

Foissner W (2006) Biogeography and dispersal of micro-organisms: a review emphasizing protists. Acta Protozool 45:111–136

Fontaneto D, Melone G (2003) Redescription of *Pleuretra hystrix*, an endemic alpine bdelloid rotifer. Hydrobiologia 497:153–160

Fontaneto D, De Smet WH, Ricci C (2006a) Rotifers in saltwater environments, re-evaluation of an inconspicuous taxon. J Mar Biol Assoc UK 86:623–656

Fontaneto D, Ficetola GF, Ambrosini R et al (2006b) Patterns of diversity in microscopic animals: are they comparable to those in protists or in larger animals? Glob Ecol Biogeogr 15:153–162

Fontaneto D, Herniou EA, Boschetti C et al (2007) Independently evolving species in asexual bdelloid rotifers. PLoS Biol 5:1–8

Friedrich C, De Smet WH (2000) The rotifer fauna of arctic sea ice from the Barents Sea, Laptev Sea and Greenland Sea. Hydrobiologia 432:73–89

Gilbert JJ (1974) Dormancy in rotifers. Trans Am Microsc Soc 93:490–513

Gislén T (1948) Aerial plankton and its conditions of life. Biol Rev 23:109–126

Gomez A, Serra M, Carvalho GC et al (2002) Speciation in ancient cryptic species complexes: evidence from the molecular phylogeny of *Brachionus plicatilis* (Rotifera). Evolution 56:1431–1444

Gray DK, Bailey SA, Duggan IC et al (2005) Viability of invertebrate diapausing eggs exposed to saltwater: implications for Great Lakes' ship ballast management. Biol Invasions 7:531–539

Green J (1972) Latitudinal variation in associations of planktonic Rotifera. J Zool, Lond 167:31–39

Havel JE, Stelzleni-Schwent J (2000) Zooplankton community structure: the role of dispersal. Verh Int Ver Limnol 27:3264–3268

Holland LE, Jenkins DG (1998) Comparison of processes regulating zooplankton assemblages in new freshwater pools. Hydrobiologia 387/388:207–214

Janiec K (1996) Short distance wind transport of microfauna in maritime Antarctic (King George Island, South Shetland Islands). Pol Polar Res17:203–211

Jenkins DG (1995) Dispersal-limited zooplankton distribution and community composition in new ponds. Hydrobiologia 313/314:15–20

Jenkins KM, Boulton AJ (2003) Connectivity in a dryland river: short-term aquatic microinvertebrate recruitment following floodplain inundation. Ecology 84:2708–2723

Jenkins DG, Underwood MO (1998) Zooplankton may not disperse readily in wind, rain, or waterfowl. Hydrobiologia 387/388:15–21

Jenkins DG, Brescacin CR, Duxbury CV et al (2007) Does size matter for dispersal distance? Glob Ecol Biogeogr 16:415–425

Jennings HS (1900) Rotatoria of the United States, with especial reference to those of the Great Lakes. Bull US Fish Comm 1899:67–104

Koste W (1978) Rotatoria. Die Rädertiere Mitteleuropas, vol 2. Borntraeger, Berlin

Koste W, Shiel RJ (1987) Rotifera from Australian inland waters. II. Epiphanidae and Brachionidae (Rotifera: Monogononta). Invertebr Taxon 7:949–1021

Koste W, Shiel RJ (1989) Classical taxonomy and modern methodology. Hydrobiologia 186/187:79–284

Lachance M-A (2004) Here and there or everywhere? Bioscience 54:884

Langley JM, Shiel RJ, Nielsen DL et al (2001) Hatching from sediment egg-bank, or aerial dispersing?-the use of mesocosms in assessing rotifer biodiversity. Hydrobiologia 446/447:203–211

Lauterborn R (1898) Vorläufige mittheilung über den Variationskreis von *Anurea cochlearis* Gosse. Zool Anz 21:597–604

Lauterborn R (1900) Der Formenkreis von *Anurea cochlearis*. Ein Beitrag zur Kenntniss der Variabilität bei Rotatorien. Verh Nat-Med Ver Heidelberg 6:412–448

Maguire BJ (1959) Passive overland transport of small aquatic organisms. Ecology 40:312

Maguire BJ (1963) The passive dispersal of small aquatic organisms and their colonization of isolated bodies of water. Ecol Monogr 33:161–185

Marchoux E (1898) Note sur un rotifère (*Philodina parasitica* n. sp.) vivant dans le tube digestif de larves aquatiques d'insectes. Compt rend Séanc Soc Biol Fil, Sér 10:749–750

McCormick PV, Cairns J (1990) Microbial colonization dynamics in temporary aquatic systems. Hydrobiologia 196:229–253

Mills S, Lunt DH, Gómez A (in press) Global isolation by distance despite strong regional phylogeography in a small metazoan. BMC Evol Biol

Nelder MP, McCreadie JW (2003) Bdelloid rotifers (Rotifera: Bdelloidea) inhabiting larval black flies (Diptera: Simuliidae) and their effect on trichomycete (Zygomycota) fungal abundance. Proc Entomol Soc Wash 105:794–796

Örstan A (1995) Desiccation survival of the eggs of the rotifer *Adineta vaga* (Davis, 1873). Hydrobiologia 313/314:373–375

Örstan A (1998) Microhabitats and dispersal routes of bdelloid rotifers. Sci Nat 1:27–36

Padilla DK, Williams SL (2004) Beyond ballast water: aquarium and ornamental trades as sources of invasive species in aquatic ecosystems. Front Ecol Environ 2:131–138

Pejler B (1977) On the global distribution of the family Brachionidae (Rotatoria). Arch Hydrobiol 53(Suppl):255–306

Pejler B (1998) History of rotifer research in northern Europe. Hydrobiologia 387/388:267–278

Pourriot R, Snell TW (1983) Resting eggs in rotifers. Hydrobiologia 104:213–224

Ricci C (1998) Anhydrobiotic capabilities of bdelloid rotifers. Hydrobiologia 387/388:321–326

Ricci C, Caprioli M (2005) Anhydrobiosis in bdelloid species, populations and individuals. Integr Comp Biol 45:759–763

Ricci C, Vaghi L, Manzini ML (1987) Desiccation of rotifers (*Macrotrachela quadricornifera*): survival and reproduction. Ecology 68:1488–1494

Ricci C, Melone G, Sotgia C (1993) Old and new data on Seisonidea (Rotifera). Hydrobiologia 255/256:495–511

Rousselet CF (1909) On the geographical distribution of the Rotifera. J Quekett Microsc Club 10:465–470

Ruttner-Kolisko A (1989) Problems in the taxonomy of rotifers, exemplified by the *Filinia longiseta–terminalis* complex. Hydrobiologia 186/187:291–298

Sandlund OT (1982) The drift of zooplankton and microzoobenthos in the river Strandaelva, western Norway. Hydrobiologia 94:33–48

Schlichting HE (1960) The role of waterfowl in the dispersal of algae. Trans Am Microsc Soc 79:160–166

Schlichting HE (1961) Viable species of algae and Protozoa in the atmosphere. Lloydia 24:81–88

Schlichting HE (1964) Meteorological conditions affecting the dispersal of airborne algae and Protozoa. Lloydia 27:64–78

Schlichting HE, Milliger LE (1969) The dispersal of microorganisms by a hemipteran, *Lethocerus uhleri* (Montadon). Trans Am Microsc Soc 88:452–454

Schröder T (2005) Diapause in monogonont rotifers. Hydrobiologia 546:291–306

Segers H (1996) The biogeography of littoral *Lecane* Rotifera. Hydrobiologia 323:169–197

Segers H (1998) An analysis of taxonomic studies on Rotifera: a case study. Hydrobiologia 387/388:9–14

Segers H (2001) Zoogeography of the Southeast Asian Rotifera. Hydrobiologia 446/447:233–246

Segers H (2003) A biogeographical analysis of rotifers of the genus *Trichocerca* Lamarck, 1801 (Trichocercidae, Monogononta, Rotifera), with notes on taxonomy. Hydrobiologia 500:103–114

Segers H (2007a) Global diversity and distribution of the rotifers (Phylum Rotifera). Hydrobiologia (in press)

Segers H (2007b) A global checklist of the rotifers (Phylum Rotifera). Zootaxa 1564:1–104

Shurin JB (2000) Dispersal limitation, invasion resistance, and the structure of pond zooplankton communities. Ecology 81:3074–3086

Sides SL (1973) Internal and external transport of algae and protozoa by sea gulls. Trans Amer Microsc Soc 92:307–311

Sørensen MV, Segers H, Funch P (2005) On a new *Seison* Grube, 1861 from coastal waters of Kenya, with a reappraisal of the classification of the Seisonida (Rotifera). Zool Stud 44:34–43

Suatoni E, Vicario S, Rice S et al (2006) An analysis of species boundaries and biogeographic patterns in a cryptic species complex: the rotifer *Brachionus plicatilis*. Mol Phylogenet Evol 18:70–74

Suren A (1990) Microfauna associated with algal mats in melt ponds of the Ross Ice Shelf. Polar Biol 10:329–335

Wallace RL, Snell TW, Ricci C et al (2006) Rotifera vol 1: biology, Ecology and Systematics. In: Segers H, Dumont HJF (eds) Guides to the identification of the microinvertebrates of the continental waters of the World 23. Kenobi productions, Ghent/Backhuys Acad Publ/The Hague 299p

Whitaker RJ, Grogan DW, Taylor JW (2003) Geographic barriers isolate endemic populations of hyperthermophilic archaea. Science 301:976–978

Diversity and geographic distribution of benthic foraminifera: a molecular perspective

Jan Pawlowski · Maria Holzmann

Originally published in the journal Biodiversity and Conservation, Volume 17, No 2, 317–328.
DOI: 10.1007/s10531-007-9253-8

Abstract The diversity and distribution of modern benthic foraminifera has been extensively studied in order to aid the paleoecological interpretation of their fossil record. Traditionally, foraminiferal species are identified based on morphological characters of their organic, agglutinated or calcareous tests. Recently, however, new molecular techniques based on analysis of DNA sequences have been introduced to study the genetic variation in foraminifera. Although the number of species for which DNA sequence data exist is still very limited, it appears that morphology-based studies largely underestimated foraminiferal diversity. Here, we present two examples of the use of DNA sequences to examine the diversity of benthic foraminifera. The first case deals with molecular and morphological variations in the well-known and common calcareous genus *Ammonia*. The second case presents molecular diversity in the poorly documented group of monothalamous (single-chambered) foraminifera. Both examples perfectly illustrate high cryptic diversity revealed in almost all molecular studies. Molecular results also confirm that the majority of foraminiferal species have a restricted geographic distribution and that globally distributed species are rare. This is in opposition to the theory that biogeography has no impact on the diversity of small-sized eukaryotes. At least in the case of foraminifera, size does not seem to have a main impact on dispersal capacities. However, the factors responsible for the dispersal of foraminiferal species and the extension of their geographic ranges remain largely unknown.

Keywords Benthic foraminifera · Molecular and morphological variation · Geographic distribution

---

Special Issue: Protist diversity and geographic distribution. Guest editor: W. Foissner.

J. Pawlowski (✉)
Department of Zoology and Animal Biology, University of Geneva, Geneva 1211, Switzerland
e-mail: jan.pawlowski@zoo.unige.ch

M. Holzmann
Department of Palaeontology, University of Vienna, Althanstrasse 14, Vienna 1090, Austria

W. Foissner et al. (eds.), *Protist Diversity and Geographical Distribution*
DOI: 10.1007/978-90-481-2801-3_7

Introduction

There are about 5,000 species of modern (living) foraminifera and more than 50,000 fossil species (Debenay et al. 1996). Almost all these species have been described based on morphological characters of their test. Compared to many other protists, biological features such as cell structures or life cycles are usually not considered in foraminiferal systematics (Pawlowski and Lee 1992). Amazingly, the majority of foraminiferal species has never been observed alive. It is a common practice to sort and identify hard-shelled foraminifera from dried sediment samples while organic-walled allogromiids are preserved in formalin or alcohol fixed samples. To recognize living foraminiferal specimens, sediment samples are regularly stained with Rose Bengal, but the effectiveness of this method is quite disputed.

During 1930 and 1950, the number of newly described foraminiferal species was rapidly increasing at an average rate of one species per day (Thalmann 1952). This was due to the extraordinary development of applied micropaleontological research and a general tendency for "splitting", i.e., describing species on the base of very subtle morphological differences, often ignoring intraspecific variations. The result was a widespread increase of synonymy in many foraminiferan taxa, creating chaos in foraminiferan nomenclature (Boltovskoy and Wright 1976). This tendency became reversed in the 1970s when experimental laboratory studies demonstrated large ecophenotypic variation in cultivated foraminifera (Schnitker 1974). Despite some critical remarks concerning ecophenotypy in foraminifera (Haynes 1992), "lumping", i.e., including a wide range of morphotypes from various geographic regions in the same morphospecies, became a dominant tendency in foraminiferal research. As a consequence, the tendency to describe new species dropped drastically. For example, only seven new recent species have been described in the Journal of Foraminiferal Research during the past 10 years.

The wide use of benthic foraminifera in palaeoecological reconstructions largely contributed to the development of ecological studies in modern benthic foraminifera (Culver and Buzas 1998). It is generally accepted that most species have distinctive depth ranges, even if these ranges are broad and change from one area to another. Biogeography was used to define the foraminiferal associations typical for particular habitats in different geographic regions (Murray 1991). Among 938 common morphospecies analysed in Murray's study, more than half show a restricted distribution from 1 to 10 biogeographic regions established by the author. Among the 25 most widely distributed species, only 20 were found in more than five regions (Table 1). Remarkably, the three most ubiquitous species (*Epistominella exigua*, *Bulimina marginata* and *Globocassidulina subglobosa*) are characteristic of bathyal and abyssal environments. Whether these species are truly ubiquitous or represent a variety of indiscriminately lumped species, as suggested by some authors (Haynes 1992), is one of the main challenges of molecular studies in benthic foraminifera.

Molecular diversity of benthic foraminifera

One of the main controversial issues in conventional morphology-based taxonomy of foraminifera is the identification of species. The limited number of morphological characters of foraminiferal tests and their pronounced variations make the distinction of some species quite arbitrary. Studies about foraminiferal diversity can sometimes be strongly influenced by the authors' tendency for lumping or splitting. The situation is particularly

Table 1 Twenty-five of the most globally distributed species of benthic forminifera (adapted from Murray 1991). Species absence shown by dark colored area

Species	Depth (m)	NE Atl	Mex	SE Atl	W Atl	Medit	Indian	W Pac	E Pac	South	Arctic
Epistominella exigua	500-7500										
Bulimina marginata	100-3800										
Globocassidulina subglobosa	50-4100										
Ammonia beccarii	0-60										
Miliammina fusca	intertidal										
Quinqueloculina seminula	0-120										
Uvigerina peregrina	30-3800										
Adercotryma glomerata	20-6200										
Cibicides lobatulus	0-2700										
Trifarina angulosa	60-1900										
Bulimina aculeata	100-1900										
Cassidulina laevigata	30-2500										
Hoeglundina elegans	2000-3800										
Nuttalides umboniferus	2900-5500										
Trochammina inflata	intertidal										
Cibicides wuellerstorfi	1300-5500										
Elphidium crispum	0-25										
Hanzawaia concentrica	20-90										
Saccammina atlantica	30-200										
Bucella frigida	0-400										
Elphidium clavatum	0-200										
Haynesina germanica	intertidal										
Islandiella islandica	50-500										
Melonis barleanum	280-3800										
Rosalina globularis	0-60										

difficult for some common species, described a long time ago, where early descriptions are uninformative and the holotypes have either been lost or have never been deposited.

During the past 10 years, molecular techniques based on analysis of DNA sequences offered new tools for the identification of foraminiferal species and studies of their intraspecific variation (Holzmann 2000; Pawlowski 2000). All these studies are based on sequences of nuclear ribosomal RNA genes. These genes bear the advantage of being easily amplified even from single-cell DNA extractions. Three rDNA regions are commonly used in foraminiferal research: the 3′ fragment of the small subunit (SSU), the internal transcribed region (ITS) and the 5′ fragment of the large subunit (LSU). Each of these fragments has its own particular rate of evolution, which may differ from one taxonomic group to another. The ITS region is the fastest evolving one and seems most appropriate for species distinction, but its use until now was rather limited (Tsuchiya et al. 2003; Schweizer et al. 2005).

Using ribosomal genes as a tool for species identification bears certain inconveniences. Foraminiferal rDNA is extremely variable in length, difficult to align and often evolves at very different rates even between closely related groups. Moreover, in some species a strong intraindividual polymorphism of rDNA copies adds a supplementary difficulty to the determination of species-specific sequences (Holzmann and Pawlowski 1996). Nuclear genes coding for actin, tubulin and ubiquitin, which have been sequenced recently for some foraminifera are too conserved to be useful for analysis at species level (Flakowski et al. 2005). We expect that more variable molecular markers will be found in the mitochondrial genome whose sequencing is still in progress.

Here, we present two examples of rDNA-based studies of benthic foraminifers' diversity. The first one describes the comparison of morphologic and molecular variations in the well-known, common shallow-water genus *Ammonia*, while the second presents the molecular diversity of a poorly known group of monothalamous foraminifera. These two examples offer abundant material for the discussion of diversity and geographic distribution in benthic foraminifera.

Molecular versus morphologic variability in Ammonia

Ammonia is widely distributed in marshes and near-shore environments around the world. The great variety of morphotypes and the lack of easily recognizable morphological characters causes difficulties in the identification (Holzmann 2000). Thirty-seven modern *Ammonia* species are listed in Ellis and Messina (1940) and supplements, the first species description dating from 1758 and the last from 1979. Of these 37 species, 26 type specimens are represented by drawings (three species including drawings of thin sections), two species are without type figures, two species were examined by scanning electron microscopy, and seven are represented by photographs. Measurements of external morphological characters are in most cases only given for the type specimens and only in one case these measurements are based on more than 100 individuals. A total of four different morphological characters had been measured in these 37 recent *Ammonia* species. Most measurements concentrate on the diameter of the test (21 species), in two species the height of the test was calculated. The diameter of the proloculus was assessed in three species and the thickness in 10 species. Two up to three of these morphological characters have been measured per species. Given the scarcity of data combined with the morphological variability in this genus it is no wonder that discussions arose about the identification of species in *Ammonia*.

Molecular studies have shown the presence of several genetically distinct types of *Ammonia* in the Mediterranean Sea, the North Atlantic and the South Pacific (Pawlowski et al. 1995; Holzmann et al. 1996; Holzmann and Pawlowski 1997; Holzmann et al. 1998; Holzmann 2000; Holzmann and Pawlowski 2000). In many places at least two different phylotypes occur together. In a comprehensive work (Hayward et al. 2004), molecular and morphological methods were combined to establish a more robust taxonomic subdivision of *Ammonia* worldwide. Thirteen phylotypes (T1–T13) could be distinguished and discriminated on the basis of morphometric analyses (Figs. 1, 2). The distinction of phylotypes is based on phylogenetic analysis of 267 partial LSU rDNA sequences, obtained from 202 living *Ammonia* specimens sampled at 30 localities from the Pacific Ocean, the Atlantic Ocean, Mediterranean Sea, Red Sea, Caribbean Sea and North Sea (Table 2, Fig. 2). The morphology of 127 sequenced specimens was recorded by SEM prior to DNA extraction and the images were utilized for morphometric analysis. Measurements or assessments of 37 external test characters were used to perform different types of analysis, suggesting that each phylotype can also be distinguished morphologically. At least 8 of the 13 phylotypes can be equated to described species. Morphometric analysis can therefore be successfully used to distinguish species in highly variable taxa if a sufficient number of specimens and morphological characters are taken into consideration.

Each phylotype (T1–T13) is monophyletic and separated by elevated genetic distances from other types. Furthermore, as no intermediate types have been observed, the different phylotypes can be regarded as distinct species (Hayward et al. 2004). This is in contrast to the popular taxonomic concept on the genus *Ammonia* that only recognizes a limited number of species with many ecophenotypes (Poag 1978; Walton and Sloan 1990). The recognition that *Ammonia* represents only a single or very few species worldwide should therefore be abandoned as a theory lacking a genetic basis. Analysis of the biogeographical patterns shows that most *Ammonia* phylotypes are characterized by a restricted distribution

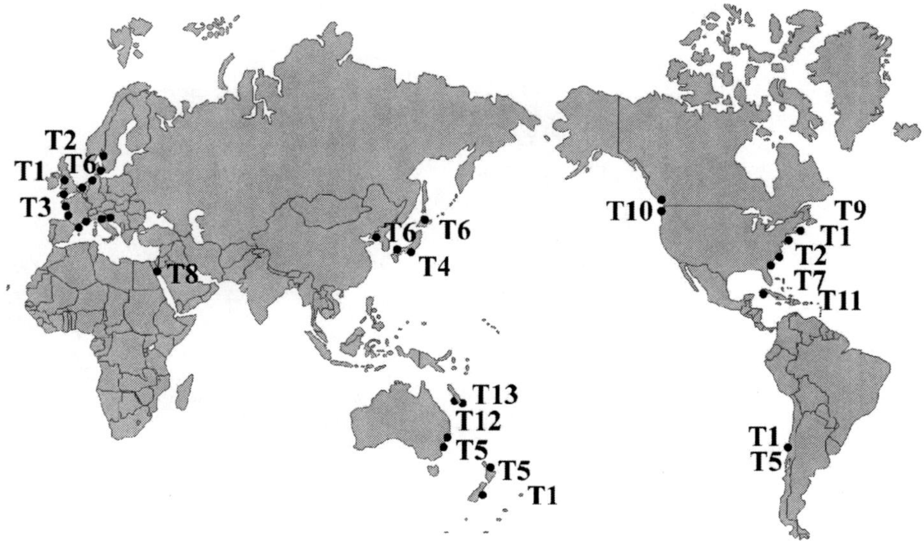

Fig. 1 Global distribution of the 13 *Ammonia* phylotypes (T1–T13) genetically identified and morphologically analysed

(Fig. 1, Table 2). Only one phylotype (T1) features a cosmopolitan distribution. Several other types are dispersed only in the northern or southern hemisphere (T2, T3, T5), some of them showing a transoceanic distribution (T2, T5). Many phylotypes are regionally restricted (T4, T7, T8, T9, T10, T11, T12, T13) while one phylotype (T6) shows a disjunct area. While transoceanic dispersal (T2, T5) could be aided by surface currents and/or transport via seabirds (Hayward and Hollis 1994), this would be difficult to accept with the disjunct distribution of T6. Human-assisted dispersal is the most likely explanation in the latter case. The genotype T6 is distributed around the coasts of China and Japan (Fig. 1) which is congruent with the habitat of *Eriocheir sinensis*, a decapod that has been introduced in the Wadden Sea at the end of the 19th century by shipping (Nehring and Leuchs 2000). Some of the ballast tank water that included *E. sinensis* could also have contained *Ammonia* individuals of the genotype T6 which since then spread out in the Wadden Sea and the adjacent Baltic Sea. Human-induced introduction of foraminifera is not an unknown phenomenon and has also played a role in the agglutinated foraminifer *Trochammina hadai* from Japan that has invaded the bay of San Francisco in the mid 1980s, most likely from ballast waters and sediments discharged from ships (Mc Gann and Sloan 1996).

The results of our studies provide just a sampling on the global diversity of the genus *Ammonia*. There are extended geographical regions that have not yet been investigated (Indian and South Atlantic Oceans, tropical and east Pacific, Southeast Asia and the East Indies). It is highly likely that the number of genetically distinct species could approach the number of formally named species (about 40; Ellis and Messina 1940 and supplements), most of which will be also distinguishable by a combination of subtle morphological characters.

Cryptic diversity in monothalamous foraminifera

In contrast to *Ammonia*, monothalamous foraminifera are a poorly known group, widely ignored by micropaleontologists. They are characterized by single-chambered organic-walled

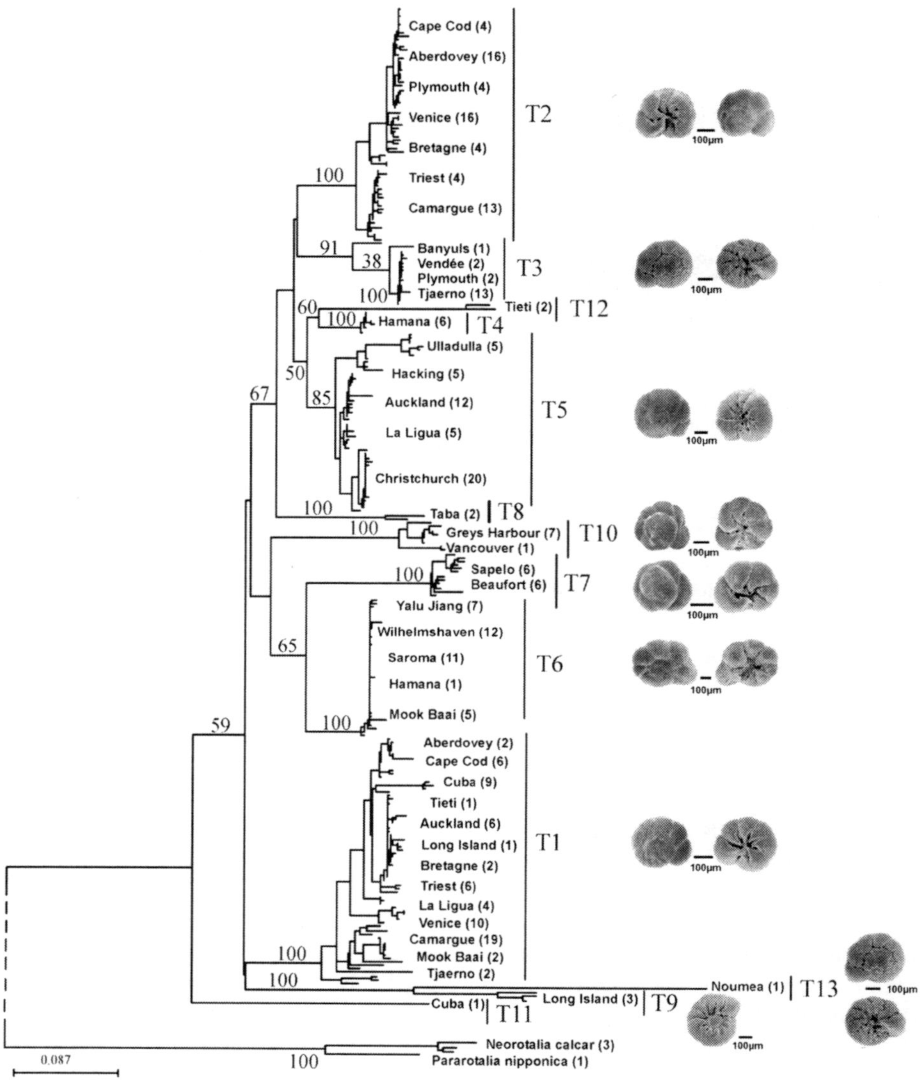

Fig. 2 Phylogenetic analysis of 267 partial LSU rDNA sequences using the Neighbour Joining method. The numbers are bootstrap percent values based on 500 resamplings. The scale bar corresponds to the number of substitutions per site. The names of sampling localities and number of sequences (in brackets) are indicated for each molecular type

or agglutinated tests that are rarely preserved in the fossil record. The fact that this group is particularly abundant in widely undersampled deep-sea and high latitude waters (Gooday et al. 2004) is also contributing to our lack of knowledge concerning its diversity. Monothalamous foraminifera are traditionally classified in the orders Allogromiida and Astrorhizida (Debenay et al. 1996). Morphological distinction of the latter orders is based on wall structure but is not confirmed by molecular studies (Pawlowski et al. 2003). SSU rDNA sequences have been used to resolve higher-level phylogenetic relationships (Pawlowski et al. 2002a, b; Pawlowski and Holzmann 2002), yet relations at lower taxonomic level remain unexplored.

Table 2 Collection localities, distribution and habitat of investigated *Ammonia* specimens

Phylotype	Number of investigated specimens	Number of partial LSU rDNA sequences	Distribution	Collection localities	Habitat
T1	52	65	Cosmopolitian	France: Camargue, Bretagne, Vendée; Italy: Lagoon of Venice, Triest	Microtidal marshes
				The Netherlands: Mook Bai; GBR: Dovey Estuary; Sweden: Tjaerno	Brackish water estuary
				USA: Cape Cod, Long Island; New Caledonia: Tieti Beach; Cuba: Playa Bailén; Chile: La Ligua; New Zealand: Waitemata Harbour	
T2	40	61	European coasts	France: Camargue, Bretagne, Vendée; Italy: Lagoon of Venice, Triest	Microtidal marshes
			Northern Atlantic coast, USA	GBR: Dovey Estuary; Sweden: Tjaerno; USA: Cape Cod	
T3	11	18	European coasts	France: Banyuls-sur-mer, Vendée; Sweden: Tjaerno	Open marine habitats, rocky shores
T4	3	6	Japanese coast	Japan: Hamana Lake	Brackish water lake
T5	45	47	Chilean coast, New Zealand, Australia	Chile: La Ligua; New Zealand: Akaroa, Pollen Island, Govemors Bay	Microtidal marshes
				Australia: Ulladulla-Burril Lake, Port Hacking-Grays Point	
T6	29	36	Wadden Sea, Baltic Sea	The Netherlands: Mook Bai; Germany: Wilhelmshavan; Japan: Lake Saroma, Hamana Lake; China: Yalu Jiang	Meso/macrotidal flats
			NE China, Japanese coast		Microtidal marshes, rocky shores
T7	6	12	Northern Atlantic coast, USA	USA: Georgia-Sapelo Island, Beaufort-North Carolina	Microtidal marshes
T8	2	2	Red Sea	Israel: Taba	Open marine habitats
T9	2	3	Northern Atlantic coast, USA	USA: Long Island	Microtidal marshes
T10	8	8	Northern Pacific coast, USA, CA	USA: Washington State, Grays Harbour; Canada: Vancouver	Microtidal marshes
T11	1	1	Caribbean Sea	Cuba: Playa Balién	Microtidal marshes
T12	2	2	Southern Pacific	New Caledonia: Tieti Beach	Brackish water estuary
T13	1	1	Southern Pacific	New Caledonia: Noumea Tjibaou	Mangroves

Our long-term survey of monothalamous foraminifera has revealed some unexpected results. Monothalamous lineages play a key role in the early evolution of foraminifera (Pawlowski et al. 2003). Their genetic diversity at different taxonomic levels by far exceeds what could be expected from morphological studies. Molecular data from material collected in Antarctica revealed an extraordinarily rich assemblage of monothalamous species. Allogromiids, athalamids and astrorhizids comprise an assemblage of more than a dozen lineages branching together at the base of the foraminiferal tree. Molecular data also show high species diversity in allogromiids (Pawlowski et al. 2002a, b, 2005). Because of the paucity of morphological characters, species distinction is particularly difficult and the majority of allogromiid genera are represented by single species descriptions (Nyholm 1974). Molecular analyses also confirmed the presence of allogromiids in freshwater and terrestrial environments (Meisterfeld et al. 2001; Holzmann and Pawlowski 2002; Holzmann et al. 2003). Very few of the genetically distinctive monothalamous taxa have been characterized morphologically and formally described or revised (Bowser et al. 2002; Gooday et al. 2004; Gooday and Pawlowski 2004; Sabbatini et al. 2004). Furthermore, a few lineages are only identified from environmental DNA extractions (Holzmann et al. 2003; Habura et al. 2004).

One of our research projects concerning monothalamous foraminifera focuses on the geographic distribution of this group, and in particular on the genetic comparison of similar morphotypes found in polar and subpolar waters of the northern and southern hemisphere. Some results of this yet unpublished study are reported here. We have compared SSU rDNA sequences of species belonging to four genera (*Micrometula*, *Psammophaga*, *Gloiogullmia* and *Hipocrepinella*) from western Svalbard (Arctic) and McMurdo Sound (Antarctic), including their representatives from the deep southern Ocean (Weddell Sea) and Arctic Ocean (Fram Strait) as well as from northern European fjords (Sweden, Scotland) wherever it was possible.

Phylogenetic analysis of our data show that within the four examined morphotypes, Arctic and Antarctic species form clearly distinctive sister clades (Fig. 3). The clades are separated by relatively large genetic distances (>5%), except in *Psammophaga* (<1%), due to either relatively rapid radiation or to an unusual slowdown of evolutionary rates in this genus. The isolates from Svalbard are closely related to those from other northern European settings. The Weddell deep-sea isolates of *Gloiogullmia* and *Micrometula* form sister groups to coastal Antarctic isolates and the Arctic deep-sea isolate of *Micrometula*, branches as sister group to the respective Antarctic clade. Interestingly, the specimens from Dunstaffnage (Scotland) either form a sister group to other northern hemisphere isolates (*Psammophaga*, *Gloiogullmia*) or to both polar clades (*Micrometula*).

Our data not only show the genetic differentiation between northern and southern populations of the examined taxa but also reveal several genetic lineages that considerably differ from each other. At present, three of the four examined genera are represented by only one described species (*Micrometula hyalostriata*, *Gloiogullmia eurystoma*, *Psammophaga simplora*). Each of these species is represented in our analyses by sequences from the area close to the type locality (Skagerrak for *M. hyalostriata*; Oslofjord *for G. eurystoma*; and Sappelo, Georgia, US for *P. simplora*). The fourth species, *Hippocrepinella hirudinea*, has been described from the Southern Ocean, and we consider our McMurdo sequences as closely related to the original type. Additionally, 12 genetically distinctive phylotypes have been revealed in our study. Remarkably, each of these types has a restricted geographic distribution. Given their apparent isolation and genetic differentiation, we may consider them as new, yet undescribed species.

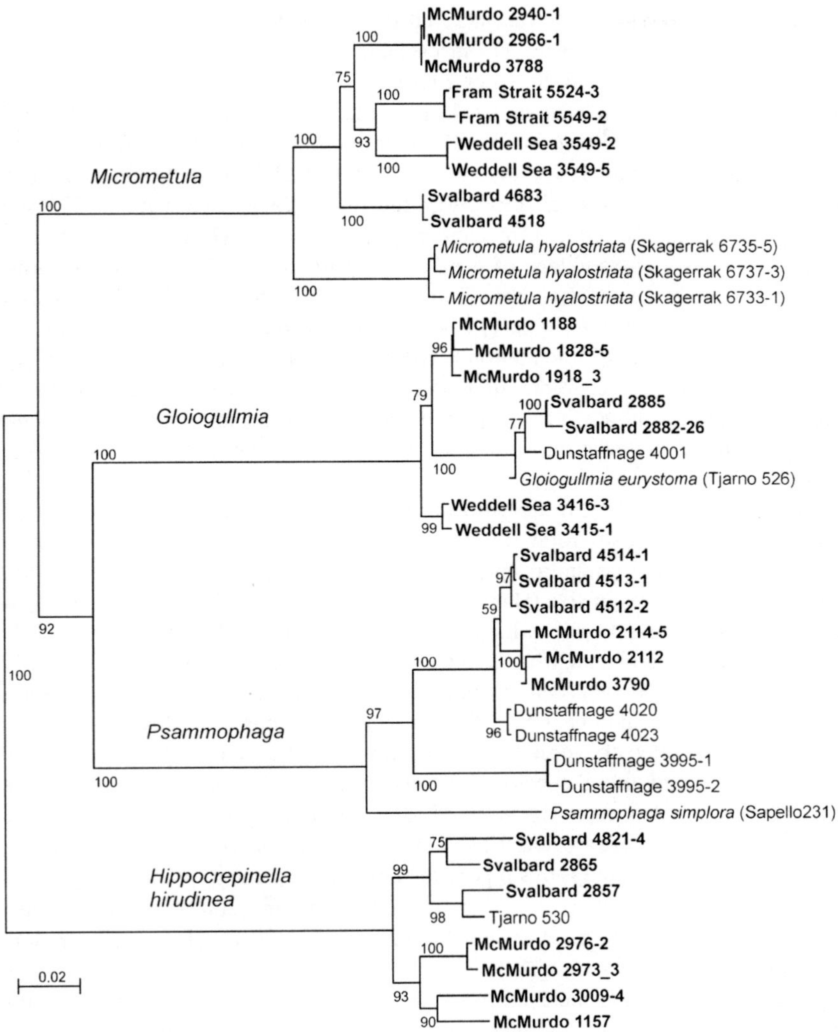

Fig. 3 Phylogenetic relations between Arctic and Antarctic monothalamous foraminifera. Sequence names indicate the locality and isolate number. Polar isolates are in bold. Species names are given to isolates from the area close to the locality of original description. The tree was obtained by the neighbour joining method with pairwise distances and 1000 bootstrap replicates

Is foraminiferan diversity different?

It has been proposed that the diversity of free-living protists is different from the diversity of larger organisms because small-sized organisms can be dispersed everywhere, and therefore the rates of allopatric speciation is low (Finlay et al. 2004). The authors assumed that small-sized organisms are generally ubiquitous and that the same species can be found wherever its preferred habitat is present (Finlay 1998). These arguments were based mostly on the study of ciliates morphospecies and have only recently been confirmed by molecular data for different ecotypes within the ciliate species *Cyclidium glaucoma* (Finlay et al. 2006).

Other results, however, point to the fact that the geographic distribution of many protist species is limited and about one third of species might be endemic in a morphological and/or genetic way (Chao et al. 2006; Foissner 2006).

Morphological and molecular studies suggest that most but not all foraminifera seem to have restricted geographic distribution. The data presented in Table 1 are based on more than 1000 studies (Murray 1991) and clearly show that globally distributed foraminiferal morphospecies represent a small proportion out of the 25 selected species. For some of them, such as *Ammonia beccarii*, molecular studies have shown that this morphospecies actually comprises an assemblage of genetically distinctive lineages. However, this does not mean that there are no ubiquitous foraminifera. As shown by molecular analyses, at least one lineage of *Ammonia* (type 1) has a global distribution. A recent molecular study shows very weak genetic differentiation between Arctic and Antarctic populations of three common species of deep-sea foraminifera (Pawlowski et al. 2007). We certainly need more molecular data to test how widely dispersed deep-sea species are. In the case of shallow-water foraminifera, however, our data indicate that most species have a limited geographic distribution.

Does size has something to do with the restricted distribution of most foraminifera? Compared to other protists, foraminifera are often larger in size and some of them particularly agglutinated polar and deep-sea species or calcareous tropical species can reach up to several centimetres in size (Haynes 1981). Yet, the majority of foraminiferal species measures from 50 to 500 μm, which is within the range of typical meiofaunal size. Undoubtedly, this is still much larger than some marine picoplanktonic algae ($\sim$2 μm), whose global distribution was demonstrated recently (Slapeta et al. 2006). The dispersal of such small organisms could be greatly facilitated by water currents. However, the example of *Ammonia* type 1 cited above shows that size might not be the main factor responsible for the dispersal of foraminiferal species. *Ammonia* specimens belonging to type 1 are within the same size range than representatives of other *Ammonia* types, and yet they are widely distributed while the others are not. What makes that particular *Ammonia* type ubiquitous is an intriguing question. Perhaps this type is the only one capable to produce dispersal forms such as the propagules observed by Alve and Goldstein (2003). Or, there are other physiological or ecological mechanisms that facilitate the dispersal of some foraminiferal species, independently of their size.

The examples presented here not only show evidence for geographic distribution of species but also confirm the importance of molecular studies for estimating the diversity of foraminifera. In both case studies, the analysis of DNA sequences revealed an extraordinarily high diversity of phylotypes at different taxonomic levels. Such high molecular diversity was found also in other foraminifera, including Soritinae (Garcia-Cuetos et al. 2006) and Glabratellidae (Tsuchiya et al. 2000, 2003). We can expect that if each of molecular types would be formally described, the number of foraminiferal species would increase at least by one factor of magnitude. The most spectacular rise of diversity is expected in the group of monothalamous foraminifera. In the much better known rotaliid genera, such as *Ammonia* or *Elphidium*, whose taxonomy is overloaded with synonyms, the number of phylotypes revealed by molecular data may approach that of described morphospecies.

Acknowledgements The authors thank T. Cedhagen, W. Majewski, S. Bowser, A. Habura and S. Korsun for stimulating discussion and help in collecting the material, and A. Gooday for comments on the manuscript. We also thank J. Fahrni, J. Guiard, D. Longet and J. S. Pawlowski for assistance. The work was supported by the Swiss NSF project 3100A0-112645 (JP) and the Austrian FWF project T270-B03 (MH).

References

Alve E, Goldstein ST (2003) Propagule transport as a key method of dispersal in benthic foraminifera (Protista). Limnol Oceanogr 48:2163–2170

Boltovskoy E, Wright R (1976) Recent foraminifera. Junk Publ., The Hague

Bowser SS, Bernhard JM, Habura A et al (2002) Structure, taxonomy and ecology of *Astrammina triangularis* (Earland) an allogromiid-like foraminifer from Explorers Cove, Antarctica. J Foram Res 32:364–374

Chao A, Li PC, Agatha S et al (2006) A statistical approach to estimate soil ciliate diversity and distribution based on data from five continents. Oikos 114:479–493

Culver SJ, Buzas MA (1998) Patterns of occurrence of benthic foraminifera in time and space. In: Donovan SK, Paul CRC (eds) The adequacy of the fossil record. Cambridge Univ. Press, Cambridge, pp 207–226

Debenay JP, Pawlowski J, Decrouez D (1996) Les foraminifers actuels. Masson, Paris

Ellis BFS, Messina AR (1940) Catalogue of foraminifera. Am Nat Mus NY Spec Publ

Finlay BJ (1998) The global diversity of protozoa and other small species. Int J Parasitol 28:29–48

Finlay BJ, Esteban GF, Fenchel T (2004) Protist diversity is different? Protist 155:15–22

Finlay BJ, Esteban GF, Brown S et al (2006) Multiple cosmopolitan ecotypes within a microbial eukaryote morphospecies. Protist 157: 377–390

Flakowski J, Bolivar I, Fahrni J et al (2005) Actin phylogeny of foraminifera. J Foram Res 35:93–102

Foissner W (2006) Biogeography and dispersal of micro-organisms: a review emphasizing protists. Acta Protozool 45:111–136

Garcia-Cuetos L, Pochon X, Pawlowski J (2006) Molecular evidence for host-symbiont specificity in soritid foraminifera. Protist 156:399–412

Gooday AJ, Pawlowski J (2004) *Conqueria laevis* gen. and sp. nov., a new soft-walled, monothalamous foraminiferan genus from the deep Weddell Sea. J Mar Biol Assoc UK 84:919–924

Gooday AJ, Holzmann M, Guiard J et al (2004) A new monothalamous foraminiferan from 100–6300 m water depth in the Weddell Sea: morphological and molecular characterization. Deep-Sea Res 2:1603–1616

Habura A, Pawlowski J, Hanes SD et al (2004) Unexpected foraminiferal diversity revealed by small-subunit rRNA analysis of Antarctic sediment. J Eukaryot Microbiol 51:173–179

Haynes JR (1981) Foraminifera. Macmillan Publ., London

Haynes JR (1992) Supposed pronounced ecophenotypy in foraminifera. J Micropaleontol 11:59–63

Hayward BW, Hollis CS (1994) Brackish foraminifera in New Zealand; a taxonomic and ecological review. Micropaleontology 40:185–222

Hayward BW, Holzmann M, Grenfell HR et al (2004) Morphological distinction of molecular types in *Ammonia*—towards a taxonomic revision of the world's most common and misidentified foraminiferal genus. Mar Micropaleontol 50:237–271

Holzmann M (2000) Species concept in foraminifera: *Ammonia* as a case study. Micropaleontology 46(Suppl 1):21–37

Holzmann M, Pawlowski J (1996) Preservation of foraminifera for DNA extraction and PCR amplification. J Foram Res 26:264–267

Holzmann M, Pawlowski J (1997) Molecular, morphological and ecological evidence for species recognition in *Ammonia* (Foraminifera, Protozoa). J Foram Res 27:311–318

Holzmann M, Pawlowski J (2000) Taxonomic relationships in the genus *Ammonia* (Foraminifera) based on ribosomal DNA sequences. J Micropaleontol 19:85–95

Holzmann M, Pawlowski J (2002) Freshwater foraminiferans from Lake Geneva: past and present. J Foram Res 32:344–350

Holzmann M, Piller W, Pawlowski J (1996) Sequence variations in large-subunit ribosomal RNA gene of *Ammonia* (Foraminifera, Protozoa) and their evolutionary implications. J Mol Evol 43:145–151

Holzmann M, Piller W, Fenner R et al (1998) Morphologic versus molecular variability in *Ammonia* spp. (Foraminifera, Protozoa) from the Lagoon of Venice, Italy. Rev Micropaleontol 41:59–69

Holzmann M, Habura A, Giles H et al (2003) Freshwater foraminiferans revealed by analysis of environmental DNA samples. J Eukaryot Microbiol 50:135–139

Mc Gann M, Sloan D (1996) Recent introduction of the Foraminifer *Trochammina hadai* Uchio in San Francisco Bay, California, USA. Mar Micropaleontol 28:1–3

Meisterfeld R, Holzmann M, Pawlowski J (2001) Morphological and molecular characterization of a new terrestrial allogromiid species: *Edaphoallogromia australica* gen. et spec. nov. (Foraminifera) from Northern Queensland (Australia). Protist 152:185–192

Murray JW (1991) Ecology and paleoecology of benthic Foraminifera. Longman, New York

Nehring S, Leuchs H (2000) Neozoen im Makrobenthos der Brackgewässer an der deutschen Nordseeküste. Lauterbornia 39:73–116

Nyholm KG (1974) New monothalamous foraminifera. Zoon 2:117–122
Pawlowski J (2000) Introduction to the molecular systematics of foraminifera. Micropaleontology 46(Suppl 1):1–12
Pawlowski J, Lee JJ (1992) The life cycle of *Rotaliella elatiana* n.sp.: a tiny macroalgavorous foraminifer from the Gulf of Elat. J Protozool 39:131–143
Pawlowski J, Holzmann M (2002) Molecular phylogeny of Foraminifera - a review. Eur J Protistol 38:1–10
Pawlowski J, Bolivar I, Fahrni J et al (1995) DNA analysis of *Ammonia beccarii* morphotypes: one or more species? Mar Micropaleontol 26:171–178
Pawlowski J, Fahrni J, Bowser SS (2002a) Phylogenetic analysis and genetic diversity of *Notodendrodes hyalinosphaira*. J Foram Res 32:173–176
Pawlowski J, Fahrni JF, Brykczynska U et al (2002b) Molecular data reveal high taxonomic diversity of allogromiid Foraminifera in Explorers Cove (McMurdo Sound, Antarctica). Polar Biol 25:96–105
Pawlowski J, Holzmann M, Berney C et al (2003) The evolution of early Foraminifera. Proc Nat Acad Sci 100:11494–11498
Pawlowski J, Fahrni JF, Guiard J et al (2005) Allogromiid Foraminifera and gromiids from under the Ross Ice Shelf: morphological and molecular diversity. Polar Biol 28:514–522
Pawlowski J, Fahrni J, Lecroq B, Longet D, Cornelius N, Excoffier L, Cedhagen T, Gooday AJ (2007) Bipolar gene flow in deep-sea foraminifera. Mol Ecol 16:4089-4096
Poag CW (1978) Paired foraminiferal ecophenotypes in Gulf Coast estuaries: ecological and paleoecological implications. Trans Gulf Coast Assoc Geol Soc 28:395–421
Sabbatini A, Pawlowski J, Gooday AJ et al (2004) *Vellaria zucchellii* n.sp., a new small monothalamous foraminifer from Terra Nova Bay, Antarctica. Antarct Sci 16:307–312
Schnitker D (1974) Ecophenotypic variation in *Ammonia beccarii* (Linné). J Foram Res 4:216–223
Schweizer M, Pawlowski J, Duijnstee IAP et al (2005) Molecular phylogeny of the foraminiferan genus *Uvigerina* based on ribosomal DNA sequences. Mar Micropaleontol 57:51–67
Slapeta J, Lopez-Garcia P, Moreira D (2006) Global dispersal and ancient cryptic species in the smallest marine eukaryotes. Mol Biol Evol 23:23–29
Thalmann HE (1952) Twenty years of "foraminiferal statistics": 1931 to 1950. Cushman Found Foram Res Contr 3–4:145–146
Tsuchiya M, Kitazato H, Pawlowski J (2000) Phylogenetic relationships among species of Glabratellidae (Foraminifera) inferred from ribosomal DNA sequences: comparison with morphological and reproductive data. Micropaleontology 46(Suppl 1):13–20
Tsuchiya M, Kitazato H, Pawlowski J (2003) Analysis of internal transcribed spacer of ribosomal DNA reveals cryptic speciation in *Planoglabratella opercularis*. J Foram Res 33:285–293
Walton WR, Sloan BJ (1990) The genus *Ammonia* Brünnich, 1772: its geographic distribution and morphologic variability. J Foram Res 20:128–156

Diversity and biogeography of testate amoebae

Humphrey Graham Smith · Anatoly Bobrov · Enrique Lara

Originally published in the journal Biodiversity and Conservation, Volume 17, No 2, 329–343.
DOI: 10.1007/s10531-007-9260-9

Abstract Testate amoebae are amoeboid protists inhabiting a test (shell). They occur globally in soils, wetlands and freshwater, especially peats and mosses. They are of ancient origin, dating from at least the Mesozoic, with possible ancestors as old as the Neoproterozoic. Approximately 2,000 taxa have been described—a number which could easily rise to 4,000 with comprehensive recording. Whilst many protists appear to be cosmopolitan as morphospecies, some of the larger testate species (exceeding 100 μm) have long been considered, controversially, to be geographically restricted. Definitive conclusions have often been confounded by gaps in distributional data and misidentification. Recent increases in recording from previously little known regions, and the rise of molecular taxonomy, have started to resolve outstanding issues—processes still far from complete. Accordingly, biogeographical studies have concentrated on "flagship" species—those which can be identified with certainty and are sufficiently recorded to determine their ecological ranges. *Apodera vas* (Certes) has been proved to be largely restricted to the Gondwanaland continents and sub-Antarctic islands, but absent from the Holartic despite the availability of much suitable habitat. An early analysis postulated a Mesozoic origin of the species and a distribution influenced by continental drift. Recent molecular evidence could imply a later origin. Either way, its current distribution is clearly influenced by the pattern of global wind currents and lack of lowland tropical habitat. By contrast a "Gondwana-tropical" group of species appears to be restricted to latitudes unaffected by glaciation. Instances of local endemism, such as restriction to a single island, are also known, which await molecular evidence for substantiation.

Special Issue: Protist diversity and geographic distribution. Guest editor: W. Foissner.

H. G. Smith (✉)
Environmental Sciences, Coventry University, Coventry, UK
e-mail: apx191@cov.ac.uk

A. Bobrov
Faculty of Soil Science, Moscow State University, Moscow, Russia

E. Lara
Wetlands Research Group, Swiss Federal Research Institute WSL, Ecublens, Lausanne, Switzerland

W. Foissner et al. (eds.), *Protist Diversity and Geographical Distribution*
DOI: 10.1007/978-90-481-2801-3_8

Keywords Biogeography · Cosmopolitanism · Diversity · Ecology · Endemism · Flagship species · Fossil testates · Testate amoebae

Introduction

Testate amoebae constitute a functional polyphyletic group of those amoeboid protozoa in which a single eukaryotic cell is enclosed within a shell or test (size range 5–300 μm) with an oral aperture, through which filose or lobose pseudopodia protrude during locomotion or feeding. They occur worldwide in a range of terrestrial, wetland and freshwater habitats, but most frequently in moist acid soils and peats, with high organic content and low nutrient turnover, and also in standing waters, lake margins and the biofilms of sewage treatment plants. They are universally associated with mosses, even in more arid conditions, such as moss cushions on tree bark and rocks.

Testate amoebae are phylogenetically divided into those with lobose and those with filose pseudopodia (Cavalier-Smith 2004). Testate lobose amoebae are Amoebozoa (Nikolaev et al. 2005) and include the larger species (>100 μm)—especially those in the families Difflugiidae, Centropyxidae, Arcellidae and Hyalospheniidae. Many species of the difflugiids and centropyxids have agglutinate tests composed of mineral particles gathered from the environment. Many large species of hyalospheniids (e.g. those in genera *Nebela* and *Heleopera*) have tests composed of idiosomes acquired from consumption of euglyphids as prey (Meisterfeld 2002a). Testate filose amoebae by contrast are closely related to the Foraminifera and Cercozoa (Longet et al. 2004) and include the families Euglyphidae and Trinematidae. These contain small species (<100 μm) with siliceous tests composed of idiosomes biosynthesised by the resident amoeba.

Testate amoebae research has contributed importantly to debate about the relevance to protists of Beijerinck's dictum "*Everything is everywhere, the environment selects*". Apologists for this paradigm have presented evidence for the cosmopolitan dispersion of all microorganisms up to ca 1 mm in size (Finlay 2002; Finlay and Fenchel 2004). Whilst their evidence is based largely on freshwater ciliate records, they have also sought to extend it to soil testates (Finlay et al. 2001; Esteban et al. 2006). However there are numerous examples of testate taxa which appear to be geographically restricted at global, regional and local scales (Foissner 2006), selected examples of which are discussed below. Larger, and thus heavier, testate species disperse less rapidly than others. Natural barriers, such as adverse winds or extended areas without appropriate habitat, slow down their progression. For a single propagule to overcome these obstacles and found a new population becomes a statistically improbable event; therefore it may take thousands or millions of years until it happens. Whilst it is true that many microbial morphospecies, including the smaller testate amoebae, tend to a cosmopolitan distribution, this process is not fixed in time. What we observe nowadays is an instantaneous (and incomplete) picture of the distribution of testate species. It is inevitable that present-day patterns will evolve over succeeding millennia as new species appear, some species spread and others become extinct.

Historical development of perceived diversity and biogeography

Whilst it is generally acknowledged that many protozoa have (at least at the morphospecies level) a cosmopolitan distribution, some species of testate amoebae have remained one of the most striking examples of microorganisms that present biogeographical patterns in their global distribution. The earliest biogeographical assessment of testates was made more than 130 years ago (Bonnet 1983). Ehrenberg concluded that large geographical areas have different faunas of testates but that the European fauna lacks uniqueness. However many protozoologists of the late 19th century were convinced that nonparasitic protozoans are essentially cosmopolitan. Penard (1902) thought that similar habitats in all parts of the world would contain similar faunal assemblages—a view which he later modified (Penard 1938). Heinis (1914) elucidated a circum-austral distribution of some species of *Nebela*. This phenomenon was further investigated by Deflandre (1928, 1936) who also noted a pantropical distribution of some species of *Arcella*. Models were continually revised through the 20th century as more locations were sampled and more species described.

An analysis by Cailleux (1978) of the publications of Decloitre noted the doubling in numbers of described species, sub-species and varieties globally: from 800 in 1952 to 1,600 in 1975. The continental distribution of taxon-richness then known was: Europe 1,031, Africa 648, Australia and Melanesia 428, North and Central America 229, Asia 220, the Arctic 220, Antarctica 89. Decloitre (1985) estimated that 1800 species, sub-species and varieties had been recorded globally. Differences between northern and southern testate faunas were noted repeatedly through the 20th century (Hoogenraad and Groot 1940, 1979; Jung 1942b; Cailleux 1978; Bonnet 1983; Smith and Wilkinson 1987; Wilkinson 2001; Meisterfeld 2002a, b; Foissner 2006), giving rise to models which have related some species' distributions to the palaeogeography of Gondwanaland and Laurasia. The high number of species recorded for Europe must surely be a reflection of the intensity of sampling, rather than true higher species-richness than the other continents. Whilst lower numbers for polar regions may be attributed to hostile environments, when the Americas, Asia and Australasia receive as much attention as Europe, it is reasonable to speculate that taxon-richness could very likely double from ca 2,000 to ca 4,000. The last 30 years have been notable for the increase in records from previously under-sampled regions—the Antarctic and the Far East (Smith 1978; Bonnet 1981; Vincke et al. 2004, 2006a; Yang et al. 2005b), thus providing a more comprehensive data base for biogeographical modelling at the global, regional and local levels.

The fossil record

Testates are certainly ancient taxa. Whilst fossil records are limited and discontinuous, tests exactly resembling those of modern species have been described from 2 and 15 Mya old sediments (Boeuf and Gilbert 1997; Foissner and Schiller 2001). Extant genera are also known from Mesozoic amber at several Holarctic sites dated at ca 100 and 220 Mya old (Poinar et al. 1993; Waggoner 1996; Schönborn et al. 1999; Schmidt et al. 2004). More recently, 'vase-shaped microfossils' with apparent testate affinities have been described from the Neoproterozoic, ca 700–750 Mya old (Corsetti et al. 2003; Porter et al. 2003). These putative testate ancestors were most likely marine, suggesting a switch to terrestrial and freshwater forms during their evolution. Their existence forms part of biologically based arguments that Neoproterozoic conditions may have been less extreme than those claimed for Snowball Earth (Corsetti et al. 2006).

There are considerable records of testates from the Quaternary sediments where they have been used as hydrological and acidity indicators (Charman 1997; Booth 2002; Mitchell et al. 2007), thus providing valuable evidence of Quaternary palaeoenvironmental changes and also of the ecology of extant species.

The whole testate taxa can thus be seen as genetically conservative, with many genera having persisted essentially unchanged (in test morphology at least) for 10's or 100's Mya. At the same time, continuing evolution at the sub-species level may well have occurred in the last few thousands of years (Grospietsch 1971).

Ecological diversity

In biogeographical research it is important to know when the absence of a species from a particular location is truly due to geographical restriction, rather than lack of appropriate habitat. However, whist there is a voluminous literature on the morphological diversity of testate species, the detailed habitat relations of many species are less well known. More is known about relations with moisture than with other environmental variables. An early scheme was proposed by Jung (1936) which classified species into six classes (FI–FVI) according to the moisture content of the habitat. The most extensive recent research on moisture relations has been by Quaternary palaeoecologists who utilize fossil testates as indicators of the moisture content and water tables of peat bogs dating from the Holocene (Charman 1997; Woodland et al. 1998; Booth 2002; Mitchell et al. 2007). This work has enabled lists of species to be determined, ordinated by moisture preference. The most hydrophilic terrestrial species include *Arcella discoides* Ehrenberg, *Apodera vas* (Certes), *Certesella certesi* (Penard) and *Difflugia bacillifera* Penard. Species of drier habitats include *Assulina muscorum* Greeff, *Corythion dubium* Taranek, *Euglypha rotunda* Wailes, *Phryganella acropodia* (Hertwig and Lesser), and *Trigonopyxis arcula* (Leidy) Penard.

The temperature relations of testate amoebae are still poorly known. A broad trend of decreasing species-richness in the southern hemisphere with increasing latitude and with declining mean January temperatures has been noted (Smith and Wilkinson 1987; Smith 1996). Similarly, testate species-richness of ten physiographic regions of Tibet and Yunnan, China, showed a strong positive correlation with the mean temperature of the warmest month (Yang et al. 2005b). However, monthly mean temperatures are an extremely crude indicator of habitat suitability within soils, wetlands and freshwater; there are doubtless many exceptions to the general trend (Vincke et al. 2006a); detailed microclimate data would be necessary to establish such a relationship with precision. In vitro experiments on the effect of temperature on growth rates and respiration of *Corythion dubium* Taranek and *Euglypha rotunda* Wailes demonstrated the ability of these species to tolerate conditions in the Antarctic, where these species are prominent in the testate communities (Smith 1973; Cowling 1983). Laybourn and Whyman (1980) found that increasing temperature promoted a higher reproductive rate in *Arcella vulgaris* Ehrenberg up to 20°C, but no further increase above that temperature. It seems likely that many testate species show some measure of facultative psychrophily, but much more research will be required to establish this as a general phenomenon of the taxon.

The eclectic nature of the feeding strategies of testate amoebae, and their influence upon other ecosystem components, have become better known in recent years. They were previously assumed to be largely bacteriophagous, since they can be routinely cultured successfully with bacteria as the sole food source. However, there is evidence that, in natural ecosystems, their feeding habits are very much more diverse; that many species will consume whatever prey items are most readily available and thus their diet may vary seasonally and with habitat. This topic has been reviewed by Schroeter (2001). His model of testate trophic relations divides testates into the smaller "panphytophagous" species, which graze on bacteria, algae, fungi (hyphae, spores and yeasts) detritus and humus particles, and the larger "predaceous" difflugid and hyalospheniid species, which additionally feed on small testates, ciliates and micrometazoa (rotifers and nematodes). Subsequent analysis of the food preferences of *Nebela* spp. in *Sphagnum* peatland by Gilbert et al. (2003) has given results consistent with Schroeter's model: some 80% of their diet consisted of micro-algae (especially diatoms) and fungi; during Summer they also consumed ciliates, rotifers and small testates.

Testate amoebae species certainly show variation in their ability as pioneer colonisers of new habitats, and thus in their position along the r-K continuum. Species of the genera *Assulina, Centropyxis, Corythion, Euglypha, Phryganella* and *Trinema* are repeatedly reported as the earliest testate species to appear in the microbial succession occurring in volcanic tephra (Smith 1985), reclaimed mining spoil (Wanner and Dunger 2002) and permeable pavement biofilms (Coupe et al. 2003). These may reasonably be seen as r-strategists in relation to other testates, though all testate species could be judged K-selected in comparison with flagellate and ciliate pioneers. Early testate colonists persist in the succession as microbial communities become more species-rich and complex, often in high numbers. Thus later colonists are in addition to, and not instead of, the pioneers (Wanner and Xylander 2005).

Intra-specific diversity

Many testate species show considerable intra-specific variation in size and shape. Observations on individuals collected in the field sometimes show gradual transition between species that can be dependant on environmental variables such as moisture or availability of test-building material (Cash et al. 1919). Test morphology can also be experimentally influenced in clonal cultures. As long ago as 1916, Jennings showed that many different morphotypes of *Difflugia corona* Wallich could be obtained by artificial selection. More recently, it has been shown that the size of the test and oral aperture in *Cyclopyxis kahli* Deflandre and *Cyclopyxis eurystoma* Deflandre can be influenced by temperature and food supply, and that these induced changes are reversible over a few generations (Wanner and Meisterfeld 1994; Wanner 1999). Interestingly, the range of morphotypic variation that could be induced artificially never exceeded that observed in wild populations. Biometric analysis of 32 natural populations of 24 species by Bobrov and Mazei (2004) revealed that maximum length exceeded minimum length by a factor of x1.2 to x2.2—a significant degree of variability within local populations, which may reasonably be supposed to have a substantial degree homogeneity of both genes and environment. A global review of 130 individuals of the hyalospheniid *Apodera vas* (Certes) by Smith and Wilkinson (2007) likewise revealed a maximum to minimum length ratio of 2.3. Thus, for identification purposes, within-species 'lumping' may be more genetically valid than 'splitting'. It is likely that many, if not all, testate taxa are characterised by an important phenotypic plasticity, that can be seen as conferring a selective advantage and may be a significant factor in enabling their long persistence through geological time.

Biogeography

Confounding factors in assembling distributional data

Considerable data have been accumulated, through the 19th and 20th centuries, of species assemblages in sampled materials from a wide range of habitats in most areas of the Earth.

Overwhelmingly, current models of testate taxonomy, phylogeny, ecology and distribution are derived from data on morphospecies descriptions based on test characteristics. Whilst identification of clades by molecular methods is not yet far advanced, there is sufficient evidence to cause wide uncertainty about the value of the criteria traditionally used for taxon discrimination in testate amoeba taxonomy. Molecular phylogenetic analysis is now being used to elucidate the relationships among species and, therefore, to evaluate the validity of the criteria used in species identification. It appears that, within the genus *Euglypha* (possibly the most abundant and widespread of testate genera) valid characters for species separation can only be observed with the electron microscope (Wylezich et al. 2001; Lara et al. 2007a). This makes them unsuitable for large scale biodiversity surveys. Within the family Hyalospheniidae, it has been suggested that the taxonomic weight of criteria used in species or genus definition may not reflect evolution within the group.

Accordingly, the classification of the whole group may have to be re-evaluated, with the result that some species might well be redefined (Lara et al. 2007b).

Even with traditional taxonomy, there are uncertainties caused by under-recording in some parts of the world, by inconsistencies amongst authors as a result of misidentification or use of synonymies, and by differing views of the validity of sub-specific nominations ('splitters' *versus* 'lumpers'). Attempts to circumvent such confounding factors have been a continuing feature of testate biogeographical research.

The "flagship" species

The uncertainties encountered in species identification have long been a pitfall in the evaluation of the geographical distribution of testate species. Lack of clear definition has confounded unequivocal conclusions about biogeography and made it difficult to interpret ancient literature. A way to circumvent these problems is to identify 'flagship' species (Tyler 1996; Foissner 2006). These are species which present a very characteristic morphology that cannot be confused with any others, and which have a restricted distribution globally, even though their habitats are not restricted to that region. To qualify as "flagship" there should not be any intermediate forms known with other species. Since they have a striking and unambiguous morphology, ancient records testifying their presence can be trusted. At present, it makes sense to concentrate attention on the distribution of these flagship species, because they are the ones for which no misidentification can be possible. It is, however, important to note that other species, which are less conspicuous, could very well also show geographically limited distributions. It is probable that these taxa will be better studied using molecular tools, such as sequencing variable genes and/or microsatellite markers.

Cosmopolitanism versus global endemism in the Hyalospheniidae

The family of testate amoebae whose biogeography has been most studied is undoubtedly the Hyalospheniidae sensu Schultze (1877). They are defined by a laterally compressed ovoid or pyrifrom test and acrostome with terminal aperture. An acrostome and pyriform shell are considered to be ancestral traits, and molecular studies have proven that this family is paraphyletic (Nikolaev et al. 2005; Lara et al. 2007b).

Apodera vas (Certes) is certainly the best known example of a testate amoeba with a geographically limited distribution (Smith and Wilkinson 2007). The shape of its test is unmistakable (Fig. 1a) and thus makes a perfect flagship species. First discovered in mosses from Tierra del Fuego by Certes (1891), it has been recorded from many other locations in the Southern hemisphere, including South Africa (Gericke 1932), Australia (Richters 1908; Meisterfeld and Tan 1998), New Zealand (Charman 1997), Kerguelen (Bonnet 1981), Iles Crozet (Vincke et al. 2004, 2006a), Marion Island (Grospietsch 1971) and South Georgia (Smith 1982; Beyens et al. 1995). It has also been reported from certain locations in the Northern hemisphere including Sumatra (Hoogenraad and Groot 1940), Nepal (Bonnet 1977), West Africa (Golemansky 1963; Bonnet 1978) and Central America (Laminger 1973a). However, it has *never* been found north of the tropical desert belt, despite the numerous studies undertaken in the Northern hemisphere (Mitchell and Meisterfeld 2005). This absence is even more remarkable in that *A. vas* is a very frequent species within its geographical range. So numerous are the records that its ecological range can be determined with certainty as occupying moist terrestrial to semi-aquatic habitats (Charman 1997) in the cool temperate to sub-Antarctic zone (Smith and Wilkinson 2007) including high montane habitats at tropical locations up to 4,000 m (Laminger 1973a; Bonnet 1980). However, it can occur in relatively unstable habitats such as forest litter, where the moisture content is variable; it must therefore have a good capacity for encystment (Bonnet 1969). The colonisation capacities of this species are illustrated by its presence on some of the most remote islands on Earth in the Southern Ocean: Kerguelen, Iles Crozet, Marion Island and South Georgia. The chance of the species having persisted on these islands in refugia during the Pleistocene here is remote indeed (Wilkinson 1990), so colonisation must have occurred during the last 10 millennia.

In an attempt to explain the continental distribution of A. *vas*, Smith and Wilkinson (1987) suggested that it appeared on the palaeo-continent of Gondwanaland in the Jurassic period. Its last common ancestor with its closest relatives, the cosmopolitan *Nebela tubulata* Brown, *N. wailesi* Deflandre and *N. lageniformis* Penard lived in Pangea more than 190 Mya ago, and *A. vas* evolved in allotropy after the separation of Gondwanaland. This hypothesis would imply that the species is very old, at least 140 Mya; it is therefore to be expected that its sister species, also found in the north, would be quite distantly related. However, molecular data currently available suggest that *A. vas* may be closely related to the cosmopolitan *N. lageniformis.* The two species present only 3% variation on a sequence of a SSU rRNA gene fragment, which includes the most variable regions of that gene. If this close similarity were confirmed by multigene analysis, it would suggest that the amount of time passed after the divergence of the two taxa might not be so large, and certainly much shorter than the separation between Gondwanaland and Laurasia (Lara et al. 2007b).

The alternative explanation is that *Apodera vas* speciated much more recently in the cool temperate Southern hemisphere (most probably South America) and the cysts were subsequently dispersed by wind to the other continents and also to the remote islands of the Southern Ocean. The strong circum-Antarctica westerly wind has been frequently cited as an effective agent for the dispersal of colonising propagules (McDowall 2005). The spread northwards of cysts as airborne particles through the Capricorn tropic, because the pattern of air currents, is highly unlikely. However, progressive land colonisation through suitable environments created by high mountain ranges, where cold and humid climates prevail (e.g. Andes, African mountains, Himalayas) is much more feasible. In contrast, crossing the Northern desert belt around the Cancer tropic (e.g. Mexican and Sahara deserts) would be much more unlikely owing to absence of suitable habitat and an adverse wind regime. This hypothesis is plausible if *Apodera vas* arose when these barriers were already present.

At least 18 other taxa of nebelid Hyalospheniidae also appear to have Gondwanaland-specific distributions (Smith and Wilkinson 1987) including *Certesella certesi* (Penard),

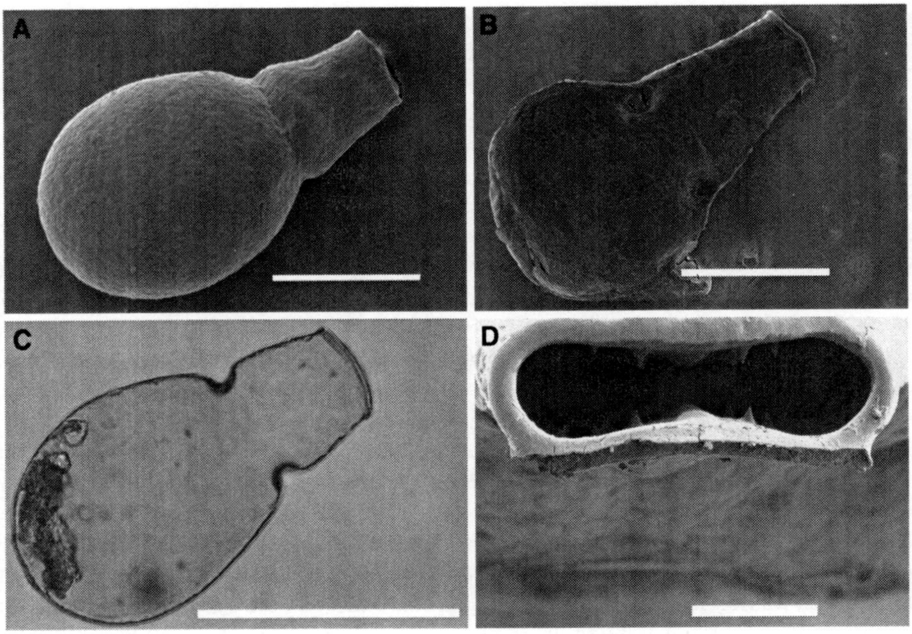

Fig. 1 Illustration of some species of Hyalospheniidae from the southern hemisphere. **A** = *Apodera vas* from Marion Island; **B** = *Certesella* sp. from Marion Island; **C** = *Alocodera cockayni* from Argentine Patagonia; **D** = A detail from the pseudostome of *Certesella murrayi* shows the typical punctuation inside the neck of the shell. Scale bars represent 50 μm (**A**, **B**, **C**) and 10 μm (**D**)

Certesella murrayi (Wailes) (Fig. 1b, d) and *Alocodera cockayni* (Wailes) (Fig. 1c), which also have characteristic, unmistakable shapes. Of these, *C. certesi* is the next most frequently recorded of the Gondwana-specific species after *A. vas* and most likely has a comparable biogeographical history. It shows an ecological difference in that its range extends to the Antarctic zone. *Certesella murrayi* and *A. cockayni are* rarer species with more disjunct distributions. *Alocodera cockayni* has been recorded several times from New Zealand as well as South America (Charman 1997), whereas *C. murrayi* may be endemic to South America (Vucetich 1978).

The existence of Northern-specific species is much more difficult to elucidate because the tropics and Southern hemisphere have been so much less sampled than Northern regions. Therefore, evidence that a species is truly absent from the South is much less likely to be conclusive. Furthermore, difficulties of access to literature can lead authors to misinterpretations. Hoogenraad and Groot (1979) cited the species *Nebela carinata* (Archer) Leidy, *N. marginata* Penard, *N. tubulosa* Penard, and *Hyalosphenia papilio* Leidy as Northern endemics, although they had been found previously in Congo by Gauthier-Lièvre (1954) and Decloitre (1965). Tentatively, the large *Nebela speciosa* Deflandre appears a good candidate for Northern endemism; its very large size (>200 μm, and even 278 μm) makes it a good flagship species and it has only been reported from North America, Germany and England (Deflandre 1936; Grospietsch 1958; Laminger 1973b; Ogden 1984). From its morphology, it appears that its closest relative might be *Nebela tubulosa* Penard (Ogden 1984).

Regional endemism

Several mid-20th century attempts were made to define regional differences in testate fauna. Jung (1942a) argued for the existence of regional faunas as a result of limited dispersal abilities of some species—particularly the forest dwelling species which have decreased ability to form cysts (Bonnet 1983). Jung (1942b) described eight new nebelid species which, like *Certesella murrayi*, appear to be restricted to South America—discoveries which are consistent with the general paradigm that the South American continent represents a biodiversity "hotspot" globally.

Decloitre (1953) sought to distinguish between the faunas of both temperate zones and the tropics, whilst Oye (1960) distinguished the faunas of the Palaearctic, the Nearctic and the Southern hemisphere. Schönborn (1966) argued that species which differ on the different Southern hemisphere continents must have evolved after the breakup of Gondwanaland. Stout (1969) compared the testate faunas of New Zealand, Australia and Europe using the family taxonomy then in favour (% figures approximate):

New Zealand: Euglyphidae 40%, Nebelidae 33%, Centropyxidae 9%, Difflugiidae 7%, Arcellidae less than in Australia and Europe.

Australia: Euglyphidae 40%, Nebelidae 25%, Centropyxidae 22%, Difflugiidae 7%, Arcellidae more than in New Zealand.

Europe: Euglyphidae 40%, Nebelidae 14%, Centropyxidae 22%, Difflugiidae 7%, Arcellidae as in Australia.

The constant figure for the Euglyphidae is notable. This family contains species almost all less than 100 μm long. These, more than species of other families, can be expected to have cosmopolitan distributions at the morphospecies level.

Bonnet (1980) proposed a "Gondwana-tropical" group of species and included in it some species of *Hoogenraadia, Planhoogenraadia, and Lamptopyxis*, whose ranges cover the tropical areas of Africa, South and Central America, Southeast Asia, New Guinea and Philippines, but lie north of circum-Austral zone. Subsequently (1983) he broadened the group by including some species of *Centropyxis, Cyclopyxis, Deharvengia, Ellipsopyxis, Plagiopyxis*, and *Protoplagiopyxis.* Korganova (1994) reported that most of the testate species found on 13 Pacific islands of Tonga and Western Samoa were cosmopolitan, but 7 species belonged to the Gondwana-tropical group. Species such as *Hoogenraadia humicola* Bonnet, *Planhoogenraadia media* Bonnet, *and Distomatopyxis couillardii* Bonnet and Gomez-Sanchez (Bobrov 2001) occupy a particular place among the faunas of the southeastern Palaearctic (Fig. 2), the ranges of *Hoogenraadia* and *Planhoogenraadia* occupying mainly the Holarctic (Bonnet and Gomez-Sanchez 1994; Todorov and Golemansky 1999; Beyens and Meisterfeld 2002). The northern boundary of these taxa's geographical ranges seems to correlate with the maximum extent of Pleistocene glaciation.

Until recently, the testate amoebae of the Far East were unknown to western scientists. Evidence is now emerging of the existence of a distinctive testate fauna with the description of four species, specific to the Far East: *Collaripyxidia dongtongiensis* (Balik and Song 2000), *Difflugia tuberspinifera* (Yang et al. 2004), *Difflugia biwae* (Yang and Shen 2005), and *Difflugia mulanensis* (Yang et al. 2005a).

Evidence of Nearctic endemism comes from the discovery of several new species from Canada. *Paraquadrula ogdeni* and *Netzelia labeosa* have been described from the arctic North West Territory (Beyens and Chardez 1997), whilst *Corythionella golemanski*, *Cyclopyxis acmodonta* and *Arcella formosa* have been described from wetland and freshwater in Ontario (Nicholls 2003, 2005).

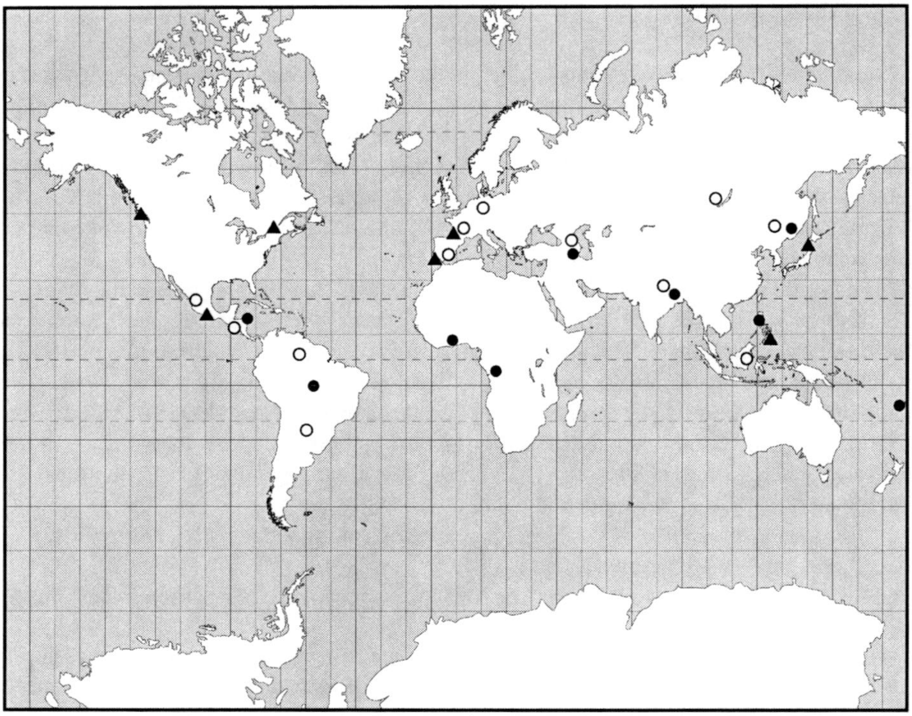

Fig. 2 The recorded global distribution of three genera of the Gondwana-tropical group of testate amoebae. •, Genus *Hoogenraadia;* ○, Genus *Planhoogenraadia;* ▲, Genus *Distomatopyxis*

A possibly unique soil environment existed in north-eastern Siberia during the late Pleistocene as shown by the studies of Bobrov et al. (2003, 2004) on Quaternary permafrost deposits on the Bykovsky Peninsula of the Laptev Sea (71° N)—an area which is now Arctic tundra, but possibly warmer during Pleistocene interglacial periods. An *Argynnia* species was found in samples dated to 45,300 + 1,200/–1050; 44,280 + 1,320/–1,120; and >41,830 years BP. Nowadays, only two species of this 15-species genus, *A. dentistoma* (Penard) and *A. vitraea* (Penard), are known to have widespread cosmopolitan distributions (Ogden and Hedley 1980). Other species have disjunct recorded distributions in Canada, Eurasia, Australia and the Antarctic (Deflandre 1936; Meisterfeld and Tan 1998). No species of this genus, except *A. dentistoma*, has been found in the present day Arctic (Beyens and Chardez 1995).

Cyclopyxis puteus Thomas (a rare and infrequent species today) was also recorded from a sample dated *ca.* 45,000 BP. This species has predominantly a Holarctic distribution in temperate and boreal forest (Bobrov 2001; Todorov 2001); however it has been recorded from Arctic tundra in Canada and Spitsbergen (Beyens and Chardez 1995). There is also a single record of a solitary specimen from the sub-Antarctic Iles Crozet (Vincke et al. 2006b). Given the large size of this species (195 μm diameter), the possibility of misidentification is remote; however this latter record appears anomalous and emphasizes the difficulty of defining the distribution patterns of rare species with certainty. It is also interesting to note the records of the *Nebela bigibbosa* Penard from the late Holocene samples, ^{14}C dated to 1,360 ± 35; 1,240 ± 60 and 1,080 ± 35 years BP (Bobrov et al.

2003, 2004). This species typically inhabits forest litter (Todorov 2002). The only previous record of it from the high Arctic is that of Penard (1903) from Spitsbergen.

A bi-polar comparison of recorded testate fauna has been made by Wilkinson (2001) which revealed that the largest testates (up to 245 μm) tended to occur in either the Arctic or the Antarctic, but not both, whilst the bi-polar species had a maximum size of 135 μm. This analysis is currently being repeated, incorporating more recent data and extending it to include the "Third Pole"—the high Himalayan mountains of Tibet and northwestern Yunnan, China, utilizing the work of Yang et al. (2005b) who has assembled a comprehensive list of 207 species across 10 physiographical regions. This tri-polar comparison is expected to yield valuable insights.

The characteristic fauna of various regions appears to be determined by an interaction of factors of a general biogeographical nature—the availability of habitats in each climatic zone, the cyst-forming and dispersal abilities of different species, and the existence of geographical barriers. It seems certain that regional endemism is a real phenomenon amongst testate species, but our understanding is so far incomplete, pending the acquisition of more comprehensive data.

Local endemism

Many cases of endemism to smaller localities, such as a single island, have been documented. Again, southern hemisphere nebelids are prominent. An often cited example is *Argynnia antarctica*, a species apparently endemic to the sub-Antarctic Marion Island, described by Grospietsch (1971), with supporting biometrical evidence. Although no molecular data are available, it appears to be recently evolved since it is likely that the whole testate fauna of Marion Island (like South Georgia and other sub-Antarctic islands) colonised the island since the end of the Pleistocene (Wilkinson 1990). It can be argued from its morphology that it evolved from the cosmopolitan *Argynnia dentistoma* (Penard), or from the Southern hemisphere endemic *A. teres* Jung, and that it has not yet had time to reach southern Africa. Other candidate species for insular endemism in the southern hemisphere include *Nebela similes* and *Nebela australis* on Tierra del Fuego (Vucetich 1972) and *Nebela subspherica* in New Zealand (van Oye 1956). An example from the Arctic is the description of *Schoenbornia smithi* from Spitsbergen by Beyens and Chardez (1997).

A less plausible example is the description by Decloitre (1964) of a new variety of *Nebela bohemica (N. bohemica* Taranek var. *adeliae*) as endemic to Terre Adélie, Antarctica. It is considerably larger than previously described specimens of *N. bohemica* (Deflandre 1936). However, in the absence of other evidence, it is valid to consider this as a phenotypic "variety" only. As in all cases of claimed endemism, it is possible to find morphologically closely related forms which have a wider distribution.

Recently reported testate species from the remote Ascension Island in the tropical mid-Atlantic (Wilkinson and Smith 2006) consist of cosmopolitan species only. This observation is consistent with the fact that this arid island was largely devoid of suitable habitats for testates until the 19th century, when imported plants with accompanying soil were artificially established there, inducing microclimatic changes (Ashmole and Ashmole 2000). It may be inferred that most of the testate fauna of Ascension Island is the result of recent arrival (since 150 years ago) and that endemic sub-speciation has not yet had time to occur.

Acknowledgements The authors wish to extend grateful thanks to Wilhelm Foissner, Thierry Heger, Edward Mitchell and David Wilkinson, whose valuable comments and guidance greatly assisted us in the preparation of the manuscript.

References

Ashmole P, Ashmole M (2002) St. Helena and Ascension Islands: a natural history. Anthony Nelson, Oswestry, UK

Balik V, Song B (2000) Benthic freshwater testate amoebae assemblages (Protozoa: Rhizopoda) from Lake Dongting, People's Republic of China, with description of a new species from genus *Collaripyxidia*. Acta Protozool 39:149–156

Beyens L, Chardez D (1995) An annotated list of testate amoebae observed in the Arctic between longitudes 27 °E and 168 °W. Arch Protistenk 146:219–233

Beyens L, Chardez D (1997) New testate amoeba taxa from the polar regions. Acta Protozool 36:137–142

Beyens L, Chardez D, Baere D et al (1995) The aquatic testate fauna of the Stromness Bay area, South Georgia. Antarct Sci 7:3–8

Beyens L, Meisterfeld R (2002) Protozoa: testate amoebae. In: Smol JP, Birks JB, Last WM et al (eds) Tracking environmental change using lake sediments. Springer, Netherlands, pp 121–153

Bobrov AA (2001) Finding of the tropical group testate amoebae (Protozoa: Testacea) at the Far East (Sikhote Alin Reserve). Biol Bull 28:401–417

Bobrov A, Mazei Y (2004) Morphological variability of testate amoebae (Rhizopoda: Testacealobosea Testaceafilosea) in natural populations. Acta Protozool 43:133–146

Bobrov AA, Siegert Ch, Andreev AA et al (2003) Testaceans (Protozoa: Testacea) in Quaternary permafrost sediments of the Bykovsky Peninsula, Arctic Yakutia. Biol Bull 30:191–206

Bobrov AA, Andreev AA, Schirrmeister L et al (2004) Testate amoebae (Protozoa: Testacealobosea and Testaceafilosea) as bioindicators in the Late Quaternary deposits of the Bykovsky Peninsula, Laptev Sea, Russia. Palaeogeogr Palaeosediment Palaeoecol 209:165–181

Boeuf O, Gilbert D (1997) Présence de Thécamoebiens du genre *Trinema* au Pliocène supérieur, découverte à Chilhac (Haute-Loire, France). C r hebd Séanc Acad Sci, Paris 325:623–627

Bonnet L (1969) Aspects généraux du peuplement thécamoebien édaphique d'Afrique intertropicale. Pub Cult Co Diam Ang 81:136–176

Bonnet L (1977) Le peuplement thécamoebien des sols du Népal et son intérêt biogéographique. Bull Soc Hist nat Toulouse 113:331–348

Bonnet L (1978) Faunistique et biogéographie des Thécamoebiens IV: Thécamboebiens des sols du massif du Nimba (Côte d'Ivorie). Bull Soc Hist nat Toulouse 114:59–64

Bonnet L (1980) Faunistique et biogéographie des Thécamoebiens. VI. Thécamoebiens de quelques sols des Philippines. Bull Soc Hist nat Toulouse 116:277–282

Bonnet L (1981) Thécamoebiens (Rhizopoda Testacea). In: Travé J (ed) Biologie des sols, vol 48. CNFRA, pp 23–32

Bonnet L (1983) Interet biogéographie et paleogéographie des Thécamoebiens des sols. Annls Stn limnol Besse 17:298–334

Bonnet L, Gomez-Sanchez M-S (1994) Thécamoebiens édaphiques (Rhizopoda, Arcellinida) à distribution géographique restreinte en Asturies et dans le sud-ouest de la France. Bull Soc Hist nat Toulouse 130: 7–14

Booth R (2002) Testate amoebae as palaeoindicators of surface-moisture changes on Michigan peatlands: modern ecology and hydrological calibration. J Paleolimnol 28:329–348

Cailleux A (1978) Reparation géographique des espèces de Thécamoebiens. C R Soc Biogeogr 472:29–39

Cash J, Wailes GH, Hopkinson J (1919) The British freshwater rhizopoda and heliozoa. IV supplement to the rhizopoda. The Ray Society, London

Certes A (1891) Protozoaires. In: Mission scientifique du Cap Horn 1882–1883, Zoology 6:1–53

Cavalier-Smith T (2004) Only six kingdoms of life. Proc R Soc London 271:1251–1262

Charman DJ (1997) Modelling hydrological relationships of testate amoebae (Protozoa: Rhizopoda) on New Zealand peatlands. J Roy Soc NZ 27:465–483

Corsetti FA, Awramik SM, Pierce D (2003) A complex microbiota from snowball Earth times: Microfossils from Neoproterozoic Kingston Peak Formation, Death Valley, USA. Proc Nat Acad Sci USA 100:4399–4404

Corsetti FA, Olcott AN, Bakemans C (2006) The biotic response to Neoproterozoic snowball Earth. Paleogeogr Paleoclim Paleoecol 232:14–130

Coupe SJ, Smith HG, Newman AP et al (2003) Biodegradation and microbial diversity within permeable pavements. Europ J Protistol 39:495–498

Cowling AJ (1983) Respiratory studies on testate amoebae from the maritime Antarctic. In: Lebrun P, André HM, de Medts A et al (eds) New trends in soil biology. Proc 8th Int Coll Soil Zoology, Louvain-la—Neuve, Belgium, p 627

Decloitre L (1953) Recherches sue les Rhizopodes thécamoebiens de l'A O F. Mem Ins Fr Afr Noire 31:1–249

Decloitre L (1964) Thécamoebiens de la XIIème Expédition Antarctique Française en Terre Adélie TAAF, Exp Pol Fr Paris

Decloitre L (1965) Contribution à la faune du Congo (Brazzaville). Mission A Dexarpentries et A Villiers III Rhizopodes Thécamoebiens. Bull Inst Fr Afr Noire 27:165–184

Decloitre L (1985) Thécamoebologie Française. Ann Soc Sc Nat Arch Toulon et du Var 37:117–122

Deflandre G (1928) Le genre *Arcella* Ehrenberg. Arch Protistenk 64:152–287

Deflandre G (1936) Etude monographique sur le genre *Nebela* Leidy. Ann Protistol 5:201–322

Esteban GF, Clarke KJ, Olmo JL et al (2006) Soil protozoa—An intensive study of population dynamics and community structure in an upland grassland. Appl Soil Ecol 33:137–151

Finlay BJ (2002) Global dispersal of free-living microbial eukaryote species. Science 296:1061–1063

Finlay BJ, Fenchel T (2004) Cosmopolitan metapopulations of free-living microbial eukaryotes. Protist 155:237–244

Finlay BJ, Esteban GF, Clarke KJ et al (2001) Biodiversity of terrestrial protozoa appears homogenous across local and global spatial scales. Protist 152:355–366

Foissner W (2006) Biogeography and dispersal of micro-organisms: a review emphasising protists. Acta Protozool 45:111–136

Foissner W, Schiller W (2001) Stable for 15 million years: scanning electron microscope investigation of Miocene euglyphid thecamoebians from Germany, with description of the new genus *Scutiglypha*. Eur J Protistol 37:167–180

Gautier-Lièvre L (1954) Les genres *Nebela, Paraquadrula* et *Pseudonebela* (Rhizopodes testacés) en Afrique. Arch Protistenk 103:241–370

Gericke JMS (1932) On two species of *Nebela* found in South Africa. South Afr J Sci Johannesburg 29:624–626

Gilbert D, Mitchell EAD, Amblard C et al (2003) Population dynamics and food preferences of the testate amoeba *Nebela tincta major—bohemica—collaris* complex (Protozoa) in a *Sphagnum* peatland. Acta Protozool 42:99–104

Golemansky V (1963) Matériaux sur la faune rhizopodique de Guinnée. Thécamoebiens du massif du Foutadjalon. Rech Africanes, NS 1:39–54

Grospietsch T (1958) Beitrag zur Rhizopodenfauna Deutschlands: Die Thekamöben der Rhön. Hydrobiologia 10:305–322

Grospietsch T (1971) Beitrag zur Ökologie der Testaceen Rhizopoden von Marion Island. In: Zinderen Bakker EM van, Winterbottom JM, Dyer RA (eds) Marion and Prince Edward Islands, report on the South African biological and geological expedition 1965–1966. AA Balkema, Cape Town, pp 411–423

Heinis F (1914) Die Moosfauna Columbiens. In: Fuhrmann O, Mayor E (eds) Voyage d'exploration scientifique en Colombie. Mem Soc Neuchateloise Sci Nat 5:675–730

Hoogenraad HR, Groot AA de (1940) Moosbewohnende Thekamoebe Rhizopoden von Java und Sumatra. Treubia 17:209–259

Hoogenraad HR, Groot AA de (1979) Die geographische Verbreitung der Süßwasser Rhizopoden. Hydrobiol Bull (Amsterdam) 13:152–171

Jennings HS (1916) Heredity, variation and the results of selection in the uniparental reproduction of *Difflugia corona*. Genetics 1:407–534

Jung W (1936) Thecamöben ursprünglicher lebender deutscher Hochmoore. Abh Landesmus Westf 7:1–87

Jung W (1942a) Illustrierte Thekamöben – Bestimmungstabellan. I. Die Systematik der Nebelinen. Arch Protistenk 95:357–390

Jung W (1942b) Südchilenische Thekamöben. Arch Protistenk 95:253–356

Korganova GA (1994) Testate amoebae (Protozoa) in island ecosystems. In: Animal populations of islands of south-west Oceania. Nauka, Moscow, pp 184–209

Laminger H (1973a) Die Testaceen (Protozoa, Rhizopoda) einiger Hochgebirgsgewässer von Mexiko, Costa Rica und Guatemala. Int Rev Ges Hydrobiol 58:273–305

Laminger H (1973b) Die Testaceen in der Umgebung der Station Büschelbach (Spessart/BRD). Hydrobiologia 41:501–513

Lara E, Heger TJ, Mitchell EAD et al (2007a) SSU rRNA reveals a sequential increase in shell complexity among the Euglyphid testate amoebae (Rhizaria: Euglyphida). Protist 158:229–237

Lara E, Heger TJ, Ekelund F et al (2007b) Ribosomal RNA genes challenge the monophyly of *Nebela* and *Hyalosphenia* (Arcellinida: Hyalospheniidae) (submitted)

Laybourn J, Whymant L (1980) Effect of diet on reproductive rate in *Arcella vulgaris* Ehrenberg (Sarcodina: Testacida). Oecologia 45:282–284

Longet D, Burki F, Flakowski J et al (2004) Multigene evidence for close evolutionary relations between *Gromia* and Foraminifera. Acta Protozool 43:303–311

McDowall RM (2005) Falkland Islands biogeography: converging trajectories in the South Atlantic Ocean. J Biogeog 32:49–62

Meisterfeld R (2002a) Order Arcellinida Kent, 1880. In: Lee JJ, Leedale GE, Bradbury P (eds) An illustrated guide to the protozoa, 2nd edn. Allen Press, Lawrence, Kansas, pp 827–860

Meisterfeld R (2002b) Testate amoebae with filopodia. In: Lee JJ, Leedale GE, Bradbury P (eds) An illustrated guide to the protozoa, 2nd edn. Allen Press, Lawrence, Kansas, pp 1054–1084

Meisterfeld R, Tan L-W (1998) First records of testate amoebae from Mount Buffalo National Park, Victoria. Vic Natural 115:231–238

Mitchell EAD, Meisterfeld R (2005) Taxonomic confusion blurs the debate on cosmopolitanism versus local endemism of free-living protists. Protist 156:263–267

Mitchell EAD, Charman DJ, Warner BG (2007) Testate amoebae (Protozoa) in ecological and paleoecological studies of wetlands: past, present and future. Biodiv Conserv (in press)

Nicholls KH (2003) *Corythionella golemanski* sp.n. a new freshwater, filose, testate rhizopod. Acta Protozool 42:75–80

Nicholls KH (2005) *Cyclopyxis acmodata* sp.n. and *Arcella formosa* n.sp.: two new species of testate rhizopods (Arcellinida, Protozoa) from threatened wetlands in Ontario, Canada. Can Field Natur 119:403–411

Nikolaev SI, Mitchell EAD, Petrov NB et al (2005) The testate lobose amoebae (order Arcellinida Kent, 1880) finally find their home within Amoebozoa. Protist 156:191–202

Ogden CG (1984) Shell structure of some testate amoebae from Britain (Protozoa, Rhizopoda). J Nat Hist 18:341–361

Ogden CG, Hedley RH (1980) An atlas of freshwater testate amoebae. British Museum (Natural History), OUP, Oxford

Oye P van (1956) On the Thecamoeban fauna of New Zealand with description of four new species and biogeografical discussion. Hydrobiologia 8:16–37

Oye P van (1960) On the Biogeography of Rhizopods and Desmids. Scientiae naturalis studia et opuscula in honorem septuagenarii W Weisbach, 61–78

Penard E (1902) Faune rhizopodique du bassin du leman. Kündig, Genève

Penard E (1903) Notice sur les Rhizopodes du Spitzbergen. Arch Protistenk 2:238–282

Penard E (1938) Les infiniment petits dans leurs manifestations vitals. Georg & Cie, Genève

Poinar GO, Waggoner BM, Bauer U-C (1993) Terrestrial soft-bodied protists and other organisms in Triassic amber. Science 259:222–224

Porter SM, Meisterfeld R, Knoll AH (2003) Vase-shaped microfossils from the Neoproterzoic Chuar Group, Grand Canyon: a classification guided by modern testate amoebae. J Palaeontol 77:409–429

Richters F. (1908) Beitrag zur Kenntnis der Moosfauna Australiens und der Inseln des Pazifischen Ozeans. Zool Jahrb 26:196–213

Schmidt AR, Schönborn W, Schäfer U (2004) Diverse fossil amoebae in German Mesozoic amber. Palaeontology 47:185–197

Schönborn W (1966) Beschalte Amöben (Testaceen). Ziemsen, Wittenberg-Lutherstadt

Schönborn W, Dörfelt H, Foissner W et al (1999) A fossilized microcenosis in Triassic amber. J Eukaryot Microbiol 46:571–584

Schroeter D (2001) Structure and function of decomposer food webs of forests along a European North-South-transect with special focus on testate amoebae (Protozoa). PhD-thesis, Department of Animal Ecology, University of Giessen

Smith HG (1973) The temperature relations and bi-polar biogeography of the ciliate genus *Colpoda*. Bull Br Antarct Surv 37:7–13

Smith HG (1978) Distribution and ecology of terrestrial protozoa of sub-Antarctic and maritime Antarctic islands. Sci Rep Br Antarct Surv 95:1–104

Smith HG (1982) The terrestrial protozoan fauna of South Georgia. Pol Biol 1:173–179

Smith HG (1985) The colonization of volcanic tephra on Deception Island by protozoa: long-term trends. Bull Br Antarct Surv 66:19–33

Smith HG (1996) Diversity of Antarctic terrestrial protozoa. Biodiv Conserv 5:1379–1394

Smith HG, Wilkinson DM (1987) Biogeography of testate rhizopods in the southern temperate and Antarctic zones. In: Trehen P (ed) 2eme Colloque sur les ecosystemes terrestres subantarctiques Paimpont, C N F R A 58:83–96

Smith HG, Wilkinson DM (2007) Not all free-living microorganisms have cosmopolitan distributions – the case of *Nebela (Apodera) vas* Certes. J Biogeog 34:1822–1831

Stout JD (1969) Significance of biogeography in ecology of soil habiting Protozoa. In: Progress in protozoology, abstracts of III international congress of protozoology, Nauka. Leningrad, pp 212–213

Todorov M (2001) Testate amoebae (Protozoa: Rhizopoda) in soil and litter of beech forests (*Fagus sylvatica* L.) from Bulgaria. Acta Zool Bulg 53:19–37

Todorov M (2002) Morphology, biometry and ecology of *Nebela bigibbosa* Penard 1890 (Protozoa: Rhizopoda). Acta Protozool 41:239–244

Todorov M, Golemansky V (1999) *Planhoogenraadia bonneti* sp.n. and *Centropyxis thailandica* sp.n. (Rhizopoda: Testacea), two new testaceans from Thailand. Acta Protozool 38:255–261

Tyler PA (1996) Endemism in freshwater algae with special reference to the Australian region. Hydrobiologia 336:1–9

Vincke S, Gremmen N, Beyens L et al (2004) The moss-dwelling testacean fauna of Île de la Possession. Pol Biol 27:753–766

Vincke S, Van der Vijver B, Gremmen N et al (2006a) The moss-dwelling testacean fauna of Strømness Bay (South Georgia). Acta Protozool 45:65–75

Vincke S, Van der Vijver B, Nijs I et al (2006b) Changes in the testacean community structure along small soil profiles. Acta Protozool 45:395–406

Vucetich MC (1972) Tres nuevas tecamebas muscicolas para la Argentina (Rhizopoda Testacea). Neotropica 18:126–128

Vucetich MC (1978) Commentarios sobre le genero *Certesella* Loeblich & Tappan, 1961 y estudio de la estereo ultraestructura tecal de tres especies austroamericanas (Rhizopoda: Testaceolobosa). Obra del Centenario del Museo de la Plata 6:305–313

Waggoner BM (1996) Bacteria and protists from Middle Cretaceous amber of Ellsworth County, Kansas. PaleoBios 17:20–26

Wanner M (1999) A review of the variability of testate amoebae: methodological approaches, environmental influence and taxonomic implications. Acta Protozool 38:15–29

Wanner M, Meisterfeld R (1994) Effect of some environmental factors on the shell morphology of testate amoebae (Rhizopoda, Protozoa). Eur J Protistol 30:91–195

Wanner M, Dunger W (2002) Primary immigration and succession of soil organisms on reclaimed opencast coal mining areas in eastern Germany. Eur J Soil Biol 38:137–143

Wanner M, Xylander WER (2005) Biodiversity development of terrestrial testate amoebae: is there any succession at all? Biol Fertil Soils 41:428–438

Wilkinson DM (1990) Glacial refugia in South Georgia? Protozoan evidence. Quaternary Newsl 62:12–13

Wilkinson DM (2001) What is the upper size limit for cosmopolitan distribution in free-living microorganisms? J Biogeography 28:285–291

Wilkinson DM, Smith HG (2006) An initial account of the terrestrial protozoa of Ascension Island. Acta Protozool 45:407–413

Woodland WA, Charman DJ, Sims PC (1998) Quantitative estimates of water tables and soil moisture in Holocene peatlands from testate amoebae. Holocene 8:261–273

Wylezich C, Meisterfeld R, Michel R et al (2001) Molecular phylogeny of amoebae. Zoology 104(Suppl IV):72

Yang J, Beyens L, Shen Y et al (2004) Redescription of *Difflugia tuberspinifera* Hu, Shen Gu *et* Gong 1997 (Protozoa: Rhizopoda: Arcellinida: Difflugiidae) from China. Acta Protozool 43:281–289

Yang J, Shen Y (2005) Morphology, biometry and distribution of *Difflugia biwae* Kawamura 1918 (Protozoa: Rhizopoda). Acta Protozool 44:103–111

Yang J, Meisterfeld R, Zhang W et al (2005a) *Difflugia mulanensis* nov.spec., a freshwater testate amoeba from Lake Mulan, China. Eur J Protistol 41:269–276

Yang J, Zhang W, Feng W et al (2005b) Geographical distribution pf testate amoebae from northwestern Yunnan and their relationship with climate. Hydrobiology 559:297–304

Diversity and geographic distribution of ciliates (Protista: Ciliophora)

Wilhelm Foissner · Anne Chao · Laura A. Katz

Originally published in the journal Biodiversity and Conservation, Volume 17, No 2, 345–363.
DOI: 10.1007/s10531-007-9254-7

Abstract About 4,500 free-living ciliate morphospecies have been described, applying an average synonymy rate of 20%. We estimate that 83–89% of the ciliate diversity is still undescribed, using the following probabilities: detailed habitat studies suggest that the described number of morphospecies must be doubled: 4,500 → 9,000; this figure has to be increased by about 50% due to species with similar interphase morphology but different resting cysts: 9,000 → 13,500; the genetic and molecular data suggest that this value must be doubled or trebled: 13,500 → 27,000 to 40,000 free-living, biological ciliate species. The knowledge on geographic distribution of ciliates heavily depends on flagship species and statistical analyses because reliable faunistic studies are rare and molecular data are still in its infancy. We present a list of 52 ciliate flagship species as a testable hypothesis, i.e., the hypothesis of restricted distribution of certain ciliate species must be refused when a considerable number of them is found in all or most biogeographic regions. Flagship species and statistical analyses consistently show Gondwanan and Laurasian ciliate communities, suggesting that the split of Pangaea deeply influenced ciliate distribution and rare species play a key role in geographic differentiation. However, there is also substantial evidence for continental, regional, and local endemism of free-living ciliates. The molecular studies usually show a high level of genetic diversity underlying ciliate morphospecies, suggesting that morphologic and molecular evolution may be decoupled in many ciliate species. Molecular studies on ciliate biogeography are at variance, possibly because most are still focusing on single molecular markers. In sum, the data indicate that

Special Issue: Protist diversity and geographic distribution. Guest editor: W. Foissner.

W. Foissner (✉)
FB Organismische Biologie, Universität Salzburg, Hellbrunnerstrasse 34, Salzburg 5020, Austria
e-mail: Wilhelm.Foissner@sbg.ac.at

A. Chao
Institute of Statistics, National Tsing Hua University, Hsin-Chu 30043 , Taiwan

L. A. Katz
Department of Biological Sciences, Smith College, Clark Science Center, Northampton, MA 01063, USA

W. Foissner et al. (eds.), *Protist Diversity and Geographical Distribution*
DOI: 10.1007/978-90-481-2801-3_9

ciliate biogeography is similar to that of plants and animals, but with an increased proportion of cosmopolites, favouring the moderate endemicity model.

Keywords Actual and estimated diversity · Cyst species · Flagship species · Floodplains · Genetic and molecular diversity · Gondwana · Laurasia · Moderate endemicity model · Pangaea

Introduction

Ciliates are unicellular, heterokaryotic organisms having a macronucleus and a micronucleus of distinctly different size and function within the same cytoplasm (Raikov 1972). The macronucleus, which is usually highly polyploid (except of the curious Karyorelictea, where it is diploid and does not divide) and divides amitotically during asexual reproduction, controls mainly somatic functions, such as RNA synthesis and ontogenesis. The diploid micronucleus is active mainly during sexual reproduction, called conjugation (Raikov 1972; Corliss 1979).

Ciliates are found in a great variety of habitats and many live on or inside of various animals, for instance, the commensalic rumen ciliates and the special species which colonize the surface of crustaceans and water beetles; some are even true parasites, for instance, *Balantidium*, which infects humans, and *Ichtyophthirius*, which causes severe disease in fish. Thus, it is not surprising that ciliates have a lot of highly specific morphologies, making species very distinct, especially since we have methods to reveal cilia and their basal bodies very selectively by scanning electron microscopy and various methods of silver impregnation (for a review, see Foissner 1991).

The distinctness of morphospecies and the comparatively large size (most between 50 μm and 200 μm) causing ciliates to lead a major role in the discussion on protist diversity and distribution (for reviews, see Finlay et al. 1996, 2004 and Foissner 1999b, 2006). Indeed, the often large (≥200 μm) flagship species have dozens of distinct features, making them easily recognizable and unmistakable (Foissner 2007). Thus, both schools on protist diversity developed on ciliates, viz., the cosmopolitic model of Finlay et al. (1996) and the moderate endemicity model of Foissner (1999b, 2004b).

The present contribution deals with free-living species because these are independent of host distribution. Further, more and better data are available than on host-bound species. Likely, less than 10% of the commensalic and parasitic ciliates have been described because most potential host species have never been investigated, except of the rumen ciliates (Foissner 1999b).

Diversity

Described diversity of free-living ciliates

An index of described ciliate species is not available. Thus, the following figures are rough estimates. Corliss (2000) suggested a total of 8,000 described ciliate morphospecies, of which about 200 are fossil tintinnids and 2,600 are commensals s.l., leaving about 5,200 extant free-living species. Since that, about 400 new species have been described, about half each in individual papers and in monographs (Song and Wang 1999; Foissner et al.

2002; Foissner and Xu 2006). Thus, we arrive at about 5,600 described free-living ciliate species.

Not all described species are valid, i.e., some are junior synonyms. Five recent monographs estimated the following synonymy rates: class Colpodea (Foissner 1993a): 22%; family Oxytrichidae (Berger 1999): 33%; suborder Urostyloidea (Berger 2006): 30%; spathidiids (Foissner and Xu 2006): 3%; aloricate Oligotrichea (Agatha, pers. comm): 10%. This gives an average synonymy rate of 20%, which is close to the 19% generic synonymy calculated by Aescht (2001). Applying this figure to the total estimate reported above, we arrive at about 4,500 valid, free-living ciliate species. Note that Berger (1999, 2006) miscalculated synonymy rates, not using the total number of species (valid and synonymous species) but only the valid ones.

Undescribed diversity of free-living ciliates

For a concise overview, we calculate the number of undescribed, free-living ciliate species ahead, providing the evidences in the following paragraphs. Applying the biological morphospecies concept, some crude calculations can be performed. The habitat studies suggest that we must at least double the described number of morphospecies: 4,500 → 9,000. This figure has to be increased by at least 50% due to the cyst species: 9,000 → 13,500. The genetic and molecular data suggest multiplying this figure by 5, at least. However, many of the "cyst species" and "ecological species" might be "genetic" and/or "molecular" species. Thus, we use conservative multiplicators of 2 and 3: 13 500 → 27,000 to 40,000 free-living, biological ciliate species, that is, 83–89% of the ciliate diversity are still undiscovered.

Habitat studies. Combining classical and modern methods, a few researchers have discovered hundreds of new ciliate morphospecies during the past 15 years, suggesting that most ciliate diversity is still unknown (Foissner 1993a, b; Petz et al. 1995; Song and Wang 1999; Foissner et al. 2002; Foissner and Xu 2006). We shall briefly discuss some recent studies, showing that our ignorance is global and concerns all main habitats (see also Cotterill et al. this issue).

The Sphagnum ponds of Simmelried in Germany (Kreutz and Foissner 2006). The Simmelried is a three hectare-sized moorland which formed after the last ice-age, that is, about 15,000 years ago. There were about 700 species of bacteria, protists, and micrometazoa which likely represent about two thirds of the morphospecies present. Many undescribed species were discovered, viz., at least 40 ciliates, 40 amoeboid organisms, and about 20 flagellates.

A statistical approach to estimate soil ciliate diversity and distribution (Chao et al. 2006). A total of 359 soil samples from Africa, Asia, Australia, South America, and Europe were investigated for ciliate diversity, using the monograph of Foissner et al. (2002) as a starting point. A total of 964 species were recorded, of which 320 were undescribed. The frequency distribution of species over samples was used for regional and global diversity estimation, applying the abundance-based coverage estimation (ACE) model. A consistent finding over all five continents was that at least half of the species diversity is still undiscovered, with a minimum of 1,928 species and a 95% confidence interval of 1600–2427 species (Table 1). This is consistent with the findings of Foissner (1997) who used a probability-based method.

Floodplain soils (Table 2). The results of Foissner (1997), Kreutz and Foissner (2006), and Chao et al. (2006) are sustained and surpassed by our data from floodplain soils, a

Table 1 Regional and global soil ciliate species diversity (from Chao et al. 2006)

Region	Subregion	Estimate of minimum species diversity	95% confidence interval	Percentage of unseen species (%)
Africa		900	(757, 1134)	41
	Kenya	217	(180, 285)	37
	Namibia	830	(685, 1078)	41
Asia		463	(319, 844)	50
Australia		865	(703, 1148)	44
Europe		706	(575, 937)	43
	Austria	446	(367, 601)	36
	Germany	552	(384, 971)	51
South America		638	(477, 971)	49
	Costa Rica	576	(377, 1100)	55
	Amazon	317	(254, 426)	46
Global		1928	(1600, 2427)	50

Table 2 Diversity and structure of floodplain soil ciliate communities

Floodplains	Total number of species	Freshwater species[a]	New or supposedly new species[b]
Danube River, Austria (2 samples from close sites)	86	28 (33%)	8 (10%)
Bukaos River, Namibia (1 sample)	90	22 (24%)	17 (19%)
Matjulu River, Krueger National Park, South Africa (1 sample)	79	32 (41%)	17 (22%)
Chobe River, Botswana (1 sample)	98	27 (27%)	26 (26%)
Niger River, Mali (2 samples from close sites)	79	20 (25%)	11 (14%)
Rio Corobici, Cost Rica (1 sample)	87	14 (16%)	5 (6%)
Amazon River, Brazil (2 samples from close sites)	112	27 (24%)	23 (21%)
Murray River, Australia (2 samples from close sites, year 1997)	110	35 (32%)	25 (23%)
Murray River, Australia (1sample taken in year 2006 from same site as in year 1997)	79	31 (39%)	15 (19%)

[a] Proportion for all described soil ciliates (643 species, Foissner 1998): 16%

[b] Overlap of new species: <5%

generally highly diverse type of ecosystem which was never investigated for protist diversity. In 13 samples from nine sites, 137 undescribed species were discovered, that is, on average 11 new species/sample (Table 2). Considering that these few samples are a glimpse when compared to the earth's floodplain diversity, there must be thousands of undescribed ciliates in floodplain soils. Thus, the above mentioned estimates of Foissner (1997) and Chao et al. (2006) are likely too conservative, possibly applying mainly to ordinary soil.

Tank bromeliads (Foissner et al. 2003). In a combined morphological, molecular, and ecological approach, Foissner et al. (2003) discovered an endemic ciliate fauna in tank bromeliads, with species reaching a length of 800 μm (Figs. 2, 3). Bromeliads occur mainly in central and South America and their tanks, which are formed by the coalescing leaf axils, form minute ponds. Altogether the tanks form a specific, above-ground ecosystem whose extensive compartmentalization obviously fosters speciation. As yet, we discovered about 50 undescribed ciliates in less than 100 samples mainly from Central America. Considering that there are about 3,000 bromelia species, many of which live in very specific habitats (e.g., Tepuis), their tanks likely contain hundreds of undescribed ciliate species.

Marine habitats. During the past 15 years, Weibo Song and his colleagues discovered about 150 undescribed, free-living and parasitic ciliate species at a single locality in China, viz., the coast near to the town of Qingdao (Song and Wang 1999; Song et al. 2003; and many individual papers, e.g., Xu et al. 2006). This matches data from Petz et al. (1995) and Dragesco (1999). Petz et al. (1995) found 46 ciliate species, of which 17 were undescribed, in Antarctic sea ice during a single cruise in the eastern Weddell Sea. Dragesco and Foissner discovered about 20 new species in the interstitial of two localities on the French coast (see Dragesco 1999 for a literature overview). Altogether, about 200 new marine ciliates have been described during the past 15 years by a few researchers, showing that the marine ciliates are as poorly known as those from freshwater and soil.

Cyst species. Most ciliates can survive adverse environmental conditions by forming a dormant stage, the so-called resting cyst. Foissner (1993a), Foissner et al. (2002) and Xu and Foissner (2005) showed that several morphologically highly similar species have different resting cysts. For instance, four very similar populations of *Epispathidium amphoriforme*, a common moss and soil ciliate, have different resting cysts, suggesting classification as different species (Foissner, unpublished). Likely, "cyst species" will increase the number of free-living ciliate species by 50% and play an important role in biogeography. Cyst morphology is increasingly used to distinguish species also in other protists. For instance, Jonckheere and Brown (2005) isolated the amoebaflagellate *Naegleria* from freshwater of Peru. Although there was little sequence difference between the new isolate and *N. pussardi* and both populations grew at 40°C, their cysts were very different and were thus used to define the new species *N. angularis*.

Genetic and molecular studies. Genetic and molecular evidences suggest that ciliates are greatly underclassified, by at least an order of magnitude, and perhaps by two orders of magnitude (Nanney et al. 1998). Dini and Nyberg (1993) reviewed the mating types of 24 species, of which 19 turned out to be a complex of morphologically highly similar taxa composed of up to 16 biological species. On average, the 24 species each consist of five biological species that do not interbreed. Taking this figure as a rough measure, the 4,500 described, valid ciliate morphospecies likely represent 22,500 biological species!

The genetic data are increasingly sustained by molecular investigations, not only in *Paramecium* (Strüder-Kypke et al. 2000; Barth et al. 2006; Hori et al. 2006) and *Tetrahymena* (Lynn and Strüder-Kypke 2006), the pets of the ciliatologists, but also in other species, such as *Carchesium polypinum* (Miao et al. 2004; Zhang et al. 2006), *Halteria grandinella* (Katz et al. 2005), *Strombidium oculatum* (Katz et al. 2005), *Stylonychia lemnae* (Schmidt et al. 2006), and *Cyclidium glaucoma* (Finlay et al. 2006). These studies consistently show a high genetic diversity underlying ciliate morphospecies that likely will be described as distinct species in the near future (Fig. 1); indeed, this has begun in both ciliates (Foissner and Berger 1999) and heterotrophic flagellates (Hausmann et al. 2006).

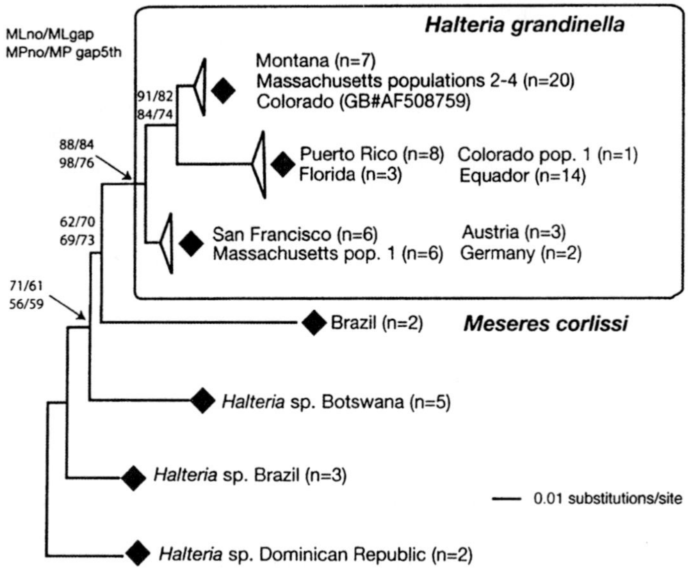

Fig. 1 *Halteria grandinella* is a very common, cosmopolitan ciliate. However, populations from different sites differ significantly in the ITS nucleotides, which even include a related species, *Meseres corlissi*. Morphologically, most populations are very similar. The figure shows a maximum-likelihood analysis of the ITS sequences. Bootstrap values are shown at nodes: upper values are maximum-likelihood with gaps removed (MLno), and maximum likelihood with gaps included (MLgap); lower values are maximum parsimony with gaps excluded (MPno), and maximum parsimony gaps treated as 5th character (MPgap5th). From Katz et al. (2005)

In only a few cases have widely-sampled isolates of ciliate morphospecies been shown to lack substantial genetic diversity, for instance, *Laboea strobila* (Katz et al. 2005).

Although most molecular studies to date have focused on only a single marker, one multimarker study of two *Paramecium* morphospecies revealed substantially different levels of variation between nuclear-encoded ITS sequences and mitochondrial cytochrome oxidase I (COI) sequences (Barth et al. 2006). Populations of *Paramecium caudatum* sampled from several sites in Europe plus a single site in China and Australia are identical at the ITS locus but show two distinct clusters for COI (Europe and China + Australia). Similarly, isolates of *Paramecium multimicronucleatum* fall into two ITS clusters but four divergent COI clusters, though there is no clear geographical pattern. Barth et al. (2006) discuss possible explanations for the observed patterns including elevated rates of mitochondrial sequence evolution compared to nuclear sequences and cryptic speciation.

The consistently high genetic diversity underlying many ciliate morphospecies suggests that morphological and molecular evolution may be decoupled in many ciliate species. Under such a scenario, morphospecies may often represent multiple genetically-isolated populations. At the same time, different morphotypes of some ciliates such as tintinnids have been shown to share identical sequences at the ITS locus (Snoeyenbos-West et al. 2002), and ecotypic variation in morphology has been demonstrated experimentally in this clade (Laval-Peuto 1981). Elucidation of the relative rates of morphological and molecular evolution in ciliates requires further analyses of multiple markers from broadly sampled morphospecies.

Ecological studies. Clonal cultures of ciliates usually reveal pronounced differences in many ecophysiological parameters, such as cell volume, growth rate, and production (for a review, see Weisse 2004). A numerical model suggests that differences in growth rates by 10% may significantly alter the clonal composition in the course of a ciliate peak (Weisse and Rammer 2006). Thus, Weisse and Rammer (2006) agree with Nanney et al. (1998) that the functional diversity of ciliates is considerably larger than it is obvious at the morphospecies level; further, they emphasize that the morphospecies concept grossly underestimates the number of species, the number of niches, and the complexity of ecosystems.

Unfortunately, the morphospecies concept is the only one which works in practice. Certainly, the biological species concept is preferable, but it is a matter of fact that more than 90% of the described species were never tested for mating, both in protists and animals; the same applies to ecological populations and clones. Thus, new ideas are needed to reconcile ecological, molecular and morphological diversity.

Geographic distribution

Flagship species

Ciliates have complex morphologies which can be clearly revealed by several methods, especially silver impregnation and scanning electron microscopy (Figs. 2–4). Thus, it is not surprising that we have sound evidences for a restricted distribution of many species. Here, for the first time, we provide a list of supposed endemics (Table 3), containing so-called flagship species which are, according to Tyler (1996) and Foissner (2006), the "ultimate proof" for protist endemism. Flagships are so showy, or so novel, making it unlikely that they would be overlooked if indeed they were widely distributed. We emphasize that the species listed in Table 3 are just a selection; they could be easily doubled! However, some doubt remains because (i) flagships could be too rare to be met in even detailed studies and (ii) reliable distribution data are rare, even for conspicuous species. For instance, *Puytoraciella dibryophrys*, an about 250 μm long colpodid described in 1979 from a temporary pond in West Africa (Foissner 1993a), has been recently found in a similar habitat of North America (Bill Bourland, personal communication). Further, we usually cannot decide whether such species are local, regional or continental endemics because large regions of the world never have been carefully investigated for ciliates (Cotteril et al., this issue). In the sense of Popper (1962), this list represents a testable hypothesis, i.e., the hypothesis of restricted distribution of certain ciliate species must be refused when a considerable number of them is found in all or most biogeographic regions.

Most of the species listed in Table 3 are likely historical (split of Pangaea) or continental endemics, for instance, *Neobursaridium gigas* and *Frontonia vesiculosa*. Others are regional or local endemics, for instance, the colepids of the ancient lakes (Lake Baikal, Lake Biwa, Lake Tanganyika). However, regional and local endemics are found not only at these famous places but also in more ordinary biotopes, as reviewed by Foissner (2007). Very recently, Stoeck et al. (2007) discovered such a species in a remote alpine lake in Germany. This ciliate is highly similar to the likely cosmopolitan *Urocentrum turbo*, both morphologically and in various molecular characteristics, but lost the ability to produce functional trichocysts. A similar case has been reported by Esteban et al. (2000), who discovered a very distinct *Lembadion* in a remote lake of Tasmania. Another example is the study by Kreutz and Foissner (2006) who found 250 ciliate species, of which 40 were

Figs. 2–4 Two undescribed ciliate flagship species in the scanning electron microscope. **2, 3**: A new, tetrahymenid ciliate from tank bromeliads. This up to 800 μm long ciliate has a functionless, minute primary oral apparatus (arrows) and a large secondary mouth (OA) used to capture large prey, e.g., rotifers. Scale bars 100 μm and 5 μm. From Foissner et al. (2003). **4**: A new colepid flagship from lake Biwa, a 4 million years old freshwater lake in Japan. This ciliate is about 80 μm long in vivo and armed with calcified plates. It is related to the widely distributed *Coleps hirtus*, but lacks the posterior spines and is genetically different by 7%. Scale bar 30 μm. From Foissner, Kusuoka and Shimano (submitted)

likely undescribed, in postglacial moorland ponds. Obviously, a considerable diversity accumulated over 15,000 years, emphasizing the great distribution capacity of microorganisms. On the other hand, some common species were lacking, for instance, the ciliate *Colpidium colpoda*, the euglenid *Phacus pleuronectes*, and rotifers of the genera *Proales* and *Floscularia*.

While a mass of undescribed species is comprehensible in amoebae, heterotrophic flagellates and ciliates, which are poorly researched, this is surprising in well-known groups, such as euglenids and chrysophytes. Thus, the authors concluded that some of the undescribed species might be regional or local endemics.

Table 3 Ciliate flagship species with very likely restricted geographic distribution

Species	Habitats	Geographic distribution[a]	References[b]
Apofrontonia dohrni	Mud from coastline puddles	Europe, Italy	Foissner (2007)
Apofrontonia lametschwandtneri	Mud from coastline puddles	South America, Venezuela	Foissner (2007)
Apofrontonia obtusa	Freshwater pond	Europe, Germany	Foissner (2007)
Baikalocoleps quadratus	Sand of open littoral	Russia, Lake Baikal	Obolkina (1995)
Bresslauides australis	Forest soil	Australia	Foissner (1993a)
Bresslauides discoideus	Soil and moss	Europe, Japan, Central America	Foissner (2006)
Bromeliophrya brasiliensis	Bromelian tanks	South America, Brazil	Foissner (2003)
Bryophyllum longisetum	Soil	Tropical Africa, Kilimanjaro	Foissner (2007)
Circinella arenicola	Sand dunes	USA, Utah	Foissner (1994)
Condylostomides etoschensis	Swamp soil	Africa (Namibia, Benin)	Foissner et al. (2002)
Corticocolpoda kaneshiroae	Tree bark	Hawaii	Foissner (1993b)
Cosmocolpoda naschbergeri	Coastal soil	Central America	Foissner (1993a)
Cyrtohymena (Cyrtohymenides) aspoecki	Floodplain soil	Austria	Foissner (2004a)
Eschaneustyla lugeri	Rainforest soil	Fiji Islands	Foissner et al. (2002)
Etoschothrix terricola	Swamp soil	Africa, Namibia (Etosha Pan)	Foissner et al. (2002)
Frontonia vesiculosa	Freshwater	Tropical Africa, South America	Dragesco and Dragesco-Kernéis (1986)
Fungiphrya strobli	Pond mud	South Africa, Table mountain	Foissner (1999a)
Heterostentor coeruleus	Marine littoral	Antarctica	Song and Wilbert (2002)
Holosticha foissneri	Sea ice	Antarctica (Weddell Sea)	Petz et al. (1995)
Jaroschia sumptuosa	Tree bark	Australia, rainforest near Cairns	Foissner (1993a)
Kentrophyllum antarcticum	Sea ice	Antarctica (Weddell Sea)	Petz et al. (1995)
Kotinia affinis	Sand of open littoral	Russia, Lake Baikal	Obolkina (1995)
Kotinia arcuata	Sand of open littoral	Russia, Lake Baikal	Obolkina (1995)
Kotinia heterolobata	Sand of open littoral	Russia, Lake Baikal	Obolkina (1995)

Table 3 continued

Species	Habitats	Geographic distribution[a]	References[b]
Krassniggia auxiliaris	Rainforest soil	Africa (Kenya) and Australia	Foissner (1993a)
Kuehneltiella muscicola	Moss	Germany	Foissner (1993a)
Kuehneltiella nambiensis	Tree bark	Africa, Namibia	Foissner et al. (2002)
Kuehneltiella terricola	Grassland soil	Australia	Foissner (1993a)
Levicoleps biwae	Littoral	Japan, Lake Biwa	Foissner et al. (submitted)
Loxocephalus foissneri	Freshwater pools	Tropical Africa, Lake Tanganyika	Dragesco and Dragesco-Kernéis (1991)
Loxodes rex	Freshwater	Tropical Africa, possibly Thailand	Foissner et al. (2002)
Luporinophrys micelae	Soil from coastline puddles	South America, Venezuela	Foissner (2005)
Macrocoleps aculeatus	Sand of open littoral	Russia, Lake Baikal	Obolkina (1995)
Macrocoleps caudatus	Sand of open littoral	Russia, Lake Baikal	Obolkina (1995)
Maristentor dinoferus	Coral reefs	Pacific Ocean	Lobban et al. (2002)
Maryna namibiensis namibiensis	Ephemeral pool	Africa, Namibia	Foissner et al. (2002)
Maryna n. costaricensis	Ephemeral pool	Central America	Foissner et al. (2002)
Neobursaridium gigas	Freshwater	Tropical Africa, South America	Dragesco and Dragesco-Kernéis (1986)
Neokeronopsis aureus	Floodplain soil	South Africa, Krueger National Park	Foissner and Stoeck (submitted)
Neokeronopsis spectabilis	Freshwater	Europe, China	Foissner and Stoeck (submitted)
Notocephalus parvulus	Marine	Antarctica (Weddell Sea)	Petz et al (1995)
Onychodromus quadricornutus	Freshwater	China, India	Berger (1999)
Planicoleps psammophilus	Littoral sand	Tropical Africa, Lake Tanganyika	Dragesco and Dragesco-Kernéis (1991)
Protospathidium namibicola	Dune sand	Africa, Namibia (Namib Desert)	Foissner et al. (2002)
Rigidothrix goiseri	Floodplain soil	Tropical Africa, Niger River	Foissner and Stoeck (2006)
Rostrophrya regis	Ephemeral pool	Tropical Africa, Cameroun	Foissner (1993a)
Saudithrix terricola	Field and floodplain soil	Saudi Arabia, China	Berger et al. (2006)
Sleighophrys pustulata	Soil from coastline puddles	South America, Venezuela	Foissner (2005)

Table 3 continued

Species	Habitats	Geographic distribution[a]	References[b]
Stentor araucanus	Lake plankton	South America	Foissner and Wölfl (1994)
Supraspathidium armatum	Swamp soil	Africa, Namibia (Etosha Pan)	Foissner et al. (2002)
Tiarinella gracilis	Sand of open littoral	Russia, Lake Baikal	Obolkina (1995)
Vermioxytricha arenicola	Sand dunes	Africa (Tunisia, Namib desert)	Foissner et al. (2002)

Note that this list contains (i) only a small selection of especially conspicuous species, (ii) includes only such flagships which have been described or redescribed with modern methods, and (iii) refers only to free-living ciliates, that is, excludes the many conspicuous species living on or inside certain hosts, e.g., on water beetles or in the digestive tract of metazoans

[a] Gives the area and/or the locality a certain species has been found. Considering our ignorance of ciliate diversity and distribution, it is likely that the range is often larger. However, we consider it as unlikely that one of these species crosses the ancient Gondwanan-Laurasian border. For instance, *Stentor araucanus*, as yet found only in various lakes of South America (Foissner and Wölfl 1994) may occur also in Australia, but not in the Holarctic

[b] For space constraints, often only reviews or monographs are cited

Last but not least, vicarious species should be mentioned, as discussed by Foissner (2007), using as examples species of the genera *Kuehneltiella* and *Apofrontonia* (Table 3). Very recently, we found other impressive examples. The first are species of the genus *Neokeronopsis*. *Neokeronopsis spectabilis* was described in the thirties of the past century and is a 300–400 μm long size-flagship, as yet found only in central Europe and China with altogether only eight records (Berger 2006). Foissner and Stoeck (submitted) discovered a new *Neokeronopsis* species in floodplain soil from South Africa. It differs from *N. spectabilis* not only in many morphological features but also in ontogenetical details, suggesting it as a representative of a distinct subgenus. The second example is the recently described *Saudithrix terricola* from field soil of Saudi Arabia (Berger et al. 2006). When the paper was in press, we found this conspicuous ciliate in soil from the Yangtse river floodplain in China, and some weeks later, Foissner discovered a new *Saudithrix* species in soil from the Chobe river floodplain in Botswana, tropical Africa. These and similar observations strongly formed Foissner's view that not everything is everywhere, both in higher and lower (micro) organisms.

Statistical studies

Large, reliable data sets on the geographic distribution of ciliates are, unfortunately, available only for terrestrial biota. Foissner (1998) and Foissner et al. (2002) investigated ciliates in over 1,000 soil samples from all biogeographic regions. Later, unpublished data were added and the whole data set analysed with new statistic tools (Chao et al. 2006). The similarity cluster resulting from the distribution of the 1,136 ciliate species clearly shows not only the separation of Laurasian from Gondwanan soil ciliate biota, but also differences within the Gondwanan sites (Fig. 5). This provides convincing statistical support for the influence of historic events on the distribution of soil ciliates, viz., the split of Pangaea and continental drift. A further study showed that geographic differentiation is related to the rare species which thus would play a key role in future studies (Chao et al. in preparation).

Fig. 5 World soil ciliate species cluster based on the classic Jaccard dis-similarity index and Ward's error sum of squares. From Chao et al. (2006). A—Australis, AR—Archinotis, H—Holarctis, N—Neotropis, P—Palaeotropis

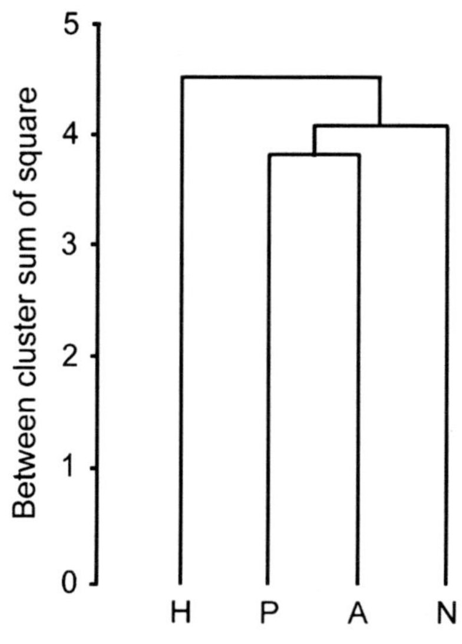

Considering the rarity of such data, we analysed the occurrence of the free-living (= freshwater, marine, and soil) species contained in the monographs and revisions of Foissner (1993a) and Berger (1999). This resulted in clusters very similar to that obtained from soil ciliates (Figs. 5–7), and suggests that the conclusions drawn above are valid for ciliates as a group.

The results of Foissner (1998), Foissner et al. (2002) and Chao et al. (2006) match the conclusions of Hillebrand et al. (2001), Green et al. (2004) and Telford et al. (2006) that similarity in species composition of various microbial assemblages generally exhibits a decreasing trend as distance increases, implying the absence of ubiquity and the existence of geographic differences.

Another possibility to test distribution patterns offers the neutral model, as shown by Řezácová and Neustupa (2007). We used this and a modified model to calculate the probability of restricted southern or northern distribution for some proposed endemics. The

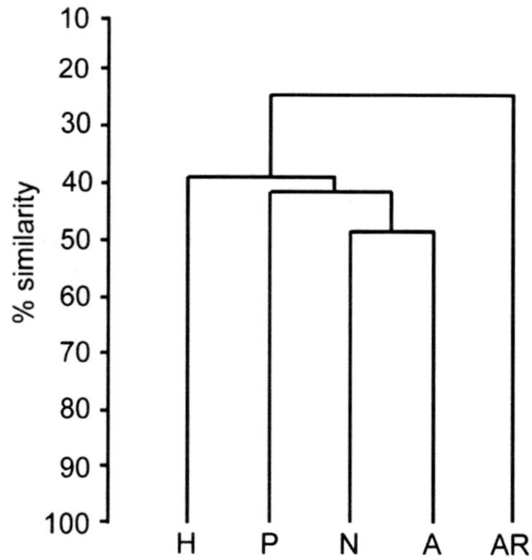

Fig. 6 The Colpodea similarity (Jaccard) cluster is based on 194 species revised/described in Foissner (1993a, 1998) and Foissner et al. (2002). Further, 25 undescribed species from sites globally are included. Thus, it is based on 219 well defined species. A—Australis, AR—Archinotis, H—Holarctis, N—Neotropis, P—Palaeotropis

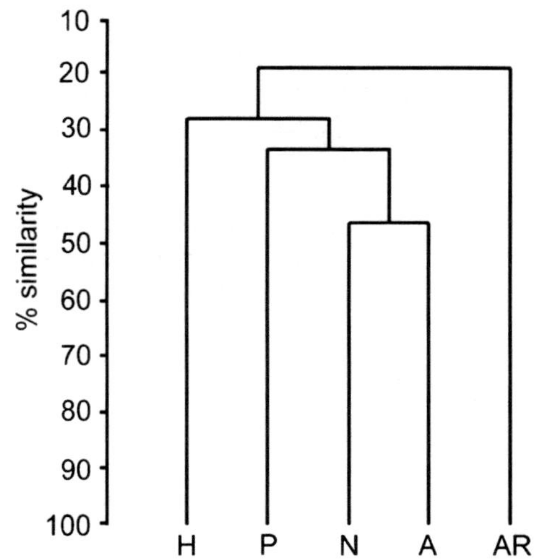

Fig. 7 The oxytrichid similarity (Jaccard) cluster is based on 191 species revised/described in Foissner (1998), Berger (1999) and Foissner et al. (2002, 2005). A—Australis, AR—Archinotis, H—Holarctis, N—Neotropis, P—Palaeotropis

Table 4 Distribution of some proposed endemics according to the neutral model, using the formula of Řezácová and Neustupa (2007)

Species	Group	Number of samples and occurrence[a]	Probability of restricted southern or northern occurrence
Nebela (Apodera) vas	Testate amoeba	100/80/220/0	0.000
Hemimastix amphikineta	Flagellate	350/42/400/0	0.000
Bresslauides discoideus	Ciliate	350/0/400/10	0.002
Rigidothrix goiseri	Ciliate	200/1/800/0	0.200
Neobursaridium gigas	Ciliate	150/6/1000/0	0.000

[a] The four numbers are: southern samples, positive southern samples, northern samples, positive northern samples. For calculation of probability, see Řezácová and Neustupa (2007). The numbers are based on original investigations (e.g., Chao et al. 2006; Foissner 1998, 2006; Foissner et al. 2002) and critical evaluation of literature data from many sources

Table 5 Distribution of the species shown in Table 4, using a formula that considers missing reports due to undersampling

Species	D = {A, Z, x}[a]	Posterior probabilities[c] for		
		Hypothesis 1	Hypothesis 2	Hypothesis 3[b]
Nebela (Apodera) vas	320/100/80	0.000	0.270	0.730
Hemimastix amphikineta	750/350/42	0.000	0.270	0.730
Bresslauides discoideus	750/400/10	0.001	0.269	0.729
Rigidothrix goiseri	1000/200/1	0.127	0.235	0.638
Neobursaridium gigas	1150/150/6	0.000	0.269	0.731

[a] The three numbers describe: A, the number of all independent reports worldwide; Z, the number of independent reports from a particular region, e.g., the northern hemisphere; x, the number of independent reports of a particular species

[b] Hypothesis 1 (H1): neutral dispersal model, i.e., individuals of a species are randomly distributed worldwide, implying the chance finding this species is the same in the northern and southern hemisphere. Hypothesis 2 (H2): the species occurs worldwide, but the chance finding it in the northern hemisphere is different from that in the southern hemisphere, i.e., missing in southern reports due to undersampling. Hypothesis 3 (H3): the species occurs only in the northern or southern hemisphere

cThe posterior probabilities (P) for each hypothesis are calculated with Bayer's theorem:

$$P(H1|D) = \frac{P(D|H1) \cdot P(H1)}{P(D|H1) \cdot P(H1) + P(D|H2) \cdot P(H2) + P(D|H3) \cdot P(H3)} \quad P(H2|D) = \frac{P(D|H2) \cdot P(H2)}{P(D|H1) \cdot P(H1) + P(D|H2) \cdot P(H2) + P(D|H3) \cdot P(H3)}$$

$$P(H3|D) = \frac{P(D|H3) \cdot P(H3)}{P(D|H1) \cdot P(H1) + P(D|H2) \cdot P(H2) + P(D|H3) \cdot P(H3)}$$

results confirm restricted distribution for all species with a probability of >95% (Tables 4, 5). The modified formula shows that undersampling plays a minor role if sample numbers are as high as in our examples (Table 5).

Molecular studies

Molecular biogeography of ciliates has just begun, and many studies to date have focused on only a single molecular marker from relatively limited geographical samples. Katz et al. (2005) found evidence of restricted southern and northern hemisphere ITS haplotypes in the cosmopolitan morphospecies *Halteria grandinella* (Fig. 1). Similarly, Schmidt et al (2006) characterized a single nucleotide polymorphism in the SSU-rDNA gene between Eurasian and North American isolates of *Stylonychia lemnae*.

Speciation as a consequence of geographic barriers has been proposed in studies of the freshwater species *Carchesium polypinum* (Miao et al. 2004) and *Halteria grandinella* as well as the tide pool species *Strombidium oculatum* (Fig. 1). The study by Miao et al. (2004) investigated the cosmopolitan ciliate *Carchesium polypinum* in China. The 18S–ITS1–5.8S rDNA separated the 19 populations analyzed into a northern and a southern phylogroup, basically matching those found in the fish fauna. More recent multilocus analysis of this taxon using inter-simple sequence repeat (ISSR) fingerprinting reveals little evidence of isolation by distance in local contemporary populations (Zhang et al. 2006). Similarly, Katz et al. (2005) argue that past geographic isolation explains the high level of genetic diversity underlying morphospecies found in habitats that are unstable over evolutionary time periods (e.g., tide pools).

For a few ciliate morphospecies, evidence has been found of high gene flow for at least some haplotypes, though these studies have also largely focused on only a single genetic marker. For example, identical haplotypes of *Strombidium oculatum* and *Laboea strobila* have been found on either side of the Atlantic Ocean (Katz et al. 2005). Similarly, there was no evidence of geographic patterns for haplotypes of *Cyclidium glaucoma*, despite the high genetic diversity underlying this morphospecies (Finlay et al. 2006). Identical (or nearly identical) haplotypes are also widespread in the freshwater species *Paramecium multimicronucleatum* and *P. caudatum* (Barth et al. 2006).

Culture-independent sampling can reveal additional insights into the biogeography of microbes. While ciliates have been sampled in numerous eukaryotic-specific surveys of microbial diversity, we are aware of only one study that focuses explicitly on ciliates (Doherty et al. submitted). These authors used primers specific for oligotrich and choreotrich ciliates to look at diversity from three Northwestern Atlantic sites sampled each in fall and spring. The bulk of haplotypes found were rare, being represented by less than two sequences in a survey of over 600 clones. Moreover, while the level of diversity was similar in all six samples, the membership within the communities varied by site and time. A few haplotypes were common to most sites, including haplotypes identical to a published sequence for *Pelagostrobilidium neptuni* and *Strombidium biarmatum*, both originally sampled from Italy (Agatha et al. 2005). Yet, the bulk of the SSU-rDNA sequences in this study had no match to sequences available from GenBank. Clearly, substantially more ciliate-specific studies are needed to assess the diversity and biogeography of uncultured ciliates.

A related emerging controversy centers on the effective population size of ciliates. Effective population size reflects the evolutionary history of populations and can be measured by the level of standing neutral variation within a species. If a strictly cosmopolitan hypothesis is true such that census population sizes are large and relatively stable over time, then one might expect ciliate populations to contain high levels of neutral variation as has been argued for *Tetrahymena thermophila* (Gerber et al. 2002; Lynch and Conery 2003) and for several species of *Paramecium* (Snoke et al. 2006). Reanalysis of the *SerH* locus used in the *Tetrahymena* studies revealed that this protein-coding gene is likely under balancing selection and therefore not a good marker for effective population size (Katz et al. 2006). In contrast, little to no silent site variation was found in an overlapping sample of natural isolates of *T. thermophila*, indicating that at least this ciliate has a low effective population size (Katz et al. 2006). This observation is consistent with the limited geographic distribution of *T. thermophila* (Foissner 2006).

Data on *Paramecium* isolates are more controversial. Observations of high levels of diversity in nuclear and mitochondrial protein coding genes from several species have been used as evidence of high effective population sizes in *Paramecium* (Snoke et al. 2006). Similarly, Barth et al. (2006) see high levels of diversity within two *Paramecium* species and recognize that, in addition to a large effective population size, alternative explanations for the observed data include undescribed syngens within this genus. Finally, Zufall et al. (2006) have demonstrated that ciliates have elevated rates of protein evolution when compared to other eukaryotes and suggest that these elevated rates may obscure estimates of effective population size from analysis of silent site variation.

Rather than two alternative explanations for the biogeography of ciliates (endemism versus cosmopolitanism), we argue that there are likely numerous patterns of distribution in ciliate morphospecies. For example, a moderate endemicity model indicates that two thirds of protists are expected to be distributed globally (Foissner 2006). Moreover, there are likely complex interactions between levels of current gene flow and genetic diversity underlying morphospecies as some of them are likely subject to low gene flow and low genetic diversity, e.g., *T. thermophila* (Katz et al. 2006); some morphospecies may have high gene flow and high genetic diversity, e.g., *P. multimicronucleatum* and *P. caudatum* (Barth et al 2006), as well as *H. grandinella* and *S. oculatum* (Katz et al. 2005) and *Cyclidium glaucoma* (Finlay et al. 2006); while still others may have high gene flow and low genetic diversity, e.g., *Laboea strobila* (Katz et al. 2005).

Acknowledgments Supported by the Austrian Science Foundation (FWF, projects P-15017 and P-19699-B17), the Taiwan National Science Council (projects NSC-94-2118-M006-001, 95-2118-M007-003), and the United States National Science Foundation (LAK DEB-0092908 and DEB-043115). The technical assistance of Mag. Birgit Peukert and Mag. Gudrun Fuss is gratefully acknowledged.

References

Aescht E (2001) Catalogue of the generic names of ciliates (Protozoa, Ciliophora). Denisia 1:1–350

Agatha S, Strüder-Kypke MC, Beran A et al (2005) *Pelagostrobilidium neptuni* (Montagnes and Taylor, 1994) and *Strombidium biarmatum* nov. spec. (Ciliophora, Oligotrichea): phylogenetic position inferred from morphology, ontogenesis and gene sequence data. Eur J Protistol 41:65–83

Barth D, Krenek S, Fokin SI et al (2006) Intraspecific genetic variation in *Paramecium* revealed by mitochondrial cytochrome c oxidase I sequences. J Eukaryot Microbiol 53:20–25

Berger H (1999) Monograph of the Oxytrichidae (Ciliophora, Hypotrichia). Monographiae Biol 78:i–xii, 1–1080

Berger H (2006) Monograph of the Urostyloidea (Ciliophora, Hypotricha). Monographiae Biol 85:i–xv, 1–1303

Berger H, Al-Rasheid KAS, Foissner W (2006) Morphology and cell division of *Saudithrix terricola* n. gen., n. sp., a large, stichotrich ciliate from Saudi Arabia. J Eukaryot Microbiol 53:260–268

Chao A, Li PC, Agatha S et al (2006) A statistical approach to estimate soil ciliate diversity and distribution based on data from five continents. Oikos 114:479–493

Corliss JO (1979) The ciliated protozoa, 2nd edn. Pergamon Press, Oxford

Corliss JO (2000) Biodiversity, classification, and numbers of species of protists. In: Raven PH, Williams T (eds) Nature and human society. The quest for a sustainable world. National Academy Press, Washington, pp 130–155

Dini F, Nyberg D (1993) Sex in ciliates. In: Jones JG (ed) Advances in microbial ecology, vol 13, Plenum Press, New York, pp 85–153

Dragesco J (1999) Revision des Geléiides (Ciliophora, Karyorelictea). Stapfia 66:1–91

Dragesco J, Dragesco-Kernéis A (1986) Ciliés libres de l'Afrique intertropicale. Faune Trop 26:1–559

Dragesco J, Dragesco-Kernéis A (1991) Free-living ciliates from the coastal area of Lake Tanganyika (Africa). Eur J Protistol 26:216–235

Esteban GF, Finlay BJ, Olmo JL et al (2000) Ciliated protozoa from a volcanic crater-lake in Victoria, Australia. J Nat Hist (London) 34:159–189

Finlay BJ, Corliss JO, Esteban G et al (1996) Biodiversity at the microbial level: the number of free-living ciliates in the biosphere. Q Rev Biol 71:221–237

Finlay BJ, Esteban GF, Fenchel T (2004) Protist diversity is different? Protist 155:15–22

Finlay BJ, Esteban GF, Brown S et al (2006) Multiple cosmopolitan ectotypes within a microbial eukaryote morphospecies. Protist 157:377–390

Foissner W (1991) Basic light and scanning electron microscopic methods for taxonomic studies of ciliated protozoa. Eur J Protistol 27:313–330

Foissner W (1993a) Colpodea (Ciliophora). Protozoenfauna 4/1:i–x, 1–798

Foissner W (1993b) *Corticocolpoda kaneshiroae* n. g., n. sp., a new colpodid ciliate (Protozoa, Ciliophora) from the bark of Ohia trees in Hawaii. J Eukaryot Microbiol 40:764–775

Foissner W (1994) Morphology and morphogenesis of *Circinella arenicola* nov. gen., nov. spec., a cephalized hypotrich (Ciliophora, Hypotrichida) from sand dunes in Utah, USA. Eur J Protistol 30: 156–170

Foissner W (1997) Global soil ciliate (Protozoa, Ciliophora) diversity: a probability-based approach using large sample collections from Africa, Australia and Antarctica. Biodivers Conserv 6:1627–1638

Foissner W (1998) An updated compilation of world soil ciliates (Protozoa, Ciliophora), with ecological notes, new records, and descriptions of new species. Eur J Protistol 34:195–235

Foissner W (1999a) Description of two new, mycophagous soil ciliates (Ciliophora, Colpodea): *Fungiphrya strobli* n. g., n. sp. and *Grossglockneria ovata* n. sp. J Eukaryot Microbiol 46:34–42

Foissner W (1999b) Protist diversity: estimates of the near-imponderable. Protist 150:363–368

Foissner W (2003) Morphology and ontogenesis of *Bromeliophrya brasiliensis* gen. n., sp. n., a new ciliate (Protozoa: Ciliophora) from Brazilian tank bromeliads (Bromeliaceae). Acta Protozool 42:55–70

Foissner W (2004a) Some new ciliates (Protozoa, Ciliophora) from an Austrian floodplain soil, including a giant, red "flagship", *Cyrtohymena* (*Cyrtohymenides*) *aspoecki* nov. subgen., nov. spec. Denisia 13:369–382

Foissner W (2004b) Ubiquity and cosmopolitanism of protists questioned. SILnews 43:6–7

Foissner W (2005) Two new "flagship" ciliates (Protozoa, Ciliophora) from Venezuela: *Sleighophrys pustulata* and *Luporinophrys micelae*. Eur J Protistol 41:99–117

Foissner W (2006) Biogeography and dispersal of micro-organisms: a review emphasizing protists. Acta Protozool 45:111–136

Foissner W (2007) Dispersal and biogeography of protists: recent advances. Jpn J Protozool 40:1–16

Foissner W, Berger H (1999) Identification and ontogenesis of the nomen nudum hypotrichs (Protozoa: Ciliophora) *Oxytricha nova* (= *Sterkiella nova* sp. n.) and *O. trifallax* (= *S. histriomuscorum*). Acta Protozool 38:215–248

Foissner W, Stoeck T (2006) *Rigidothrix goiseri* nov. gen., nov. spec. (Rigidotrichidae nov. fam.), a new "flagship" ciliate from the Niger floodplain breaks the flexibility-dogma in the classification of stichotrichine spirotrichs (Ciliophora, Spirotrichea). Eur J Protistol 42:249–267

Foissner W, Wölfl S (1994) Revision of the genus *Stentor* Oken (Protozoa, Ciliophora) and description of *S. araucanus* nov. spec. from South American lakes. J Plankton Res 16:255–289

Foissner W, Xu K (2006) Monograph of the Spathidiida (Ciliophora, Haptoria) Vol. I: Protospathidiidae, Arcuospathidiidae, Apertospathulidae. Monographiae Biol 81:i–ix, 1–485

Foissner W, Agatha S, Berger H (2002) Soil ciliates (Protozoa, Ciliophora) from Namibia (Southwest Africa), with emphasis on two contrasting environments, the Etosha Region and the Namib Desert. Denisia 5:1–1459

Foissner W, Strüder-Kypke M, van der Staay GWM et al (2003) Endemic ciliates (Protozoa, Ciliophora) from tank bromeliads: a combined morphological, molecular, and ecological study. Eur J Protistol 39:365–372

Foissner W, Berger H, Xu K et al (2005) A huge, undescribed soil ciliate (Protozoa: Ciliophora) diversity in natural forest stands of Central Europe. Biodivers Conserv 14:617–701

Gerber CA, Lopez AB, Shook SJ et al (2002) Polymorphism and selection at the SerH immobilization antigen locus in natural populations of *Tetrahymena thermophila*. Genetics 160:1469–1479

Green J, Holmes AJ, Westoby M et al (2004) Spatial scaling of microbial eukaryote diversity. Nature 432:747–753

Hausmann K, Selchow P, Scheckenbach F et al (2006) Cryptic species in a morphospecies complex of heterotrophic flagellates: the case study of *Caecitellus* spp. Acta Protozool 45:415–431

Hillebrand H, Watermann F, Karez R et al (2001) Differences in species richness patterns between unicellular and multicellular organisms. Oecologia 126:114–124

Hori M, Tomikawa I, Przyboś E, Fujishima M (2006) Comparison of the evolutionary distances among syngens and sibling species of *Paramecium*. Mol Phylogent Evol 38:697–704

de Jonckheere JF, Brown S (2005) Description of a new species with a remarkable cyst structure in the genus *Naegleria: Naegleria angularis* sp. n. Acta Protozool 44:61–65

Katz LA, McManus GB, Snoeyenbos-West LO et al (2005) Reframing the "everything is everywhere" debate: evidence for high gene flow and diversity in ciliate morphospecies. Aquat Microb Ecol 41:55–65

Katz LA, Snoeyenbos-West OLO, Doerder FP (2006) Unusual patterns of molecular evolution at the SerH surface antigen locus in *Tetrahymena thermophila*: implication for estimates of effective population size. Mol Biol Evol 23:608–614

Kreutz M, Foissner W (2006) The *Sphagnum* ponds of Simmelried in Germany: a biodiversity hot-spot for microscopic organisms. Protozool Monogr 3:1–267

Laval-Peuto M (1981) Construction of the lorica in Ciliata Tintinnina. In vivo study of *Favella ehrenbergii*: variability of the phenotypes during the cycle, biology, statistics, biometry. Protistologica 17:242–279

Lobban CS, Schefter M, Simpson AGB et al (2002) *Maristentor dinoferus* n. gen., n. sp., a giant heterotrich ciliate (Spirotrichea: Heterotrichida) with zooxanthellae, from coral reefs on Guam, Mariana Islands. Mar Biol 140:411–423

Lynch M, Conery JS (2003) The origins of genome complexity. Science 302:1401–1404

Lynn D, Strüder-Kypke MC (2006) Species of *Tetrahymena* identical by small subunit rRNA gene sequences are discriminated by mitochondrial cytochrome *c* oxidase I gene sequences. J Eukaryot Microbiol 53:385–387

Miao W, Yu Y, Shen Y et al (2004) Intraspecific phylogeography of *Carchesium polypinum* (Peritrichia, Ciliophora) from China, inferred from 18S-ITS1-5.8S ribosomal DNA. Sci China, C Life Sci 47:11–17

Nanney DL, Park C, Preparata R et al (1998) Comparison of the sequence differences in a variable 23S rRNA domain among sets of cryptic species of ciliated Protozoa. J Eukaryot Microbiol 45:91–100

Obolkina LA (1995) New species of the family Colepidae (Prostomatida, Ciliophora) from Lake Baikal. Zool Zh 74:3–19 (in Russian)

Petz W, Song W, Wilbert N (1995) Taxonomy and ecology of the ciliate fauna (Protozoa, Ciliophora) in the endopagial and pelagial of the Weddell Sea, Antarctica. Stapfia 4:1–223

Popper K (1962) The logic of scientific discovery. Harper and Row, New York

Raikov IB (1972) Nuclear phenomena during conjugation and autogamy in ciliates. In: Chen T-T (ed) Research in protozoology vol 4. Pergamon Press, New York, pp 146–289

Řezácová M, Neustupa J (2007) Distribution of the genus *Mallomonas* (Synurophyceae) – ubiquitous dispersal in microorganisms evaluated. Protist 158:29–37

Schmidt SL, Ammermann D, Schlegel M et al (2006) *Stylonychia lemnae* strains from North American sites differ in a single nucleotide within the small subunit rDNA from Eurasian strains. J Eukaryot Microbiol 53:308–309

Snoeyenbos-West OLO, Salcedo T, McManus GB et al (2002) Insights into the diversity of choreotrich and oligotrich ciliates (Class: Spirotrichea) based on genealogical analyses of multiple loci. Int J Syt Evol Microbiol 52:1901–1913

Snoke MS, Berendonk TU, Barth D et al (2006) Large global effective population sizes in *Paramecium*. Mol Biol Evol 23:2474–2479

Song W, Wang M (1999) New name list of marine ciliates in China. In: Song W (ed) Progress in protozoology. Qingdao Ocean University Press, Qingdao, China, pp 65–76

Song W, Wilbert N (2002) Faunistic studies on marine ciliates from the Antarctic benthic area, including descriptions of one epizoic form, 6 new species and 2 new genera (Protozoa: Ciliophora). Acta Protozool 41:23–61

Song W, Zhao Y, Xu K et al (2003) Pathogenic protozoa in mariculture. Science Press, Beijing (in Chinese)

Stoeck T, Bruemmer F, Foissner W (2007) Evidence for local ciliate endemism in an alpine anoxic lake. Microb Ecol (in press)

Strüder-Kypke MC, Wright AG, Fokin SI et al (2000) Phylogenetic relationships of the genus *Paramecium* inferred from small subunit rRNA gene sequences. Mol Phylogenet Evol 14:122–130

Telford RJ, Vandvik V, Birks HJB (2006) Dispersal limitations matter for microbial morphospecies. Science 312:1015

Tyler PA (1996) Endemism in freshwater algae with special reference to the Australian region. Hydrobiologia 336:1–9

Weisse T (2004) Pelagic microbes – protozoa and the microbial food web. In: ÓSullivan P, Reynolds CR (eds) The lakes handbook, vol I. Blackwell, Oxford, pp 417–460

Weisse T, Rammer S (2006) Pronounced ecophysiological clonal differences of two common freshwater ciliates, *Coleps spetai* (Prostomatida) and *Rimostrombidium lacustris* (Oligotrichida), challenge the morphospecies concept. J Plankton Res 28:55–63

Xu D, Song W, Lin X et al (2006) On two marine oligotrich ciliates, *Spirostrombidium agathae* n.sp. and *S. schizostomum* (Kahl, 1932) n. comb. from China, with a key to the identification of seven well-characterized *Spirostrombidium* spp. (Ciliophora: Oligotrichida). Acta Protozool 45:433–442

Xu K, Foissner W (2005) Descriptions of *Protospathidium serpens* (Kahl, 1930) and *P. fraterculum* n.sp. (Ciliophora, Haptoria), two species based on different resting cyst morphology. J Eukaryot Microbiol 52:298–309

Zhang W-J, Yang J, Yu Y-H et al (2006) Population genetic structure of *Carchesium polypinum* (Ciliophora: Peritrichia) in four Chinese lakes inferred from ISSR fingerprinting: high diversity but low differentiation. J Eukaryot Microbiol 53:358–363

Zufall RA, McGrath C, Muse SV et al (2006) Genome architecture drives protein evolution in ciliates. Mol Biol Evol 23:1681–1687

The "*Tetrahymena pyriformis*" complex of cryptic species

Ellen M. Simon · David L. Nanney · F. Paul Doerder

Originally published in the journal Biodiversity and Conservation, Volume 17, No 2, 365–380.
DOI: 10.1007/s10531-007-9255-6

Abstract Cryptic species are common among protists and have long been known in ciliates. The ciliate genus *Tetrahymena* contains a large group of morphologically indistinguishable species referred to as the '*T. pyriformis*' complex. These species include those reproductively isolated by mating type as well as asexual species characterized by the absence of the germinal micronucleus. This paper examines the molecular diversity of the species and describes the biogeography of '*T. pyriformis*' species. Most species are globally distributed, though the best studied species, *T. thermophila*, is confined to North America and gives evidence of population structure in local populations. Selfers and asexual species are common and arise from sexual species, a possible exploitation of nuclear dimorphism. It is argued that the cryptic species likely have different ecological roles and that the biodiversity of *Tetrahymena* in particular, and ciliates in general, is underestimated.

Keywords *Tetrahymena* · Biodiversity · Ciliates · Cryptic species · Population structure · Species problem

Abbreviations

COX1 Cytochrome oxidase 1 subunit
CT Connecticut
D2 A hypervariable region of LSU
FL Florida
IL Illinois

Special Issue: Protist diversity and geographic distribution. Guest editor: W. Foissner.

E. M. Simon · D. L. Nanney
Department of Animal Biology, University of Illinois at Urbana-Champaign,
505 South Goodwin Avenue, Urbana, IL 61801, USA

F. P. Doerder (✉)
Department of Biological, Geological and Environmental Sciences, Cleveland State University,
2121 Euclid Avenue, Cleveland, OH 44115, USA
e-mail: f.doerder@csuohio.edu

W. Foissner et al. (eds.), *Protist Diversity and Geographical Distribution*
DOI: 10.1007/978-90-481-2801-3_10

LSU Large ribosomal subunit
MA Massachusetts
ME Maine
MI Michigan
PA Pennsylvania
NH New Hampshire
NW PA North West Pennsylvania
RRNA Ribosomal RNA
VT Vermont

Introduction

Small ciliates now known as *Tetrahymena* were described and studied under various names (e.g., *Leucophyrs* and *Glaucoma*) well before the genus *Tetrahymena* was established by Furgason (1940). The generic name distinguishes over 40 species from other similar ciliates: tetrahymenas have four ciliated membranes in their buccal (oral) apparatus, one undulating membrane and three adoral membranelles. In common with other ciliates, all species have a transcriptionally active macronucleus, but only the breeding species possess the germinal micronucleus necessary for conjugation and sexual recombination. It is likely that nuclear dimorphism plays important roles in ciliate speciation.

The ability of tetrahymenas to grow rapidly in axenic media (without other organisms as food; Lwoff 1923) quickly led to their wide use as "animal" models in biochemical and physiological research (Hill 1972; Elliott 1973a). As investigators collected more tetrahymenas the name *T. pyriformis* soon came to include a wide variety of similar, often subtly different "strains", most of which lacked the micronucleus necessary for genetic analysis. In the absence of markers to distinguish among strains, inevitable mixup and mislabeling occurred. In the 1950s Elliott and coworkers brought some order to the micronucleate strains by demonstrating that they separate into biological species based on mating reactions (reviewed in Elliott 1973b). In the mid-1950s Nanney and Allen domesticated what later would be called *Tetrahymena thermophila* and made the inbred strains available to investigators who quickly exploited nuclear dimorphism to make fundamental discoveries (Gall 1986; Asai and Forney 2000). Its genome has been sequenced (Eisen et al. 2006), and today *T. thermophila* is the most commonly used *Tetrahymena*.

This article focuses on the diversity and biogeography of the '*T. pyriformis*' cryptic species. It was Sonneborn (1937) who discovered that morphologically indistinguishable cells of '*P. aurelia*' discriminate among themselves by forming conjugating (sexual) pairs only with a complementary type. Within twenty years of his discovery, it was well established that reproductively isolated groups identified by mating type were indeed species that differed not only in modes of mating type determination but also had significant genetic differences (reviewed in Sonneborn 1957). Cryptic species are now found in a wide variety of protists (Slapeta et al. 2005; Creach et al. 2006; also see Fawley et al. 2006; reviewed by Foissner 2006; Hausmann et al. 2006; Scheckenbach et al. 2006; Slapeta et al. 2006). Among ciliates, cryptic species also appear to be particularly common. In their reviews Przybos (1986) and Dini and Nyberg (1993) mention cryptic species in *Aspidisca, Colpidium, Euplotes, Glaucoma, Oxytricha, Paramecium, Pseudourostyla*, *Stylonychia*, *Tetrahymena*, *Tokophrya*, and *Uronychia*. More recently, there are indications of cryptic species in *Paramecium duboscqui* (Fokin et al. 1999), *Paramecium jenningsi* (Maciejewska 2006), *Stylonychia mytilus/lemnae* (Haentzsch et al. 2006; Schmidt et al. 2006), *Uronychia*

(Chen et al. 2003), *Euplotes octocarinatus* (Mollenbeck 1999), *Euplotes crassus-minuta-vannus* (Petroni et al. 2002), and *Sterkiella histriomuscorum* (Foissner and Berger 1999). This list of ciliate morphospecies with cryptic species is almost certainly incomplete as the vast majority of the variously estimated 3,000–8,000 ciliate species are known only by morphology and have not been examined by genetic or molecular methods that would identify cryptic species. In addition, new ciliates are constantly being discovered by classical techniques (Chao et al. 2006) and by PCR-based methods (Lopez-Garcia et al. 2001; Moreira and Lopez-Garcia 2002; Massana et al. 2004; Slapeta et al. 2005; Stoeck et al. 2006). Most of the latter have not been examined microscopically or cultured in the laboratory.

While breeding analysis is definitive in cryptic species identification, it does not work for asexual (amicronucleate) strains which normally do not mate, and it is of little use with selfers, clones showing intraclonal conjugation. Both of these are common in *Tetrahymena*, raising questions about the nature of speciation in this genus. DNA sequences (ribosomal and mitochondrial barcodes) allow identification of most of these species without reference to living strains. As might be expected, the systematics of *Tetrahymena* species is still work in progress, and even with new molecular information, not all workers are in agreement. Nevertheless, it is timely to examine what is known regarding the cryptic species of '*T. pyriformis*' and to place the results in the larger question of the evolution and role of cryptic species.

Tetrahymena and the '*T. pyriformis*' morphospecies

Originally the species *T. pyriformis* (Ehrenberg, 1830) Lwoff 1947 included all morphologically indistinguishable tetrahymenas, including those with (sexual) and without (asexual) micronuclei (Corliss 1973). When electrophoretic studies on isozymes (Borden et al. 1973a, b) showed that sexual strains could be distinguished independently of mating and that asexual strains had differences similar to those of the sexual species, binomial names were assigned to all '*T. pyriformis*' (Nanney and McCoy 1976). A logical but unfortunate decision resulting in continued confusion was to retain the name *T. pyriformis* for one group of amicronucleate strains. This is why, in keeping with Frankel's (2000) suggestion, we use the form '*T. pyriformis*' in quotation marks and roman type when referring to the morphospecies. In our usage, the '*T. pyriformis*' complex includes all species previously known under that name as varieties, syngens or phenosets (Nanney and McCoy 1976). Present evidence indicates that *Tetrahymena* is a monophyletic genus, though this has been disputed (Jerome and Lynn 1996).

Table 1 lists the species in the genus *Tetrahymena*. The '*T. pyriformis*' morphospecies appear in groups, five of which comprise the species that appear on five branches of the phylogenetic tree (Figure 2 in Nanney et al. 1998) based on the sequences of a portion of the D2 LSU rRNA. These are indicated by a syngen number (1–20) and/or a phenoset letter (A–E) or a strain designation. Sixteen unnamed new species or subspecies in these groups indicated by numerals are included for completeness. A separate publication will describe these species. Table 1 also lists a group of miscellaneous *Tetrahymena* species.

Attempts have been made to distinguish the cryptic '*T. pyriformis*' species by cortical features (Nanney 1967), tolerance to metals (Nyberg 1974), tolerance to higher temperatures (Nyberg 1981a), and multivariate analyses (Gates and Berger 1974). However, the overlap among the species for all these criteria prevents clear discrimination except for species falling at the extreme ends of the spectra. This is consistent with the highly conserved

Table 1 Species and biogeography of *Tetrahymena*

Code[a]	Species and references[a]	ATCC[a]	Strain[b]	Micronucleus[c]	Biogeographical region[d]				
					HA NA	HA EU	PT	NT	AU
T. thermophila group									
TTH	*T. thermophila* (Nanney and McCoy 1976)	*	1	+	29	–	–	–	–
TMA	*T. malaccensis* (Simon et al. 1985)	*	16	+	–	–	2	–	–
TEL	*T. elliotti* (Nanney and McCoy 1976)	*	20, B	+, –	47	5	3	7	–
283	T. n. sp. 1	*205076	B-(A)	–	2	–	–	–	–
T. pyriformis group									
TPY	*T. pyriformis* (Nanney and McCoy 1976)	*	A#	–	7	4	1	–	–
	Tse *T. setosa*[e] (Holz and Corliss 1956; McCoy 1975)	30782	#	+, –	12	1	–	–	–
128	T. n. sp. 2	*205038	A-(A)	–	15	–	–	–	–
137	T. n. sp. 3	205063	A-(B)	–	1	–	–	–	–
TUR	*T. leucophrys*[f] (Williams et al. 1984)	50069		–	1	–	–	–	
TSI	*T. silvani*[f] (Simon et al. 1985)	*50084		+	–	–	2–	–	
V2S	*T. vorax*[fg] (Kidder 1940)	30421	#	+	4	–	–	–	–
T. tropicalis group									
TTR	*T. tropicalis*[h] (Nanney and McCoy 1976)	*	9, C, E	+, –	21	2	4	15	–
D61	T. n. sp. 4	205044	9 (A)	–	–	–	–	1	–
DR6	T. n. sp. 5	205045	9 (B)	–	–	–	–	1	–
70	T. n. sp. 6	205046	9 (C)	–	1	–	–	–	–
KAK	T. n. sp. 7	205047	9 (D)	–	2	–	–	–	–
T. borealis group									
TBO	*T. borealis* (Nanney and McCoy 1976)	*	3, D	+, –	54	1	–	–	–
TCA	*T. canadensis* (Nanney and McCoy 1976)	*	7	+, –	18	–	1	–	–
	Tro *T. rostrata*[e,g,i] (Kahl 1926; Corliss 1952b)	30770	#	+, –	4	2	–	–	3
M51	T. n. sp. 8	50402	MP 51	–	–	–	1	–	–
SV3	T. n. sp. 9	50413	SHV 3	–	1	–	–	–	–
TMI	*T. mimbres* (Meyer and Nanney 1987)	30330	G	–	2	–	–	–	–

Table 1 continued

Code[a]	Species and references[a]	ATCC[a]	Strain[b]	Micronucleus[c]	Biogeographical region[d]				
					HA NA	HA EU	PT	NT	AU
T. americanis group									
TAM	Tam T. n. ssp. 10	*30315	2	+	96	–	4	11	–
	Tap T. n. ssp. 11	*205004	19	+	3	–	–	2	–
	Tau *T. australis* (Nanney and McCoy 1976)	*	11	+, –	27	5	7	2	6
	The *T. hegewischi* (Nyberg 1981b)	*	5	+	19	–	–	–	–
	Thy *T. hyperangularis* (Nanney and McCoy 1976)	*	10	+	6	4	–	–	–
	Tna *T. nanneyi* (Simon et al. 1985)	*	15	+	5	–	–	–	–
	Tni *T. nipissingi* (Nyberg 1981b)	*	14	+	2	–	–	–	–
	Tpi T. n. ssp. 12	*30364	6	+	17	3	–	2	–
	Tpo T. n. ssp. 13	*30278	8	+	16	–	–	–	–
	Tpe T. n. ssp. 14	*205034	18	+	1	10	–	2	–
	Tso *T. sonneborni* (Nyberg 1981b)	*	13	+	7	4	–	–	–
TAS	*T. asiatica* (Simon et al. 1985)	*	17	+	–	–	2	–	–
TCP	*T. capricornis* (Nanney and McCoy 1976)	*	12	+	–	–	–	–	10
TPT	*T. patula*[fh] (Corliss 1952a)	*50064	#	+, –	1	1	–	–	–
241	T n. sp. 15	50403	WLK 5	+	2	–	–	–	–
Tcs	*T. cosmopolitanis* (Nanney and McCoy 1976)		4	+	11	3	1	–	–
Tem	*T. empidokyrea* (Jerome et al. 1996)			+	18	–	–	–	–
Tsh	*T. shanghaiensis* (Feng et al. 1988)	205039	S1	+	–	–	1	–	–
Miscellaneous group									
TLI	*T. limacis* (Warren 1932; Kozloff 1946)	30771	#	+, –	2	–	–	–	–
TPA	*T. paravorax*[ef] (Corliss 1957)	205177	#	+, –	–	1	–	1	–
TCO	*T. corlissi*[e,g] (Thompson 1955)	50086	#	+	2	–	–	–	–
BOT	*Lambornella clarki*[j] (Corliss and Coats 1976)		#	+	+	–	–	–	–
20F	T. sp. UK 20f			–	–	1	–	–	–
U28	T. sp.UK 17, 28			–	–	2	–	–	–
TCU	*T. caudata*[e] (Simon et al. 1985)		*	+	–	–	4	–	–
KZ2	T. sp.		Kacz 2	+	–	2	–	–	–
L2B	T. sp.		LO 2h	?	–	–	1	–	–
T12	T. n. sp. 16	205048	TI 12	+	–	–	1	–	–

Table 1 continued

Code[a]	Species and references[a]	ATCC[a]	Strain[b]	Micronucleus[c]	Biogeographical region[d]				
					HA NA	HA EU	PT	NT	AU
Tch	*T. chironomi* (Corliss 1960)		#	+	+	+	–	–	–
Tdi	*T. dimorpha*[f] (Batson 1983)			+	–	+	–	–	–
Trt	*T. rotunda* (Lynn et al. 1981)			+	1	–	–	–	–
Tsa	*T. sialidos* (Batson 1985)			+	–	+	–	–	–
Tfa	*T. farleyi* (Lynn et al. 2000)			–	1	–	–	–	–

[a] Codes consisting of capital letters and/or numerals identify species/strains which appear on branches of the D2 LSU rDNA phylogenetic tree (see Table 1 and Fig. 1 in Nanney et al. 1998). Indented codes indicate identical sequences. The miscellaneous group consists of species not on those branches and species for which LSU-D2 sequences are not available. "T. n. sp." or "T. n. ssp." designate 16 new species or subspecies to be described elsewhere. American Type Culture Collection accession numbers are included for all new species and subspecies. An asterisk indicates that ATCC has two or more strains

[b] Arabic numerals identify the "syngen" number (1–20) of reproductively isolated groups of micronucleate strains. Capital letters identify groups of (mostly) amicronucleate strains called "phenosets" (A–E) that shared very similar isozymic phenograms and were assigned specific binomials by (Nanney and McCoy 1976). When amicronucleate strains with sequences nearly identical to those of named species appeared we also used capital letters to identify them; they are in ( ). A # identifies a species recognized by Corliss (1973). See text regarding *T. pyriformis*. The list is available at http://www.life.uiuc.edu/nanney/index.html

[c] Indicates differences among strains regarding presence or absence of the germinal micronucleus

[d] Number of strains isolated from biogeographical regions. HA NA, Holarctic North America; HA EU, Holarctic Europe (Germany, France, United Kingdom, Poland; PT, Paleotropis (Malaysia; So. China; Thailand); NT, Neotropis (Yucatan, Panama, Colombia, Brazil, Dominican Republic, Puerto Rico); AU, Australia. Numerical data were assembled by EMS from three sources: (1) strains collected by or donated to the Nanney lab since the 1950s; (2) Elliott's (1973) data and (3) abstracts and unpublished sources. These numbers should not be used for quantitative purposes. Geographic locations marked with a "+" are from literature cited in the species designation. The >1,000 strains of *T. thermophila* collected in PA and elsewhere in North America by FPD are not enumerated here

[e] Caudal cilium present

[f] Dimorphic. Under appropriate environmental conditions small microstome cells transform into large macrostomes. Except for *T. dimorpha* these are carnivorous. The original *T. leucophrys* strain was obtained from Turtox Supply Co. from an unknown geographic source. A strain with D2 LSU rRNA identical to *T. leucophrys* has been collected by FPD from Pennsylvania

[g] Reproductive cysts provide an alternative to binary fission as a means of reproduction

[h] Strains of *T. furgasoni* and *T. tropicalis* have identical LSU-D2 regions and agree on 9 of the 13 enzyme systems examined and therefore the former is considered a junior synonym of the latter

[i] May form resting cysts

[j] Associated with tree hole mosquitoes, transferred to the genus *Tetrahymena* (Corliss 1973) then returned to *Lambornella* (Corliss and Coats 1976)

morphology that includes ultrastructural details (Elliott and Kennedy 1973) and the number and size of micronuclear chromosomes (Ray and Elliott 1954). The cell structure of '*T. pyriformis*' morphospecies (see Figures 3 and 5 in Frankel 2000) characteristic of the complex is a very successful form.

The structural homogeneity of *Tetrahymena* cells contrasts sharply with the heterogeneity of their molecular components. The differences have been reviewed before (Nanney 1982; chapters 4 to 9 in Gall 1986; and chapters 3 and 8 in Asai and Forney 2000) and include nucleotide base composition, genome and mitochondrial DNA reannealing kinetics, restriction enzyme analysis of the rDNA palindromes, expression of esterase enzymes, and molecular weights and charges of cortical, ciliary and ribosomal proteins. Secretion of porphoryn and other pigments by cells in exhausted media is characteristic of certain species, e.g., by *T. pigmentosa* (Nanney and McCoy 1976) and is indicative of differences in metabolic pathways. Antibodies to the cell surface immobilization antigen (i-ag) are species specific though intraspecific polymorphism is too great to be of practical utility (Loefer et al. 1958; Margolin et al. 1959). The self-splicing intron of the LSU rRNA discovered by Zaug and Cech (1980) in *T. thermophila* is present in only five of 12 '*T. pyriformis*' species surveyed and appears to have been independently acquired following speciation (Sogin et al. 1986). '*Tetrahymena pyriformis*' species show variability of DNA rearrangement during macronuclear development (Huvos 1995) and differences in location or sequence of internally eliminated micronuclear sequences (Huvos 2007).

Collectively, these observations unequivocally show that despite the highly conserved cell morphology, '*T. pyriformis*' morphospecies have undergone considerable molecular diversification. On a quantitative level the diversity rivals that of the vertebrates. The *Tetrahymena* species are possibly very ancient, perhaps tens of millions of years (Nanney et al. 1998; Van Bell 1985a). Identity of ecological niche in view of such large differences seems highly unlikely.

As might be expected, nucleotide sequences are of increasing utility in identifying '*T. pyriformis*' species. These include the histone H4 proteins (Sadler and Brunk 1992), the histone H3II/H4II intergenic region (Brunk et al. 1990), the SSU rRNA (Jerome and Lynn 1996), 5S and 5.8S rRNA (Van Bell 1985a; b), and telomerase RNA (Ye and Romero 2002). The D2 LSU rRNA sequences distinguish among most species, though eight of the reproductively isolated species in the *americanis* group (Table 1) are identical. Mitochondrial COX1 (cytochrome oxidase 1) sequences readily discriminate among species, including those with identical D2 LSU regions (Lynn and Struder-Kypke 2006), and therefore are the most useful DNA barcodes. Significantly, we have found no examples of differences in the D2 LSU rRNA region sequences among strains of any one species including the *T. americanis* or *T. pigmentosa* subspecies. Similarly, no differences in the D2 LSU rRNA region were observed in over 60 *T. thermophila* isolates from PA and New England (Doerder, unpubl.).

It is increasingly clear from PCR-based surveys that many, perhaps most, protists have escaped detection because they cannot be cultured in the laboratory. This apparently applies to ciliates as well (Stoeck et al. 2006). It is possible that the number of *Tetrahymena* species is underestimated because some of them fail to grow when removed from nature. Both DLN and FPD have observed that some *Tetrahymena*-like cells isolated either directly from the sample or from samples enriched with peptone fail to divide or fail to form thriving cultures. Perhaps the laboratory food source, *Klebsiella pneumoniae*, is not preferred or perhaps the medium containing this bacterium becomes toxic. It would be worthwhile to perform a PCR-based survey on water samples to detect these potential tetrahymenas, crudely estimated as 5–10% of *Tetrahymena*-like isolates.

Origin of selfers and amicronucleate species

Selfers and amicronucleate strains are especially common among *Tetrahymena* natural isolates, the latter representing 30–70% of isolates in many collections. They pose special problems as neither can be identified as to species by mating reactions. The origins of both may be related to nuclear dimorphism and errors of macronuclear development. Selfers exhibit intraclonal conjugation in the absence of complementary mating types, perhaps by switching among or expressing multiple mating types, and amicronucleate cells isolated from nature have never been observed to mate. Among cells cultured in the laboratory, selfers sometimes appear in progeny of crosses, but these usually resolve into stable mating types. Many wild selfers are persistent selfers, though like laboratory selfers, some do resolve into mating types (Simon and Meyer 1992). With one exception (Kaney and Speare 1983), amicronucleate cells arising from micronucleate cells in the laboratory inevitably die due to simultaneous loss of the oral apparatus (Nanney 1957; Allen 1967). An essential relationship between the micronucleus and oral function is known in many ciliates. Perhaps in nature the amicronucleate condition is made possible by the chance acquisition by the macronucleus of the essential micronuclear function at conjugation when during macronuclear development micronuclear sequences normally eliminated are accidentally retained (Kaney and Speare 1983; Karrer et al. 1984). It is a matter of speculation as to why *Tetrahymena* has so many selfer strains and amicronucleate species, though inbreeding by selfers and the ability of asexual amicronucleates to take advantage of beneficial mutations by macronuclear assortment may play a role.

Dini and Nyberg (1993) suggested that selfers and amicronucleates may be derivatives of sexual species. Molecular evidence now confirms that a species can contain amicronucleate and selfing forms in addition to strains with stable mating types. Wild selfing strains of *T. australis* have been reported (Simon and Meyer 1992; Nanney et al. 1998), as have selfing strains of *T. elliotti* (Simon and Meyer 1992), one of the original amicronucleate species (Nanney and McCoy 1976). Some micronucleate strains of *T. elliotti* behave as typical breeding species with stable mating types (Simon and Meyer 1992). *Tetrahymena elliotti*, selfing and non-selfing with micronuclei and asexual without micronuclei, have been independently found in ponds of NW PA. All PA strains are identical in both the D2 LSU rRNA sequences as well as in a segment of the more rapidly evolving COX1 gene. Amicronucleate *T. thermophila* also have been found coexisting with micronucleate counterparts in NW PA (Doerder, unpubl). All five isolates were identical in the D2 LSU rRNA sequence to micronucleate strains. The COX1 sequences of all five amicronucleates also were identical but differed at 36/400 positions from inbred *T. thermophila* (one amino acid change). This number of substitutions approaches the number of differences between species (e.g., *T. thermophila* versus *T. elliotti*) and contrasts to the 4–5 (silent) nucleotide differences within the micronucleate strains of *T. thermophila*. The greater number of substitutions suggests that amicronucleate *T. thermophila* strains are much older than amicronucleate *T. elliotti*. Increasing use of DNA barcodes is likely to assign more selfers and amicronucleate strains to particular species. It is possible that some amicronucleates are descendants of extinct micronucleate species.

Biogeography of *Tetrahymena*

Historically the morphospecies '*T. pyriformis*' was described as ubiquitous. After the phenomenon of mating was brought under control (Elliott and Nanney 1952) members of both laboratories collected strains from fresh water habitats in North and Central America, Columbia, Mexico, Europe, Australia, New Zealand, Pacific Islands and Asia (Elliott 1973b), North America, Dominican Republic, Puerto Rico, Asia (Nyberg 1981; Simon et al. 1985; Doerder et al. 1995; Nanney et al. 1998). No attempts were made to do systematic surveys of the areas. Only in recent years have selfers and amicronucleates been retained for analysis. Now that the '*T. pyriformis*' species are named and many strains of each have been isolated, it is appropriate to inquire again as to biogeography. The question is important from the viewpoint of the tetrahymena speciation and as well as contributing to the debate as to whether all microbes are globally distributed (Finlay 2002; Finlay et al. 2004; Finlay and Fenchel 2004; Foissner 2006).

Table 1 provides a quasi-quantitative summary of the biogeography of '*T. pyriformis*' morphospecies. Elliott (1973) reported that three of the 12 syngens known at that time, 1, 7, and 8, later named *T. thermophila, T. canadensis* and *T. pigmentosa* (subspecies Tpo, Table 1), had been found only in North America and syngens 11 and 12, *T. australis* and *T. capricornis*, only in Australia. Other syngens had broader distributions. Combined data assembled over the past 40 years (Table 1) show that *T. thermophila* and *T. pigmentosa* (Tpo) are still apparently limited to North America, as are *T. hegewischi* and the amicronucleate strain 128 (n. sp. 2). *Tetrahymena capricornis* is still limited to Australia. *Tetrahymena borealis, T. hyperangularis* and *T. sonneborni* are limited to the Holarctic. Other species and subspecies of the complex for which more than eight strains have been collected are present in 2–3 biogeographical zones. Three species, syngens 9 and 11, *T. tropicalis* and *T. australis*, as well as *T. elliotti*, have been found from every major region of the world where collections have been made. For *T. elliotti* both micronucleate and amicronucleate forms have been collected from North America and East Asia. A systematic world-wide survey would provide useful information.

How members of the '*T. pyriformis*' complex became so widely distributed is a matter of speculation. Most '*P. aurelia*' species are also widely distributed. As these ciliates are found only in fresh water, and resistant cysts are not found in either morphospecies, it is difficult to imagine how they could be dispersed over oceanic distances on bodies of insects or waterfowl or by wind currents. It is possible that episodic glaciation played a significant role in distributing species. Shifting wetlands and temporary glacial lakes following retreat of the last glacier may have particularly influenced present distributions of northerly populations. It is perhaps worth noting in this regard that various tetrahymenas and their close relative *Lambornella* can be histophagous and have been isolated from various larval and adult aquatic invertebrates including chironomids (Corliss 1960; Golini and Corliss 1981), mosquitoes (Muspratt 1945; Corliss 1961; Egerter and Anderson 1985), blackflies (Lynn et al. 1981), and slugs (Kozloff 1957; Corliss et al. 1962; Brooks 1968). Tetrahymenas have also been reported from fish (Hoffman et al. 1975) and (!) dog urine (Lynn et al 2000). Except for the latter, these hosts may also serve as dispersal agents. A contribution by human activity also is likely.

Genetic and molecular information is increasingly important in assessing biogeography and population structure. *Tetrahymena thermophila* is the only *Tetrahymena* for which sufficient information is available for comment. This species has been collected from natural and man-made ponds, originally in MA, VT, and MI, and later in IL, PA, NH, ME, and FL. Crosses among wild isolates from distant locations are inter-fertile (Nyberg 1981a), and indeed, crosses of wild isolates to highly inbred laboratory strains also are fertile (Saad and Doerder 1995). This suggests that present populations have not been separated for long periods or that there is gene flow. The number of mating types appears to be fixed at seven, and, usually, all seven are found in any extensive collection (Doerder et al. 1996; Doerder unpubl; Simon unpubl). There is emerging evidence of population structure. There are numerous *mat* (mating type) alleles specifying different frequencies of mating type (Nanney et al. 1955; Phillips 1968; Doerder et al. 1995, 1996), so many so that it appears that some ponds may have unique *mat* alleles. Molecular identification of the *mat* gene is greatly needed. Alleles at the *Ser*H i-ag locus have been sequenced, and it appears that for this locus too there are pond-specific alleles (Gerber et al. 2002; Doerder unpubl).

Mitochondrial COX1 sequences are widely used to examine population substructure. In *T. thermophila* mitochondria are inherited clonally and are not exchanged at conjugation (Roberts and Orias 1973). In *T. thermophila*, three COX1 haplotypes have been found in NW PA populations (41 isolates), and five have been found among New England isolates (5 isolates; details to be published elsewhere). Only one of these is in common, suggesting regional differences and a possible New England origin for NW PA populations. The distribution of haplotypes in eight NW PA ponds is shown in Fig. 1. The most common haplotype PA2 is present in all ponds, but haplotype PA1 is present only in ponds 2–7 clustered within a 5 km radius. These data, the presence of multiple *mat* alleles, and the presence of multiple '*T. pyriformis*' species, e.g., *T. thermophila* and *T. elliotti*, suggest that these ~40–60 year old NW PA ponds each were colonized by multiple cells. The source of colonists could be surrounding natural sources (e.g., beaver ponds), but human activity (e.g., fish stocking) cannot be ruled out. PA ponds more distant from ponds 1–8 have repeatedly tested negative for tetrahymenas, though ponds sampled for the first time in 2006 in both west and east PA have yielded *T. thermophila* as well as other putative '*T. pyriformis*' strains. Among 54 (mostly natural) ponds in MA, CT, and NH sampled in 2006 (1–6 samples per pond), only 6 yielded *T. thermophila*; an additional 33 yielded tetrahymena-like ciliates. Clearly, *T. thermophila* is not a common species, but it is often locally abundant. Much more field work remains to be done.

Sympatry of species often suggests different ecological roles. Contrary to classical ideas of speciation, there is no evidence of competitive exclusion among the species of '*T. pyriformis*'. We cite two examples. Of 53 samples, obtained by the Nanney lab in one collecting trip, only 16 (30%) yielded clones of a single biological species, 29 contained strains of two species, six yielded three species, one, four species and one, five species. In NW PA ponds in which *T. thermophila* is the dominant species ~10% of isolates are other species of *Tetrahymena*. *Tetrahymena elliotti*, a relative of *T. thermophila*, is the most

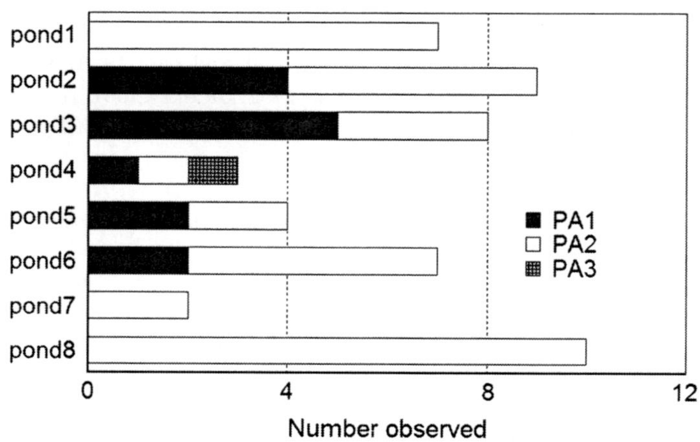

Fig. 1 Distribution of PA1, PA2 and PA3 COX1 haplotypes in eight ponds in NW PA

frequently found sympatric species. Other, rarer, species include *T. pyriformis*, *T. vorax*, *T. leucophrys* and at least one tetrahymena for which no match is found in the sequence databases. If multiple species can exist sympatrically only if they occupy different niches, the niches must be extremely small, i.e., specialized. Perhaps there are differences in preferred food sources, as has been suggested for certain cryptic species of *Euplotes* (Dini and Nyberg 1999).

The species problem and ecological dilemmas

Protists are critical members of communities and ecosystems. Their roles are diverse and in most cases poorly understood. The emerging consensus is that morphologically indistinguishable cryptic species are common, perhaps quite common, and, that they are molecularly diverse and likely have different ecological roles. Sonneborn's (1957) analysis of the species problem emphasized the fact that a definition of a biological species that is based solely on reproductive isolation ignores the asexual species common among protists. In Sonneborn's opinion the use of the same term, e.g., species, for groups of sexual and asexual organisms "requires genetic differences of the same kind and magnitude" in both. In a recent review of the species problem, Schlegel and Meisterfeld (2003) argue that sufficient 'objective characters', including molecular characters, now exist to apply Sonneborn's criterion more broadly among the protists. The data and references presented in this article show that this criterion is increasingly satisfied, especially by sequence information and other molecular data. The assignment of amicronucleate tetrahymenas (e.g., amicronucleate *T. thermophila*) to breeding species by molecular means and the finding of micronucleate strains (e.g., *T. elliotti*) previously known only from amicronucleate strains verify the utility of molecular approaches to species identification. It is perhaps ironic that whereas Sonneborn's criterion was met with resistance (Schloegel 1999), its validation by the finding of numerous molecular differences among members of numerous cryptic species, sexual and asexual, has now become the principle by which investigators who find molecular differences among populations postulate the existence of cryptic species.

If ciliate morphospecies are as rich in cryptic species as present information suggests, and even if they are not, the mechanisms of ciliate speciation deserve considerably more attention. Both competition and predation are widely considered as the two major drivers of adaptive radiation (Chase et al. 2002). Unfortunately, next to nothing is known about the specific food preferences most ciliates, including *Tetrahymena* species. In the laboratory and presumably in natural habitats tetrahymenas eat bacteria, possibly unknown species; possibly they also consume microdetritus and may be facultative histophages. Independent evolution of histophagy in the genus (Struder-Kypke et al. 2001) is consistent with evolution of diverse food preferences. The laboratory selection of *Tetrahymena* strains based on their ability to grow on a single species of bacterium or in an enriched axenic medium obscures nutritional differences among species. Tetrahymenas are preyed upon by other ciliates, including members of their own genus, various invertebrate larvae, including mosquitoes, and are likely consumed by numerous other invertebrates and small vertebrates. Some of these predators might also contribute to dispersal from pond to pond as hypothesized above. Differences in temperature and metal tolerance also suggest niche partition. Unfortunately, as with food preferences, there are no systematic field data on species stratification and distribution in ponds or streams, though analysis of sampling records suggests that some species may have preferences (Doerder and Simon unpubl).

The molecular diversity of the '*T. pyriformis*' morphospecies, though now of utility in species identification, obviously does not explain what brought about speciation, or why there are so many species, including selfers and amicronucleates. Possibly nuclear dimorphism has a significant contributing role. The shift from outbreeding to inbreeding as implied by selfing suggests that there are strong, perhaps recurrent evolutionary pressures regarding speciation and life history strategies in this group. Autogamy in other ciliates suggests the same kinds of pressures. It is worth noting in this context that tetrahymenas have yet another way to achieve functional homozygosity, a mechanism independent of sex. The phenomenon of macronuclear assortment (Doerder et al. 1992) involves the random distribution of alleles during macronuclear division during binary fission such that eventually one allele or the other is lost in a lineage. This occurs at every heterozygous locus, effectively recombining macronuclear genes in the absence of meiosis and generating up to 2^N combinations of alleles (where N is the number of heterozygous loci). In this way a single germinal genotype can generate diversity to explore varied habitats, a possible reproductive advantage in non-uniform habitats. Assortment also would allow lineages to express new macronuclear mutations. While these mutations would not be transmissible at conjugation, they likely are advantageous in the reproductive success of amicronucleate strains. Nuclear dualism likely has other as yet unstudied consequences for speciation.

Cryptic species raise the special concern that biodiversity is underestimated. This is an important factor not only in ecology and evolution but also practically in terms of policy and management decisions, say, for instance, when ciliates are used as indicators of water quality (Foissner 1988). As this review makes abundantly clear, the morphospecies '*T. pyriformis*' consists of molecularly diverse species that very likely do not have exactly the same ecological role. The problem of species identification posed by cryptic species is not new. Indeed, over 20 years ago Corliss and Daggett (1983) treated this problem at some length with emphatic recommendations. We agree that it is not practical for survey ecologists and other field biologists to do mating tests to distinguish cryptic species within a morphospecies. To use the term "complex" as we and others have done to encompass the array of cryptic species within a morphospecies is not an ideal solution. The terms '*P. aurelia*' and '*T. pyriformis*' each encompass too much diversity. For proper interpretation of results, especially from a comparative viewpoint, it is critical that complete strain designations are used and that the source and date from which each was obtained are documented. This bears repeated emphasis because not only in past were strains misidentified, the lack of proper identification still occurs today (we refrain from citing offending papers). Now that molecular methods are more widely and cheaply available, DNA barcodes, particularly mitochondrial COX1 sequences (Lynn and Struder-Kypke 2006) are especially useful.

Acknowledgments This work was supported by grants from NIH and NSF, personal funds, and the Graduate College of Cleveland State University. FPD is particularly grateful to David Nanney and Ellen Simon for their 40 years of advice and encouragement.

References

Allen SL (1967) Cytogenetics of genomic exclusion in *Tetrahymena*. Genetics 55:797–882

Asai DJ, Forney JD (eds) (2000) *Tetrahymena thermophila*. Academic Press, San Diego

Batson BS (1983) *Tetrahymena dimorpha* sp. nov. (Hymenostomatida: Tetrahymenidae), a new ciliate parasite of Simuliidae (Diptera) with potential as a model for the study of ciliate morphogenesis. Phil Trans R Soc Lond 301:345–363

Batson BS (1985) A paradigm for the study of insect-ciliate relationships:*Tetrahymena sialidos* sp. nov. (Hymenostomatida: Tetrahymenidae), parasite of larval *Sialis lutaria* (Linn.) (Megaloptera: Sialidae). Phil Trans R Soc Lond 310:123–144

Borden D, Whitt GS, Nanney DL (1973a) Electrophoretic characterization of classical *Tetrahymena pyriformis* strains. J Protozool 20:693–700

Borden D, Whitt GS, Nanney DL (1973b) Isozymic heterogeneity in *Tetrahymena* strains. Science 181:279–280

Brooks WM (1968) Tetrahymenid ciliates as parasites of the gray garden slug. Hilgardia 39:205–276

Brunk CF, Kahn RW, Sadler LA (1990) Phylogenetic relationships among *Tetrahymena* species determined using the polymerase chain reaction. J Mol Evol 30:290–297

Chao A, Li PC, Agatha S et al (2006) A statistical approach to estimate soil ciliate diversity and distribution based on data from five continents. Oikos 114:479–493

Chase JM, Abrams PA, Grover JP et al (2002) The interaction between predation and competition: a review and synthesis. Ecol Lett 5:302–315

Chen ZG, Song WB, Warren A (2003) Species separation and identification of *Uronychia* spp. (Hypotrichia:Ciliophora) using RAPD fingerprinting and ARDRA riboprinting. Acta Protozool 42:83–90

Corliss JO (1952a) Comparative studies on holotrichous ciliates in the *Colpidium-Glaucoma-Leucophyrs-Tetrahymena* group. I. General considerations and history of strains in pure culture. Trans Am Microsc Soc 71:159–184

Corliss JO (1952b) Le cycle autogamique de *Tetrahymena rostrata*. C R Acad Sci Paris 235:399–402

Corliss JO (1957) *Tetrahymena paravorax* n. sp., the first caudal-ciliated member of the genus referable to the *vorax-patula* complex. J Protozool 4:13

Corliss JO (1960) *Tetrahymena chironomi* sp. nov., a ciliate from midge larvae, and the current status of facultative parasitism in the genus *Tetrahymena*. Parasitology 50:111–153

Corliss JO (1961) Natural infection of tropical mosquitoes by ciliated protozoa of the genus *Tetrahymena*. Trans Roy Soc Trop Med Hyg 55:149–152

Corliss JO (1973) History, taxonomy, ecology, and evolution of species of *Tetrahymena*. In: Elliott AM (ed) Biology of *Tetrahymena*. Dowden. Hutchinson & Ross, Stroudsburg, pp 1–55

Corliss JO, Coats DW (1976) A new cuticular cyst-producing tetrahymenid ciliate, *Lambornella clarki* n. sp., and the current status of ciliatosis in culicine mosquitoes. Trans Am Microsc Soc 95:725–739

Corliss JO, Daggett PM (1983) *Paramecium* aurelia and *Tetrahymena pyriformis*—current status of the taxonomy and nomenclature of these popularly known and widely used ciliates. Protistologica 19:307–322

Corliss JO, Smith AC, Foullres J (1962) A species of *Tetrahymena* from the British garden slug *Milax budapestensis*. Nature 196:1008–1009

Creach V, Ernst A, Sabbe K et al (2006) Using quantitative PCR to determine the distribution of a semicryptic benthic diatom, *Navicula phyllepta* (Bacillariophyceae). J Phycol 42:1142–1154

Dini F, Nyberg D (1993) Sex in ciliates. In: Jones JG (ed) Advances in microbial ecology. Plenum Press, New York, pp 85–154

Dini F, Nyberg D (1999) Growth rates of marine ciliates on diverse organisms reveal ecological specializations within morphospecies. Microb Ecol 37:13–22

Doerder FP, Deak JC, Lief JH (1992) Rate of phenotypic assortment in *Tetrahymena thermophila*. Dev Genet 13:126–132

Doerder FP, Gates MA, Eberhardt FP et al (1995) High frequency of sex and equal frequencies of mating types in natural populations of *Tetrahymena thermophila*. Proc Natl Acad Sci USA 92:8715–8718

Doerder FP, Arslanyolu M, Saad Y et al (1996) Ecological genetics of *Tetrahymena thermophila*: mating types, i-antigens, multiple alleles and epistasis. J Eukaryot Microbiol 43:95–100

Egerter DE, Anderson JR (1985) Infection of the western treehole mosquito, *Aedes sierrensis* (Diptera: Culcidae) with *Lambornella clarki* (Ciliophora: Tetrahymenidae). J Invert Path 46:296–304

Eisen JA, Coyne RS, Wu M et al (2006) Macronuclear genome sequence of the ciliate *Tetrahymena thermophila*, a model eukaryote. PloS Biol 4:1620–1642

Elliott AM (1973a) Biology of *Tetrahymena*. Dowden, Hutchinson & Ross, Stroudsburg, PA

Elliott AM (1973b) Life cycle and distribution of *Tetrahymena*. In: Elliott AM (ed) Biology of *Tetrahymena*. Dowden, Hutchinson & Ross, Stroudsburg, PA, pp 259–286

Elliott AM, Kennedy JR (1973) Morphology of *Tetrahymena*. In: Elliott AM (ed) Biology of *Tetrahymena*. Dowden, Hutchinson & Ross, Stroudsburg PA, pp 57–87

Elliott AM, Nanney DL (1952) Conjugation in *Tetrahymena*. Science 116:33–34

Fawley MW, Dean ML, Dimmer SK et al (2006) Evaluating the morphospecies concept in the Selenastraceae (Chlorophyceae, Chlorophyta). J Phycol 42:142–154

Feng XL, Sun Q, Cao TG et al (1988) The S1 strain of *Tetrahymena* from Shanghai - *Tetrahymena shanghaiensis* sp-nov. Acta Zool Sin 34:42–51

Finlay BJ (2002) Global dispersal of free-living microbial eukaryote species. Science 296:1061–1063

Finlay BJ, Fenchel T (2004) Cosmopolitan metapopulations of free-living microbial eukaryotes. Protist 155:237–244

Finlay BJ, Esteban GF, Fenchel T (2004) Protist diversity is different? Protist 155:15–22

Foissner W (1988) Taxonomic and nomenclatural revision of Sladecek list of ciliates (Protozoa, Ciliophora) as indicators of water quality. Hydrobiologia 166:1–64

Foissner W (2006) Biogeography and dispersal of microorganisms: a review emphasizing protists. Acta Protozool 45:111–136

Foissner W, Berger H (1999) Identification and ontogenesis of the nomen nudum Hypotrichs (Protozoa: Ciliophora) *Oxytricha nova* (= *Sterkiella nova* sp n.) and *O. trifallax* (= *S. histriomuscorum*). Acta Protozool 38:215–248

Fokin S, Stoeck T, Schmidt H (1999) *Paramecium duboscqui* Chatton, Brachon, 1933. Distribution, ecology and taxonomy. Eur J Protistol 35:161–167

Frankel J (2000) Cell biology of *Tetrahymena thermophila*. Methods Cell Biol 62:27–125

Furgason WH (1940) The significant cytostomal pattern of the *Glaucoma-Colpidium* group, and a proposed new genus and species, *Tetrahymena geleii*. Arch Protistenk 94:224–266

Gall J (1986) The molecular biology of ciliated protozoa. Academic Press, Orlando

Gates MA, Berger J (1974) A biometric study of three stains of *Tetrahymena pyriformis* (Ciliatea: Hymenostomatida). Can J Zool 52:1167–1183

Gerber CA, Lopez AB, Shook SJ et al (2002) Polymorphism and selection at the *Ser*H immobilization antigen locus in natural populations of *Tetrahymena thermophila*. Genetics 160:1469–1479

Golini VI, Corliss JO (1981) A note on the occurrence of the hymenostome ciliate *Tetrahymena* in chironomid larvae (Diptera: Chironomidae) from the Laurentian Great Lakes. Trans Am Microsc Soc 100:89–93

Haentzsch M, Schmidt SL, Bernhard D et al (2006) A PCR-based method to distinguish the sibling species *Stylonychia mytilus* and *Stylonychia lemnae* (Ciliophora, Spirotrichea) using isocitrate dehydrogenase gene sequences. J Eukaryot Microbiol 53:343–347

Hausmann K, Selchow P, Scheckenbach F et al (2006) Cryptic species in a morphospecies complex of heterotrophic flagellates: the case study of *Caecitellus* spp. Acta Protozool 45:415–431

Hill DL (1972) The biochemistry and physiology of *Tetrahymena*. Academic Press, New York

Hoffman GL, Landolt M, Camper JE et al (1975) A disease of freshwater fishes caused by *Tetrahymena corlissi* Thompson, 1955, and a key for identification of holotrich ciliates of freshwater fishes. J Parasitol 61:217–223

Holz GG Jr, Corliss JO (1956) *Tetrahymena setifera* n. sp., a member of the genus *Tetrahymena* with a caudal cilium. J Protozool 3:112–118

Huvos PE (1995) Developmental DNA rearrangements and micronucleus-specific sequences in five species within the *Tetrahymena pyriformis* species complex. Genetics 141:925–936

Huvos PE (2007) Extensive changes in the locations and sequence content of developmentally deleted DNA between *Tetrahymena thermophila* and its closest relative, *T. malaccensis*. J Eukaryot Microbiol 54:73–82

Jerome CA, Lynn DH (1996) Identifying and distinguishing sibling species in the *Tetrahymena pyriformis* complex (Ciliophora, Oligohymenophorea) using PCR/RFLP analysis of nuclear ribosomal DNA. J Eukaryot Microbiol 43:492–497

Jerome CA, Simon EM, Lynn DH (1996) Description of *Tetrahymena empidokyrea* n.sp., a new species in the *Tetrahymena pyriformis* sibling species complex (Ciliophora, Oligohymenophorea), and an assessment of its phylogenetic position using small-subunit rRNA sequences. Can J Zool 74:1898–1906

Kahl A (1926) Neue und wenig bekannte Formen der holotrichen und heterotrichen Ciliaten. Arch Protistenk 55:197–438

Kaney AR, Speare VJ (1983) An amicronucleate mutant of *Tetrahymena thermophila*. Exp Cell Res 143:461–467

Karrer K, Stein-Gavens S, Allitto BA (1984) Micronucleus-specific DNA sequences in an amicronucleate mutant of *Tetrahymena*. Dev Biol 105:121–129

Kidder GW (1940) Growth studies on ciliates IV The influence of food on the structure and growth or *Glaucoma vorax* sp. nov. Biol Bull 78:9–23

Kozloff EN (1946) The morphology and systematic position of a holotrichous ciliate parasitizing *Deroceras agreste* (L.). J Morphol 79:445–465

Kozloff EN (1957) A species of *Tetrahymena* parasitic in the renal organ of the slug *Deroceras reticulatum*. J Protozool 4:75–79

Loefer JB, Owen RD, Christensen E (1958) Serological types among thirty-one strains of the ciliated protozoan *Tetrahymena pyriformis*. J Protozool 5:209–217

Lopez-Garcia P, Rodriguez-Valera F, Pedros-Alio C et al (2001) Unexpected diversity of small eukaryotes in deep-sea Antarctic plankton. Nature 409:603

Lwoff A (1923) Sur la nutrition des infusoires. C R Acad Sci Paris 176:928–930

Lynn DH, Molloy D, LeBrun R (1981) *Tetrahymena rotunda* n. sp. (Hymenostomatida: Tetrahymenidae), a ciliate parasite of the homolymph of *Simulium* (Diptera: Simuliidae). Trans Am Microsc Soc 100:134–141

Lynn DH, Struder-Kypke MC (2006) Species of *Tetrahymena* identical by small subunit rRNA gene sequences are discriminated by mitochondrial cytochrome c oxidase I gene sequences. J Eukaryot Microbiol 53:385–387

Lynn DH, Gransden SG, Wright A-DG et al (2000) Characterization of a new species of the ciliate *Tetrahymena* (Ciliophora: Oligohymenophorea) isolated from the urine of a dog: first report of *Tetrahymena* from a mammal. Acta Protozool 39:289–294

Maciejewska A (2006) Sibling species within *Paramecium jenningsi* revealed by PCR-RFLP. Acta Protozool 45:387–393

Margolin P, Loefer JB, Owen RD (1959) Immobilizing antigens of *Tetrahymena pyriformis*. J Protozool 6:207–215

Massana R, Balague V, Guillou L et al (2004) Picoeukaryotic diversity in an oligotrophic coastal site studied by molecular and culturing approaches. FEMS Microbiol Ecol 50:231–243

McCoy JW (1975) Updating the Tetrahymenids. III. Natural variation in *Tetrahymena setosa* nov. comb. Acta Protozool 14:253–262

Meyer EB, Nanney DL (1987) Isozymes in the ciliated protozoan *Tetrahymena*. In: Rattazzi MC et al (eds) Isozymes in the biological and medical sciences. A. R. Liss, New York, pp 61–101

Mollenbeck M (1999) Genetic relationship of 32 cell lines of the *Euplotes octocarinatus* species complex revealed by random amplified polymorphic DNA (RAPD) fingerprinting. Mol Ecol 8:1971–1979

Moreira D, Lopez-Garcia P (2002) The molecular ecology of microbial eukaryotes unveils a hidden world. Trend Microbiol 10:31–38

Muspratt J (1945) Observations on the larvae of tree-hole breeding *Culicini* (Diptera: Culicidae) and two of their parasites. J Entomol Soc S Afr 8:13–20

Nanney DL (1957) Inbreeding degeneration in *Tetrahymena*. Genetics 42:137–146

Nanney DL (1967) Comparative corticotype analyses in *Tetrahymena*. J Protozool 14:690–697

Nanney DL (1982) Genes and phenes in *Tetrahymena*. BioScience 32:783–788

Nanney DL, Caughey PA, Tefankjian A (1955) The genetic control of mating type potentialities in *Tetrahymena pyriformis*. Genetics 40:668–680

Nanney DL, McCoy JW (1976) Characterization of the species of the *Tetrahymena pyriformis* complex. Trans Am Microsc Soc 95:664–682

Nanney DL, Park C, Preparata R et al (1998) Comparison of sequence differences in a variable 23S rRNA domain among sets of cryptic species of ciliated protozoa. J Eukaryot Microbiol 45:91–100

Nyberg D (1974) Breeding systems and resistance to environmental stress in ciliates. Evolution 28:367–380

Nyberg D (1981a) Fertility is not a function of geographic distance in *Tetrahymena*. J Hered 72:94–96

Nyberg D (1981b) Three new "biological" species of *Tetrahymena* (*T. hegewischi* n. sp., *T. sonneborni* n. sp., *T. nipissingi* n. sp.) and temperature tolerance of members of the "*pyriformis*" complex. J Protozool 28:65–69

Petroni G, Dini F, Verni F et al (2002) A molecular approach to the tangled intrageneric relationships underlying phylogeny in *Euplotes* (Ciliophora, Spirotrichea). Mol Phylogenet Evol 22:118–130

Phillips RB (1968) Mating-type alleles in Illinois strains of *Tetrahymena pyriformis*, syngen 1. Genet Res (Cambridge) 11:211–214

Przybos E (1986) Species structure in ciliates. Folia Biol 34:103–132

Ray C Jr, Elliott AM (1954) Chromosome number of four varieties of *Tetrahymena*. Anat Rec 20:228

Roberts CT Jr, Orias E (1973) Cytoplasmic inheritance of chloramphenicol resistance in *Tetrahymena*. Genetics 73:259–272

Saad Y, Doerder FP (1995) Immobilization antigen variation in natural isolates of *Tetrahymena thermophila*. Eur J Protistol 31:45–53

Sadler LA, Brunk CF (1992) Phylogenetic relationships and unusual diversity in histone H4 proteins within the *Tetrahymena-pyriformis* complex. Mol Biol Evol 9:70–84

Scheckenbach F, Wylezich C, Mylnikov AP et al (2006) Molecular comparisons of freshwater and marine isolates of the same morphospecies of heterotrophic flagellates. Appl Environ Microbiol 72:6638–6643

Schlegel M, Meisterfeld R (2003) The species problem in protozoa revisited. Eur J Protistol 39:349–355

Schloegel JJ (1999) From anomaly to unification: Tracy Sonneborn and the species problem in protozoa, 1954–1957. J Hist Biol 32:93–132

Schmidt SL, Bernhard D, Schlegel M et al (2006) Fluorescence in situ hybridization with specific oligonucleotide rRNA probes distinguishes the sibling species *Stylonychia lemnae* and *Stylonychia mytilus* (Ciliophora, Spirotrichea). Protist 157:21–30

Simon EM, Meyer EB (1992) Suicide is not the inevitable outcome of "perpetual" selfing in tetrahymenines collected from natural habitats. Dev Genet 13:47–52

Simon EM, Meyer EB, Preparata RM (1985) New wild *Tetrahymena* from southeast Asia, China and North America, including *T. malaccensis*, *T. asiatica*, *T. nanneyi*, *T. caudata*, and *T. silvana* n. spp. J Protozool 32:183–189

Slapeta J, Moreira D, Lopze-Garcia P (2005) The extent of protist diversity:insights from molecular ecology of freshwater eukaryotes. Proc R Soc Lond 272:2073–2081

Slapeta J, Lopez-Garcia P, Moreira D (2006) Global dispersal and ancient cryptic species in the smallest marine eukaryotes. Mol Biol Evol 23:23–29

Sogin ML, Ingold A, Karlok M et al (1986) Phylogenetic evidence for the acquisition of ribosomal RNA introns subsequent to the divergence of some of the major *Tetrahymena* groups. EMBO J 5:3625–3630

Sonneborn TM (1937) Sex, sex inheritance and sex determination in *Paramecium aurelia*. Proc Natl Acad Sci USA 23:378–385

Sonneborn TM (1957) Breeding systems, reproductive methods, and species problems in protozoa. In: Mayr E (ed) The species problem. AAAS, Washington, pp 155–324

Stoeck T, Hayward B, Taylor GT et al (2006) A multiple PCR-primer approach to access the microeukaryotic diversity in environmental samples. Protist 157:31–43

Struder-Kypke MC, Wright AD, Jerome CA et al (2001) Parallel evolution of histophagy in ciliates of the genus *Tetrahymena*. BMC Evol Biol 1:5

Thompson JC Jr (1955) Morphology of a new species of *Tetrahymena*. J Protozool 2:12

Van Bell CT (1985a) 5S and 5.8S ribosomal RNA evolution in the suborder *Tetrahymena* (Ciliophora:Hymenostomatida). J Mol Evol 22:231–236

Van Bell CT (1985b) The 5S and 5.8S ribosomal RNA sequences of *Tetrahymena thermophila* and *T. pyriformis*. J Protozool 32:640–644

Warren R (1932) On a ciliate protozoon inhabiting the liver of a slug. Ann Natal Mus 7:1–53

Williams NE, Buhse HE Jr, Smith MG (1984) Protein similarities in the genus *Tetrahymena* and a description of *Tetrahymena leucophrys* n. sp. J Protozool 31:313–321

Ye AJ, Romero DP (2002) Phylogenetic relationships amongst tetrahymenine ciliates inferred by a comparison of telomerase RNAs. Int J Syst Evol Microbiol 52:2297–2302

Zaug AJ, Cech TR (1980) In vitro splicing of the ribosomal RNA precursor in nuclei of *Tetrahymena*. Cell 19:331–338

Diversity and geographic distribution of desmids and other coccoid green algae

Peter F. M. Coesel · Lothar Krienitz

Originally published in the journal Biodiversity and Conservation, Volume 17, No 2, 381–392.
DOI: 10.1007/s10531-007-9256-5

Abstract Taxonomic diversity of desmids and other coccoid green algae is discussed in relation to different species concepts. For want of unambiguous criteria about species delimitation, no reliable estimations of global species richness can be given. Application of the biological species concept is seriously hampered by lack of sexual reproduction in many species. Molecular analyses demonstrated cases of close affiliation between morphologically highly different taxa and, contrary, examples of little relationship between morphologically similar taxa. Despite the fact that desmids and chlorococcal algae, because of their microbial nature, can be readily distributed, cosmopolitan species are relatively scarce. The geographic distribution of some well-recognizable morphospecies is discussed in detail. Of some species a recent extension of their area could be established, e.g., in the desmids *Micrasterias americana* and *Euastrum germanicum*, and in the chlorococcaleans *Desmodesmus perforatus* and *Pediastrum simplex*.

Keywords · Chlorococcal algae · Desmids · Diversity · Geographic distribution · Green algae

Introduction

This review focuses on the diversity and geographic distribution of some groups of green algae showing a coccoid level of organization but belonging to different taxonomic units. According to modern systematic views, the desmids (Desmidiales) are placed in the

---

Special Issue: Protist diversity and geographic distribution. Guest editor: W. Foissner.

P. F. M. Coesel (✉)
Institute for Biodiversity and Ecosystem Dynamics, University of Amsterdam, P.O. Box 94062, Amsterdam 1090 GB, The Netherlands
e-mail: coesel@science.uva.nl

L. Krienitz
Leibniz-Institute of Freshwater Ecology and Inland Fisheries, Alte Fischerhütte 2, 16775 Stechlin-Neuglobsow, Germany

divison Charophya ('Streptophyta'), class Zygnemophyceae (Lewis and McCourt 2004). Desmids are coccoid and have a striking morphology characterized by two symmetrical halves (semicells). They comprise both solitary and colonial taxa. The other coccoid green algae studied here (in the following text designated as 'chlorococcal algae'), were formerly artificially classified under the Chlorococcales *sensu lato* and are now grouped in several orders of Chlorophyceae, Trebouxiophyceae and Prasinophyceae (Krienitz et al. 2003; Lewis and McCourt 2004). These orders mainly contain solitary or colonial algae with a spherical, ellipsoidal or needle-shaped morphology. We selected such diverse groups of green micro-algae in order to show the different state of the art in research on diversity and geography of these tiny protists.

Geographically the best studied group of green algae is that of the desmids, due to their often appealing appearance. Samples from far abroad revealed a lot of astonishing, exotic forms. Reliable knowledge of geographical distribution patterns, of course, is confined to those taxa that cannot be confused with any other ones. Fortunately, among the desmids quite a number of such taxa may be designated. Particularly the genus *Micrasterias* is marked by a high percentage of well recognizable species the distribution of which is confined to relatively small parts of the world (Figs. 1–5). In a previous paper, Coesel (1996) distinguished 10 desmid floral regions: Indo-Malaysia/Northern Australia, Equatorial Africa, Tropical South and Central America, North America, Extratropical South America, Eastern Asia, Southern Australia and New Zealand, South Africa, Temperate Eurasia and, finally, the circumpolar and high mountain regions. The number of species supposed to be endemic to any of those regions roughly speaking decreases from over a hundred to less than 10 and presumably goes hand in hand with the total desmid species diversity to be encountered.

As compared to the desmids, the study of diversity and distribution of chlorococcal algae presents more difficulties because of a high degree of morphological uniformity ('green balls') on the one hand and an extreme phenotypic variability of colony structure and cell wall equipment such as bristles, spines, ornamentations and incrustations on the other. Therefore, the conventional morphological species concept does not reflect the real diversity. Most of the classical phycologists describing chlorococcal algal species from tropical and polar regions (for details see Komárek and Fott 1983) did not collect the material themselves. They examined fixed samples taken from scientific travellers and their species descriptions are usually difficult to link to modern, DNA-based views of chlorococcal taxonomy.

Both for desmids and chlorococcal algae it holds that experimental systematic studies, including ecophysiological tests and molecular sequence analyses on 'exotic' species are badly needed.

Species concepts in desmids and chlorococcal algae

Problems concerning species definition in microalgae considerably hamper the interpretation of algal biodiversity and biogeography. The morphologic species concept considers species as groups of morphologically identical or similar organisms (Futuyma 1998). The biologic species concept defines species as groups of interbreeding populations which are reproductively isolated from other groups (Mayr 1942). However, both species concepts give problems when applied to algal species. In this context, John and Maggs (1997) come to the conclusion that at present no operational species concept is available for eukaryotic algae.

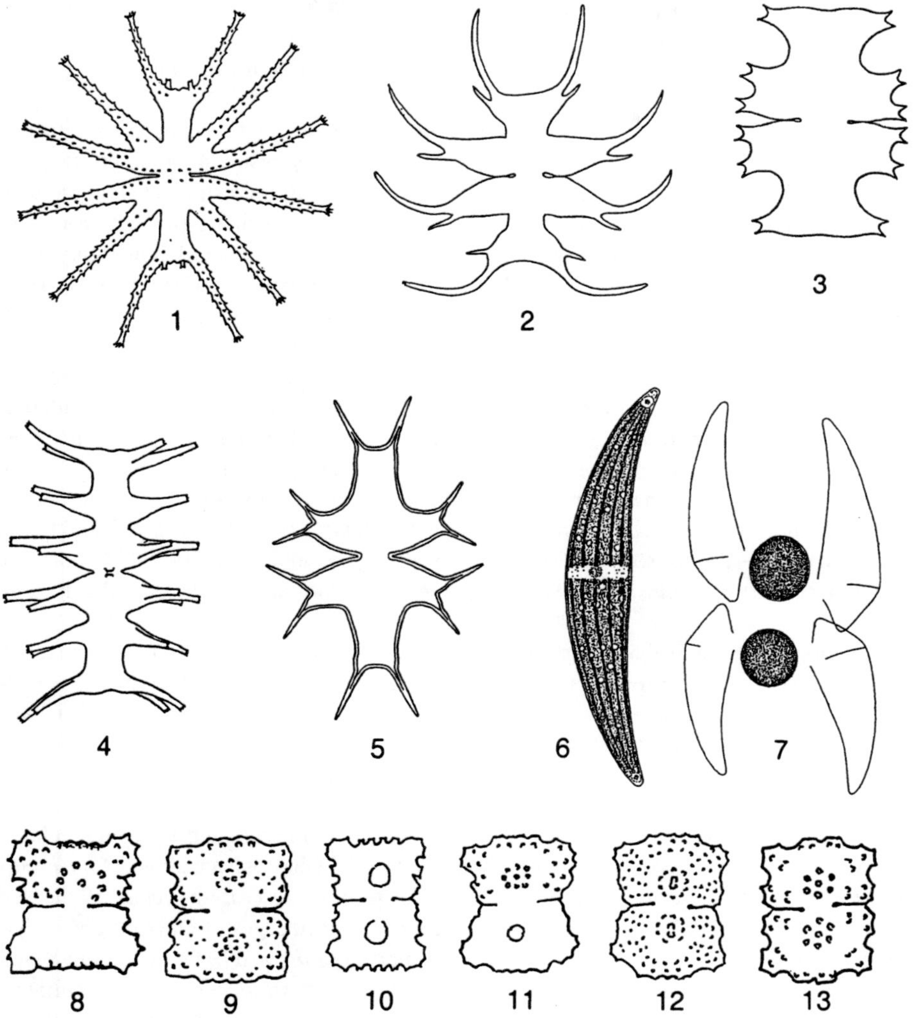

Figs. 1-13 Desmids with limited geographical distribution (1–5) and cryptic (6, 7) and synonymous (8–13) species. **1**: *M. hardyi* (S.E. Australia; after Krieger 1939); **2**: *M. sudanensis* (tropical Africa; after Thomasson 1960); **3**: *M. depauperata* (American continents; after Krieger 1939); **4**: *M. muricata* (North America; after Krieger 1939); **5**: *M. ceratophora* (S.E. Asia and N. Australia: after Scott and Prescott 1961). **6, 7**: *Closterium ehrenbergii* (vegetative cell and sexual reproduction stage), a morphospecies consisting of at least 12 different biological species (original). **8–13**: Six *Cosmarium* species which are probably synonymous **8**: *C. wallichii* West and West (after Wallich 1860, as *Euastrum clepsydra* Wall.); **9**: *C. seelyanum* Wolle (after Wolle 1884); **10**: *Cosmarium nobile* (Turner) Krieg. (after Turner 1892, as *Euastrum nobile* Turner); **11**: *C. naivashensis* Rich (after Rich 1932); **12**: *C. divergens* Krieg. (after Krieger 1932); **13**: *C. subnobile* Hinode (after Hinode 1966)

Traditional, morphology-based desmid taxonomy is overloaded with synonyms and suffers from a high rate of splitting (see Diversity chapter below). Unfortunately, sexual reproduction—essential for applying the biologic species concept—is a relatively rare phenomenon in this algal group (of many species no sexual stages are known at all). In addition, desmids are haploid organisms, so most mutations are immediately expressed. Consequently, by predominant lack of sexual reproduction (so possible exchange of genes during meiotic cell division) genotypically determined morphological variation is not wiped out. Therefore, exclusively clonal reproduction (particularly in euplanktic species) may readily result in the formation of desmid microspecies, just like in apomictically reproducing macrophytes (Coesel and Joosten 1996).

One of the possible reasons that sexual reproduction stages (zygospores) are usually encountered only incidentally is that but few clonal populations are homothallic (i.e., self-fertile). In experiments, out of some 120 randomly selected desmid strains belonging to 16 genera and over 80 species, only three showed homothallic sexual reproduction (Coesel and Teixeira 1974). On the other hand, mating experiments between different clones, often originating from different sites, revealed the phenomenon of heterothallism in, e.g., *Closterium ehrenbergii* Ralfs (Ichimura 1981; Coesel 1988), *Closterium strigosum* Bréb. (Watanabe and Ichimura 1978), *Pleurotaenium mamillatum* West (Ling and Tyler 1976) and *Micrasterias thomasiana* Archer (Blackburn and Tyler 1987). Especially in *Closterium ehrenbergii* (Figs. 6, 7), many mating types have been demonstrated, that is, populations which mutually show (almost) complete sexual isolation (Ichimura 1981). Such mating types can be considered syngens or biological species. Often, but not necessarily, mating types of one and the same morphospecies slightly differ in morphology, ecology and/or geographical distribution (Ichimura and Kasai 1990; Ichimura et al. 1997). No doubt, such sibling species—also to be traced from DNA analyses (Denboh et al. 2003)—will occur in many more morphospecies, and it is clear that they will substantially increase the diversity of this algal group.

Also in chlorococcal algal taxonomy, the morphologic species concept is burdened with a high degree of misinterpretation. For example, one and the same morphotype, such as the globular 'green ball', has evolved in different phylogenetic lineages. On the other hand, highly diverse morphotypes can belong to one and the same lineage (Luo et al. 2005). This ambivalent interpretation of morphotypes was documented on the archetypical green ball *Chlorella* and relatives. In freshwater and brackish habitats several distinct lineages of spherical '*Chlorella*-like' green algae have been found and designated as separate clades: *Mychonastes/Pseudodictyosphaerium*, now placed in the Chlorophyceae; and *Choricystis/Nanochlorum/Chlorella*, now placed in the Trebouxiophyceae (Krienitz et al. 1999; Hepperle and Krienitz 2001). In marine habitats, several other lineages of globular green algae have been reported, e.g., in the prasinophytes *Pycnococcus provasolii* Guillard, now placed in the Pseudoscourfieldiales (Daugbjerg et al. 1995); and *Ostreococcus tauri* Courties et Chrétiennot-Dinet, now placed in the Mamiellales (Courties et al. 1998). Further members of the *Chlorella* morphotype await taxonomic treatment and a transfer to other lineages of the Chlorophyta (Komárek and Fott 1983). The multiple origin of '*Chlorella*-like' algae may be explained by an adaptive advantage of the coccoid morphotype in ecosystems (Potter et al. 1997). On the other hand, sequence analyses revealed that several morphologically different algae, that were grouped in different lineages, such as *Closteriopsis*, *Actinastrum*, *Dictyosphaerium*, *Didymogenes* and *Micractinium* cluster close to globular 'true' *Chlorella* sensu stricto species (Fig. 14).

In numerous chlorococcal algal species sexual reproduction is unknown. Therefore, the biologic species concept is not applicable. Fortunately, molecular analyses offer an alternative to get insight in their relationships. After the introduction of DNA sequencing and phylogenetic analyses, the systematics of algae is going through a dramatic phase of change. The recent situation is marked by the quest for a compromise between the conventional (artificial) and the phylogenetic system. This upheaval will have significant consequences on biodiversity and biogeography of algal taxa. Some examples of these changing views will be given in the following two chapters.

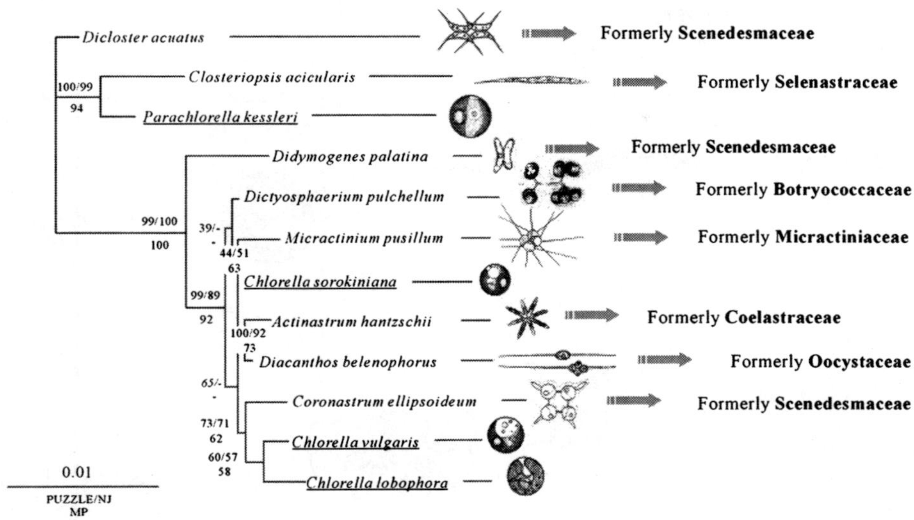

Fig. 14 Phylogenetic tree of Chlorellaceae derived from 18S rRNA gene sequences. The true, spherical *Chlorella* species, *C. vulgaris*, *C. sorokiniana* and *C. lobophora*, are intermixed by morphologically different taxa which were formerly placed within other families of coccoid green algae. *Chlorella kessleri* is transferred to the genus *Parachlorella* which is closely related to species of the needle-shaped *Closteriopsis* and *Dicloster* (according to Krienitz et al. 2004 and Luo et al. 2005)

Diversity

According to Gerrath (1993), estimations of global desmid species richness highly vary, but for the most part will be too high in view of the fair number of current synonyms. To give here a single example of possible synonymy: *Cosmarium seelyanum* Wolle, *C. wallichii* West and West, *C. divergens* Krieg., *C. naivashensis* Rich, *C. nobile* (Turner) Krieg., and *C. subnobile* Hinode are possibly the same species (Figs. 8–13), which occur in the literature also under the name of *C. abruptum* Lundell. Minor morphologic differences often are within the phenotypic variability of a species, and are thus not of taxonomic significance. Many forms, at any time described as separate varieties or even species, appear to be interconnected by transitional forms, so most likely belong to one and the same taxon, e.g, the common species *Staurastrum furcatum* (Ralfs) Bréb. and *Staurastrum aciculiferum* (West) Andersson (Figs. 15–17). However, such causes for overestimation of species diversity are counterbalanced by reasons for underestimation. Different species the vegetative cells of which are very much alike in practice usually will not be recognized. Certain small-sized, smooth-walled *Cosmarium* species can only be identified reliably when zygospores are encountered (Figs. 18–23). As zygospores usually are met only incidentally, new species can be expected on account of zygospore morphology. Apart from possible under- and overestimations of species diversity, related to insufficient observations or failures in knowledge of relevant literature, a reliable assessment of species diversity is also hindered by the experience that polymorphism in a 'species' is not always phenotypic but also may have got a genetic base, like in sibling species.

Taking into account all these complications it is fully understandable that estimations of total desmid species richness in (rather recent) literature range from 1,500 to 12,000

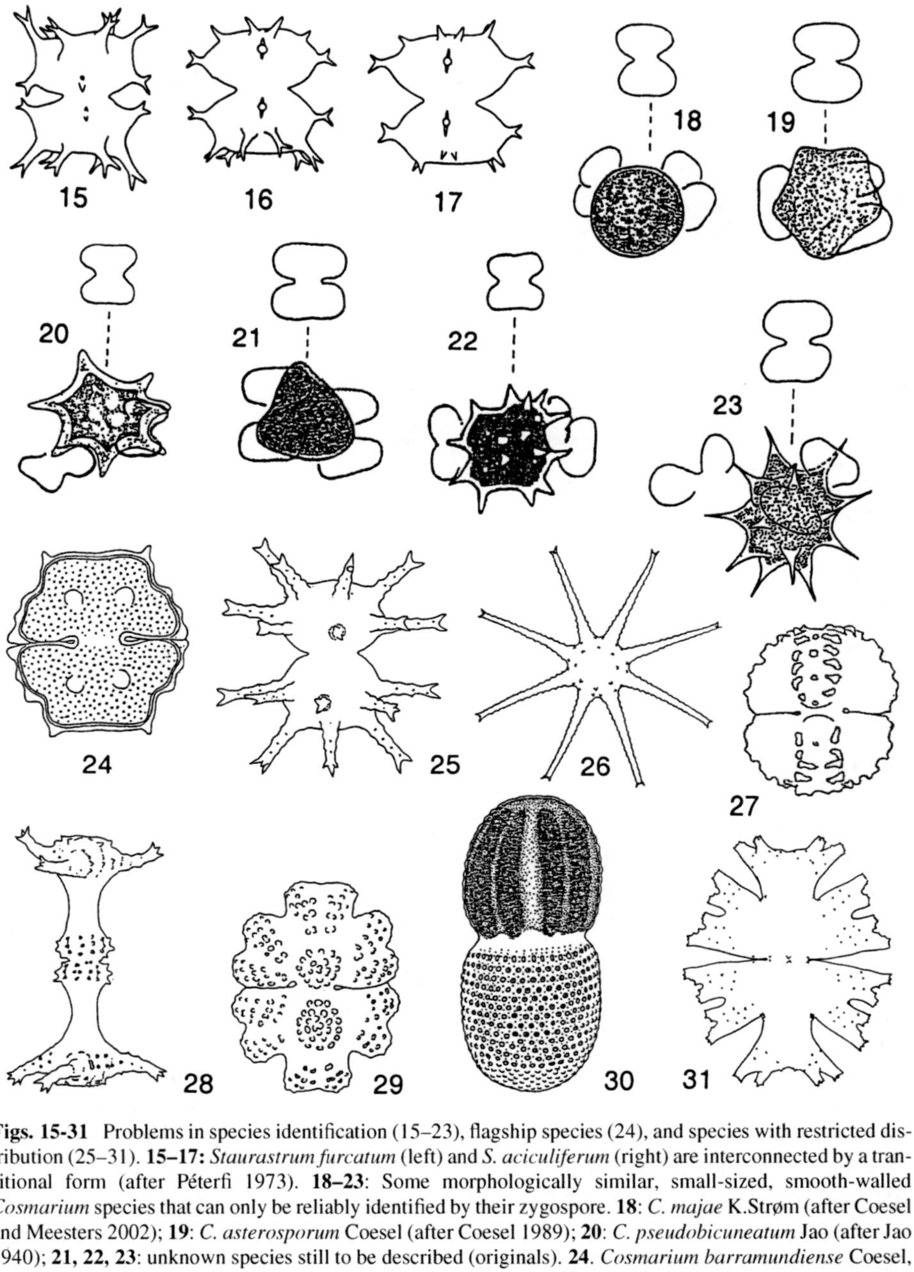

Figs. 15-31 Problems in species identification (15–23), flagship species (24), and species with restricted distribution (25–31). **15–17:** *Staurastrum furcatum* (left) and *S. aciculiferum* (right) are interconnected by a transitional form (after Péterfi 1973). **18–23**: Some morphologically similar, small-sized, smooth-walled *Cosmarium* species that can only be reliably identified by their zygospore. **18**: *C. majae* K.Strøm (after Coesel and Meesters 2002); **19**: *C. asterosporum* Coesel (after Coesel 1989); **20**: *C. pseudobicuneatum* Jao (after Jao 1940); **21, 22, 23**: unknown species still to be described (originals). **24**. *Cosmarium barramundiense* Coesel, a recently described flagship species from northern Australia (after Coesel 2004). **25–28**: *Staurastrum* species that have, within Europe, a distinctly atlantic-subarctic distribution (originals). **25**: *S. arctiscon*; **26**: *S. ophiura* (in top view); **27**: *S. maamense*; **28**: *S. elongatum*. **29, 30**: Two desmids (*Euastrum germanicum*, *Cosmarium striolatum*) with a predominantly continental distribution within Europe (originals). **31**: *Micrasterias americana*, a fast expanding desmid species in The Netherlands (original)

(Gerrath 1993). Gerrath (1993) thinks that there are approximately 3,000 'good' desmid species worldwide. When extrapolating the number of morphospecies distinguished in an ongoing inventory of European Staurastra, the first author comes to a comparable number. Anyhow, it is remarkable that hardly any new desmid flagship species are found. A recent, rather extensive investigation in northern Australia revealed but a single 'brand-new' species, quite different from all taxa described before (Fig. 24). This could be an indication that the number of desmid species endemic to a relatively small geographical area is limited, which for a group of readily to be transported micro-organisms is not really surprising.

The most recent comprehensive monograph on chlorococcal algae (Komárek and Fott 1983) contains about 1,200 species and subspecific taxa. Estimations of the real number of species are extremely vague because of the pending situation in species concepts. To illustrate this, the former genus *Scenedesmus* (now split into the genera *Scenedesmus*, *Desmodesmus*, *Acutodesmus*) may be used as an example. As a result of the high variability in morphologic characters (shape and organization of coenobia, spines, incrustations, cell wall ornamentations), more than 1,300 (morpho)species and subspecies have been desribed (Hegewald and Silva 1988). Studies using unialgal cultures to estimate the morphological variability revealed a severe overestimation of species number (Hegewald et al. 1990; Trainor 1998; Hegewald 1999). These observations were supported by molecular studies (Hegewald 2000). On the other hand, combined studies on fine structure and gene sequence (ITS2) of 22 clones identified as *Desmodesmus costato-granulatus* (Skuja) Hegewald indicated a higher diversity than expected. According to this interdisciplinary approach, it was split into five species (Vanormelingen et al. 2007).

Another example of the ambivalent situation in estimation of species diversity is *Botryococcus braunii* Kützing considered to be a 'well-known' microplanktont of inland waters, showing typical, large colonies like bunches of grape. Komárek and Marvan (1992) collected 74 populations of *Botryococcus*-like algae worldwide, studied their morphologic characteristics, and found 13 different morphotypes which fulfilled the species status according to the commonly used criteria in chlorococcal algal taxonomy. Unfortunately, molecular studies on these algae are deficient. The only known study on the phylogenetic placement of *Botryococcus* is supporting a polyphyletic origin (Senousy et al. 2004).

Also taxon delimitation within the Selenastraceae, a family of needle-shaped and lunate chlorococcal algae, has experienced considerable changes in recent times. Komárková-Legnerová (1969) and Marvan et al. (1984) provided revisions of the Selenastraceae based on morphotypes. Hindák (1984) described several new species which are of intermediate morphology with respect to described species, e.g., *Monoraphidium intermedium* Hindák as an intermediate taxon of *M. griffithii* (Berk.) Kom.-Legn. and *M. obtusum* (Korsh.) Kom.-Legn. The first molecular phylogenetic study on the Selenastraceae (Krienitz et al. 2001) revealed an intermixing of common members of the genera *Ankistrodesmus*, *Monoraphidium* and *Selenastrum* which contradicts the traditional way of circumscription of genera and species in this family. Finally, Fawley et al. (2005) discovered cryptic species in the Selenastraceae. It was found that isolates of one and the same morphotype can differ in 18S rDNA sequences, whereas isolates with identical or similar 18S rDNA sequence can exhibit different morphologies. These results give further arguments for the necessity of interdisciplinary work in algal systematics and diversity.

Geographic distribution

Examples of peculiar, well-defined distribution patterns in desmids were already shown by Donat (1926) and refined by Heimans (1969). Most striking is the occurrence of a number of flagship species that, within Europe, are characterized by a marked atlantic-subarctic distribution: *Staurastrum elongatum* Barker, *S. maamense* Archer, *S. arctiscon* (Ralfs) Lundell, *S. cerastes* Lundell, *S. ophiura* Lundell, *S. brasiliense* Nordst., and *S. sexangulare* (Bulnh.) Lundell (Figs. 25–28). As we have to do with aquatic organisms, it is difficult to imagine which climatic factor(s) could be responsible for such a remarkable distribution pattern. Likely it is some ecological parameter, linked to the nearness of seawater, that is decisive for their occurrence. The above-mentioned species are also known from the North American continent, but their distribution over there seems to be less distinct (Prescott et al. 1982). Anyhow, even in the atlantic and subarctic regions of Europe none of these species is really common. From The Netherlands, out of the seven above-mentioned species, *S. elongatum*, *S. ophiura*, *S. arctiscon*, *S. brasiliense* and *S. cerastes* have been reported, but only from one or a few sites and only in the first half of the last century. Some of them were regularly found during a longer period, e.g., *S. ophiura* between 1912 and 1930. From the fact that none of these species succeeded in enlarging its regional area, it is suggested that ecological demands rather than climatic factors or dispersal abilities are limiting.

In Europe, versus atlantic-subarctic species, also continental desmid species can be distinguished. Striking examples are *Cosmarium striolatum* (Nägeli) Archer [synonymous with *C. tesselatum* (Delponte) Nordst.] and *Euastrum germanicum* (Schmidle) Krieg. (Figs. 29, 30). Both *Cosmarium striolatum* and *Euastrum germanicum* are widely distributed on the continent (Heimans 1969; Coesel 1978), but are absent from Great Britain (Brook and Williamson 1991). In this case, the lack might be attributed to the isolated position of the British Islands. A few decades ago, these species were extremely rare in The Netherlands, too. Yet, in recent years Dutch records, particularly of *Euastrum germanicum*, are remarkably increasing in number. Obviously, both species are advancing in western direction and it may be only a question of (relatively little) time that they reach England.

Possibly, the fast increasing number of Dutch records of *Euastrum germanicum* in the last decade has to do with increasing average year temperatures. In this context, also the expansion of another conspicuous Dutch desmid species in The Netherlands has to be stressed, i.e., of *Micrasterias americana* Ralfs (Fig. 31). Although The Netherlands have been intensively inventoried for desmids already from the beginning of the 20th century, this species was not recorded before 1952 (Heimans 1969). Since then the number of Dutch records steadily increased, but it is only in the last decade that it has become one of the most common *Micrasterias* species of The Netherlands, also found in disturbed habitats. This latter phenomenon, though, might be an indication of a changed genetic constitution enabling the filling of another (larger) ecological niche.

In contrast to the desmids, the chlorococcal algae are generally supposed to be ubiquitous and to have a cosmopolitan distribution. As such they would serve as a good example of Beijerinck's metaphor, taken up by Baas-Becking and finally accentuated by Fenchel et al. (1997), Finlay (2002) and Fenchel and Finlay (2004): 'in micro-organisms, everything is everywhere, the environment selects'. This statement has evoked a heated discussion focusing on the species concepts. In recent times, numerous articles have been published which contradict the ubiquity hypothesis (reviewed by Foissner 2006 and Logares 2006). For micro-algae this statement is still under disputation (Coleman 2002, Finlay and Fenchel 2002). In diatoms, Hillebrandt et al. (2001) revealed a decreasing similarity of

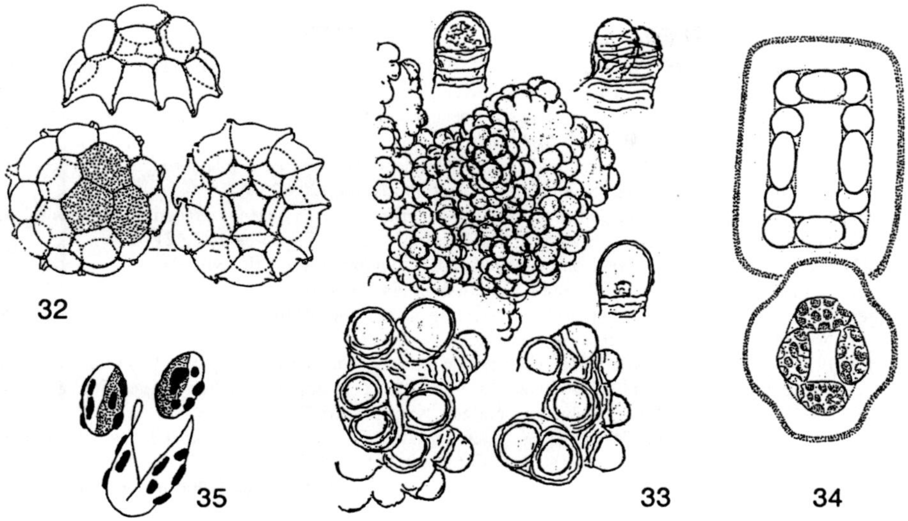

Figs. 32-35 Some examples of chlorococcal algae with limited geographic distribution. **32**: *Pediastrum marvillense* Théréz. and Couté (Antarctic; after Thérézien and Couté 1977); **33**: *Botryococcus fernandoi* Komárek and Marvan (Ethiopia; after Komárek and Marvan 1992); **34**: *Makinoella tosaensis* Okada (S.E. Asia; after Okada 1949); **35**: *Amphikrikos variabilis* Krienitz (Namibia; after Krienitz 1998)

species composition with increasing geographic distance, and thus they reject strict ubiquity of unicellular taxa. Coleman (2001) found local adaptation and endemism in phytoflagellates of the genera *Pandorina* and *Volvulina*.

As for the chlorococcal algae, there are indications of endemism both in classical, morphological and in modern, molecular approaches. Wille (1924) described the genus *Soropediastrum* which contains two species only found on the Kergueles. From the same Antarctic region *Pediastrum marvillense* Thérèz. and Couté (Fig. 32) was discovered, which also seems to be endemic (Komárek and Jankovská (2001). Detailed studies on chlorococcal algae of Cuba demonstrated in 20% of the taxa slight morphological differences in comparison to the original descriptions based on material from the temperate zone (Komárek 1983; Comas 1996). The latter authors found 21 taxa only recorded from Cuba. Several morphospecies of *Botryococcus* studied by Komárek and Marvan (1992) are only known from a few localities, e.g., *B. fernandoi* Komárek and Marvan (Fig. 33). The large oocystacean *Makinoella tosaensis* Okada (Fig. 34) has been reported only from Japan and Korea (Hegewald et al. 1999). *Amphikrikos variabilis* Krienitz (Fig. 35), with a distinct pattern of cell wall incrustation, has been found only in swamps and rivers of Namibia (Krienitz 1998).

Though, there are also indications that several species enlarged their distribution area considerably during the last decades ('invading species'), e.g., the formerly 'tropical/subtropical' species *Desmodesmus perforatus* (Lemmerm.) Hegewald and *Pediastrum simplex* Meyen. Nowadays, these species are encountered regularly in the temperate zone (Jeon and Hegewald 2006; Geissler and Kies 2003).

Slapeta et al. (2006) performed molecular analyses in the morphospecies *Micromonas pusilla* Butcher, a marine, picoplanktic prasinophyte. This morphospecies appeared to be a complex of morphologically indistinguishable phylogenetic lineages, representing cryptic species. Although some of these entities were shown to have a global, oceanic distribution, a more restricted distribution of other ones could not be precluded.

References

Blackburn SI, Tyler PA (1987) On the nature of eclectic species—a tiered approach to genetic compatibility in the desmid *Micrasterias thomasiana*. Br phycol J 22:277–298

Brook AJ, Williamson DB (1991) A check-list of desmids of the British Isles. Occasional publication 28, Freshwater Biological Association, Ambleside

Coesel PFM (1978) Taxonomical, geographical and ecological notes on *Euastrum mononcylum* var. *germanicum* Schmidle (Chlorophyta, Desmidiaceae). Arch Protistenk 120:436–445

Coesel PFM (1988) Biosystematic studies on the *Closterium moniliferum/ehrenbergii* complex (Chlorophyta, Conjugatophyceae) in western Europe. II. Sexual compatibility. Phycologia 27:421–424

Coesel PFM (1989) Taxonomic notes on Dutch desmids. Cryptogam Algol 10:181–193

Coesel PFM (1996) Biogeography of desmids. Hydrobiologia 336:41–53

Coesel PFM (2004) Some new and otherwise interesting desmid species from Kakadu National Park (Northern Australia). Quekett J Microsc 39:779–782

Coesel PFM, Joosten AMT (1996) Three new planktic *Staurastrum* taxa (Chlorophyta, Desmidiaceae) from eutrophic water bodies and the significance of microspecies in desmid taxonomy. Algol Stud 80:9–20

Coesel PFM, Meesters J (2002) Signalen van sieralgen. Natura 99:68–70

Coesel PFM, Teixeira RMV (1974) Notes on sexual reproduction in desmids. II. Experiences with conjugation experiments in uni-algal cultures. Acta Bot Neerl 23:603–611

Coleman AW (2001) Biogeography and speciation in the *Pandorina/Volvulina* (Chlorophyta) superclade. J Phycol 37:836–851

Coleman AW (2002) Microbial eukaryote species. Science 297:337

Comas AG (1996) Las Chlorococcales dulciacuícolas de Cuba. Biblthca phycol 99:1–192

Courties C, Perasso R, Chretiennot-Dinet MJ et al (1998) Phylogenetic analysis and genome size of *Ostreococcus tauri* (Chlorophyta, Prasinophyceae). J Phycol 34:844–849

Daugbjerg N, Moestrup O, Arctander P (1995) Phylogeny of genera of Prasinophyceae and Pedinophyceae (Chlorophyta) deduced from molecular analysis of the rbcL gene. Phycol Res 43:203–213

Denboh T, Ichimura T, Hendrayanti D et al (2003) The *Closterium moniliferum-ehrenbergii* (Charophyceae, Chlorophyta) species complex viewed from the 1506 group I intron and ITS2 of nuclear rDNA. J Phycol 39:960–977

Donat A (1926) Zur Kenntnis der Desmidiaceen des norddeutschen Flachlandes. Pflanzenforschung 5. Fischer, Jena

Fawley MW, Dean ML, Dimmer SK et al (2005) Evaluating the morphospecies concept in the Selenastraceae (Chlorophyceae, Chlorophyta). J Phycol 42:142–154

Fenchel T, Esteban GF, Finlay BJ (1997) Local versus global diversity of microorganisms: cryptic diversity of ciliated protozoa. Oikos 80:220–225

Fenchel T, Finlay BJ (2004) The ubiquity of small species: patterns of local and global diversity. BioScience 54:777–784

Finlay BJ (2002) Global dispersal of free-living microbial eukaryote species. Science 296:1061–1063

Finlay BJ, Fenchel T (2002) Microbial eukaryote species. Science 297:337

Foissner W (2006) Biogeography and dispersal of micro-organisms: a review emphasizing protists. Acta Protozoologica 45:111–136

Futuyma DJ (1998) Evolutionary biology. Sinauer Assoc, Massachusetts

Geissler U, Kies L (2003) Artendiversität und Veränderungen in der Algenflora zweier städtischer Ballungsgebiete Deutschlands: Berlin und Hamburg. Nova Hedwigia, Beih 126:1–777

Gerrath JF (1993) The biology of desmids: a decade of progress. In: Round FE, Chapman DJ (eds) Progress in phycological research 9. Biopress, Bristol, pp 79–192

Hegewald E (1999) Polymorphismus und Variabilität in der Grünalgengattung *Scenedesmus* Meyen, 1829 (Chlorophyta, Chlorococcales). Cour Forsch-Inst Senckenberg 215:123–128

Hegewald E (2000) New combinations in the genus *Desmodesmus* (Chlorophyceae, Scenedesmaceae). Algol Stud 96:1–18

Hegewald E, Hindák F, Schnepf E (1990) Sudies on the genus *Scenedesmus* Meyen (Chlorophyceae, Chlorococcales) from South India, with special reference to the cell wall ultrastructure. Beih Nova Hedwigia 99:1–75

Hegewald E, Schnepf E, Jeon SL (1999) Report on *Makionella tosaensis* Okada (Chlorophyta, Oocystaceae). Algae 14:87–90

Hegewald E, Silva P (1988) Annotated catalogue of *Scenedesmus* and nomenclaturally related genera including original desriptions and figures. Biblthca phycol 80:1–587

Heimans J (1969) Ecological, phytogeographical and taxonomic problems with desmids. Vegetatio 17:50–82

Hepperle D, Krienitz L (2001) Systematics and ecology of chlorophyte picoplankton in German inland waters along a nutrient gradient. Internat Rev Hydrobiol 86:269–284

Hillebrandt H, Watermann F, Karez R et al (2001) Differences in species richness patterns between unicellular and multicellular organisms. Oecologia 126:114–124

Hindák F (1984) Studies on the chlorococcal algae (Chlorophyceae). III. VEDA, Slovak Acad Science, Bratislava

Hinode T (1966) Desmids from the northern district of Tokushima Prefecture (3). J Jap Bot 41:305–316

Ichimura T (1981) Mating types and reproductive isolation in *Closterium ehrenbergii* Meneghini. Bot Mag Tokyo 94:325–334

Ichimura T, Kasai F (1990) Mating systems and speciation in haplontic unicellular algae, desmids. In: Kawano S (eds) Biological approaches and evolutionary trends in plants. Academic Press, London, pp 309–332

Ichimura T, Kasai F, Coesel PFM (1997) Geographical and ecological distribution of highly polyploid populations of the *Closterium ehrenbergii* species complex (Chlorophyta). Phycologia 36:157–163

Jao CC (1940) Studies on the freshwater algae of China. IV. Subaerial and aquatic algae from Nanyoh, Hunan. II. Sinensia 11:241–361

Jeon SL, Hegewald E (2006) A revision of the species *Desmodesmus perforatus* and *D. tropicus* (Scenedesmaceae, Chlorophyceae, Chlorophyta). Phycologia 45:567–584

John DM, Maggs CA (1997) Species problems in eukaryotic algae: a modern perspective. In: Claridge MF, Dawah HA, Wilson MR (eds) Species: the units of biodiversity. Chapman and Hall, New York, pp 83–107

Komárek J (1983) Contribution to the chlorococcal algae of Cuba. Nova Hedwigia 37:65–180

Komárek J, Fott B (1983) Chlorophyceae (Grünalgen) Ordnung: Chlorococcales. In: Huber-Pestalozzi G (ed) Das Phytoplankton des Süßwassers 7. Teil, 1. Hälfte. Schweizerbart, Stuttgart

Komárek J, Jankovská V (2001) Review of the green algal genus *Pediastrum*; implication for pollenanalytical research. Biblthca phycol 108:1–127

Komárek J, Marvan P (1992) Morphological differences in natural populations of the genus *Botryococcus* (Chlorophyceae). Arch Protistenk 141:65–100

Komárková-Legnerová J. (1969) The systematics and ontogenesis of the genera *Ankistrodesmus* Corda and *Monoraphidium* gen. nov. In: Fott B (ed) Studies in phycology. Academia Praha, pp 262–292

Krieger W (1932) Die Desmidiaceen der Deutschen Limnologischen Sunda-Expedition. Arch Hydrobiol, Suppl 11:129–230

Krieger W (1939) Die Desmidiaceen Europas mit Berücksichtigung der aussereuropäischen Arten. Rabenhorst's Kryptogamen-Flora 13, 1, 2. Akad Verlagsges, Leipzig

Krienitz L (1998) *Amphikrikos variabilis* sp. nova (Chlorophyta), a common species of inland waters of Namibia. Algol Stud 91:1–10

Krienitz L, Hegewald E, Hepperle D et al (2003) The systematics of coccoid green algae: 18S rRNA gene sequence data versus morphology. Biologia 58:437–446

Krienitz L, Hegewald E, Hepperle D et al (2004) Phylogenetic relationship of *Chlorella* and *Parachlorella* gen. nov. (Chlorophyta, Trebouxiophyceae). Phycologia 43:529–542

Krienitz L, Takeda H, Hepperle D (1999) Ultrastructure, cell wall composition, and phylogenetic position of *Pseudodictyosphaerium jurisii* (Chlorococcales, Chlorophyta) including a comparison with other picoplanktonic green algae. Phycologia 38:100–107

Krienitz L, Ustinova I, Friedl T et al (2001) Traditional generic concepts versus 18S rRNA gene phylogeny in the green algal family Selenastraceae (Chlorophyceae, Chlorophyta). J Phycol 37:852–865

Lewis LA, McCourt RM (2004) Green algae and the origin of land plants. Am J Bot 91:1535–1556

Ling HU, Tyler PA (1976) Meiosis, polyploidy and taxonomy of the *Pleurotaenium mamillatum* complex (Desmidiaceae). Br phycol J 11:315–330

Logares RE (2006) Does the global microbiota consist of a few cosmopolitan species? Ecologia Austral 16:85–90

Luo W, Krienitz L, Pflugmacher S et al (2005) Genus and species concept in *Chlorella* and *Micractinium* (Chlorophyta, Chlorellaceae): genotype versus phenotypical variability under ecosystem conditions. Verh Int Ver Limnol 29:170–173

Marvan P, Komárek J, Comas A (1984) Weighting and scaling of features in numerical evaluation of coccal green algae (genera of the Selenastraceae). Algol Stud 37:363–399

Mayr E (1942) Systematics and the origin of species. Columbia Univ Press, New York

Okada Y (1949) *Makinoella tosaensis*, a new genus of the Oocystaceae. Jap J Bot 24:166–168

Peterfi LS (1973) Studies on Romanian Staurastra. I. Variability and taxonomy of *Staurastrum spinosum* (Brébisson) Ralfs. Nova Hedwigia 24:121–144

Potter D, Lajeunesse TC, Saunders GW et al (1997) Convergent evolution masks extensive biodiversity among marine coccoid picoplankton. Biodiv Conserv 6:99–107

Prescott GW, Bicudo CEM, Vinyard WC (1982) A Synopsis of North American Desmids. II, 4. Univ Nebraska Press, Lincoln, London

Rich F (1932) Contributions to our knowledge of the freshwater algae of Africa. Trans Roy Soc South Africa 20:149–188

Scott AM, Prescott GW (1961) Indonesian desmids. Hydrobiologia 17:1–13

Senousy HH, Beakes GW, Hack E. (2004) Phylogenetic placement of *Botryococcus braunii* (Trebouxiophaceae) and *Botryococcus sudeticus* isolate UTEX 2629 (Chlorophyceae). J Phycol 40:412–423

Slapeta J, López-García P, Moreira D (2006) Global dispersal and ancient cryptic species in the smallest marine eukaryotes. Mol Biol Evol 23:23–29

Thérézien Y, Couté A (1977) Algues d'eau douce des Iles Kerguelen et Crozet (à l'exclusion des Diatomées). CNFRA 43:1–91

Thomasson K (1960) Notes on the plankton of Lake Bangweulu. 2 Nova Acta Regiae Soc Sci Upsal 17:1–43

Trainor FR (1998) Biological aspects of Scenedesmus (Chlorophyceae) phenotypic plasticity. Nova Hedwigia, Beih 117:1–367

Turner WB (1892) The freshwater algae (principally Desmideae) of East India. Kongl Svensk Vetensk Acad Handl 25:1–187

Vanormelingen P, Hegewald E, Braband A et al (2007) The systematics of a small spineless *Desmodesmus* taxon, *D. costato-granulatus* (Sphaeropleales, Chlorophyceae), based on ITS2 rDNA sequence analyses and cell wall morphology. J Phycol 43:378–396

Wallich GC (1860) Desmidiaceae of Lower Bengal. Ann Mag Nat Hist 3:184–197, 273–285

Watanabe M, Ichimura T (1978) Biosystematic studies of the *Closterium peracerosum-strigosum-littorale* complex. II. Reproductive isolation and morphological variation among several populations from the northern Kanto area in Japan. Bot Mag Tokyo 91:1–10

Wille N (1924) Süsswasseralgen von der deutschen Südpolar-Expedition auf dem Schiff "Gauss". 1.–2. Teil. Deutsche Südpolar-Expedition 8:373–445

Wolle F (1884) Desmids of the United States and list of American pediastrums. Moravian Publ Off, Bethlehem, P.A

The diversity and distribution of diatoms: from cosmopolitanism to narrow endemism

Pieter Vanormelingen · Elie Verleyen · Wim Vyverman

Originally published in the journal Biodiversity and Conservation, Volume 17, No 2, 393–405.
DOI: 10.1007/s10531-007-9257-4

Abstract It has been claimed that microbial taxa will not exhibit endemism because their enormous populations remove dispersal as an effective constraint on geographical range. Here we review evidence that challenges this ubiquity hypothesis for the most speciose group of microbial eukaryotes, the diatoms. Detailed taxonomic inventories using fine-grained morphological characteristics, molecular markers, and crossing experiments have revealed that the geographic distribution of diatoms ranges from global to narrow endemic. Records of human-mediated introductions of exotic species further provide a strong indication that geographic dispersal was limiting in the past. Finally, recent studies have revealed that diatom community structure and diversity are influenced by geographical factors independent of environmental conditions. Diatom communities are thus regulated by the same processes that operate in macro-organisms, although possibly to a different degree, implying that dispersal limitation is significant and the endemism observed in isolated areas is real. These results underscore the pressing need to (1) continue research into diatom biology, ecology and the factors driving diatom species diversity and geographic distributions, and (2) protect relatively isolated areas against further introductions of exotic species.

Keywords Biogeography · Diatoms · Dispersal · Diversity · Endemism · Macroecology · Microorganisms · Ubiquity hypothesis

Introduction

The diversity and taxonomic composition of local communities result from a balance between processes operating on a regional scale, such as allopatric species formation and geographic dispersal, which both add species to communities, and processes capable of

---

Special Issue: Protist diversity and geographic distribution. Guest editor: W. Foissner.

P. Vanormelingen · E. Verleyen · W. Vyverman (✉)
Laboratory Protistology & Aquatic Ecology, Gent University, Krijgslaan 281 S8, 9000 Gent, Belgium
e-mail: Wim.Vyverman@UGent.be

promoting local extinction, including predation, competitive exclusion, and stochastic variation (Ricklefs 1987). While it is generally accepted that these processes shape the distribution and community diversity of macroscopic organisms (Leibold et al. 2004; Cox and Moore 2005), considerable controversy has arisen about their importance for microscopic organisms (<1 mm; Whitfield 2005; Green and Bohannan 2006; Martiny et al. 2006). According to advocates of the ubiquity hypothesis (Finlay 2002), the vast population sizes of micro-organisms drive ubiquitous dispersal (Baas-Becking 1934) and make local extinction virtually impossible (Fenchel and Finlay 2004). Geographic isolation is therefore absent and as a result, allopatric speciation should be rare or non-existent, which would explain the perceived low global morphospecies diversity of microbial eukaryotes (Finlay 2002). Evidence supporting the ubiquity hypothesis comes from high local to global morphospecies ratios for various groups of photosynthetic and heterotrophic protists (Finlay and Clarke 1999; Finlay and Fenchel 2004) and the maintenance of consistent patterns of local abundance or rarity on a global scale (Finlay and Clarke 1999; Finlay 2002). In contrast, various recent studies suggest that micro-organisms, like macro-organisms, do display restricted geographic ranges, as recently outlined in the moderate endemicity model (Chao et al. 2006; Foissner 2006). Evidence supporting this model includes observations on the allopatric divergence of cyanobacteria and archaea in hot springs (Papke et al. 2003; Whitaker et al. 2003), foraminifera occurring in the polar oceans (Darling et al. 2004, 2007), and fungi on a continental scale (Taylor et al. 2006). In addition, a positive correlation between habitat size and bacterial diversity (Bell et al. 2005; Reche et al. 2005), distinct soil ciliate faunas on different continents (Chao et al. 2006), and distance-decay relationships in microbial eukaryotes (Green et al. 2004) have been observed.

Diatoms are one of the most successful contemporary groups of photosynthetic eukaryotic micro-organisms. They inhabit almost any kind of aqueous environment, and can also occur as endosymbionts in dinoflagellates and foraminifers (Round et al. 1990). They have global significance in biogeochemical cycles, and provide 20–25% of globally fixed carbon and atmospheric oxygen (Mann 1999). Their hallmark is an intricately shaped and ornamented silica cell wall called the frustule, which provides a wealth of ultrastructural characters on which diatom classification is largely based (Round et al. 1990). As for most other groups of microscopic organisms however, remarkably little is as yet known about diatom biology, ecology, and the factors driving diatom species diversity and geographic distributions (Mann 1999; Chepurnov et al. 2004).

Below, we review diatom species diversity and biogeography. We start with a concise description of the current state of diatom species taxonomy, and how this influences our understanding of diatom species diversity and its spatial structuring. Next, we consider three lines of research which, taken together, draw a more subtle picture of diatom diversity and biogeography than that proposed by the ubiquity hypothesis. These are (1) assessments of the geographic distribution of diatom genera and species based on dedicated, fine-grained morphological investigations, sometimes in combination with molecular and crossing data; (2) well-documented accounts of human-mediated introductions, and (3) estimates of the relative influence of spatial and environmental factors in determining diatom community structure and diversity on multiple geographic scales. Finally, we highlight important areas for future research on diatom biogeography.

Problems with diatom species taxonomy

As in many protist groups, diatom species taxonomy is notoriously messy (see Mann 1999 for a detailed review). This is mainly because taxon delineation has been (and still is) almost exclusively based on morphological features of the siliceous cell wall, without a full understanding of the underlying causes of morphological variation patterns. As a result, the description of new species has sometimes been based on morphological differences between allopatric populations which are purely phenotypic (Cox 1995). This can lead to the perception that species have restricted geographic distribution patterns. For example, the marine planktonic diatoms *Thalassiosira gravida* Cleve and *T. rotula* Meunier were originally considered to be two distinct species, the former being restricted to high latitudes and the latter to lower latitudes (Hasle 1976). Culture experiments showed however that valve morphology changes from typical *T. rotula* to typical *T. gravida* with decreasing temperature (Syvertsen 1977). Other examples of such taxonomic redundancy are given by Mann (1999). In addition, the gradual reduction in size associated with the diatom life cycle, can lead to pronounced changes in valve morphology (Round et al. 1990). During the 20th century, these findings led to a strong tendency to regard most morphological variability as purely phenotypic. The widely (i.e., globally) used European freshwater diatom floras of Hustedt (1927–1966) and Krammer and Lange-Bertalot (1986–1991) were strong advocates of this idea, and applied wide species boundaries. In addition, the use of these floras in non-European localities led to 'force-fitting' of diatom taxa to 'European' names (Tyler 1996). Lumping and force-fitting had severe consequences for our understanding of the ecology, diversity, and distribution of diatoms, as it not only stretched the morphological boundaries of many species, but also reinforced the idea that most diatom species have cosmopolitan distributions and are ecologic generalists (Kociolek and Spaulding 2000).

During the last decades however, it has become clear that subtle, discontinuous morphological variation patterns, which were hitherto assumed to be of no taxonomic significance, are instead generally correlated with variation in reproductive, molecular-genetic, physiological, and ecological characters (see review by Mann 1999 and references therein; Behnke et al. 2004; Beszteri et al. 2005; Créach et al. 2006; Lundholm et al. 2006; Amato et al. 2007). These studies suggest that many diatom species contain several subtly distinct, semi-cryptic entities that are worth taxonomic recognition at the species level, and that, as a consequence, diatom species diversity has been severely underestimated rather than overestimated. The actual global diversity of diatoms may therefore be an order of magnitude higher than the current number of described species, with up to 200,000 extant species (Mann and Droop 1996). Almost at the same time as these findings, the lumping trend among 'alpha taxonomists' has been halted and even reversed. In the Iconographia Diatomologica series (edited by Lange-Bertalot) alone, >1400 new diatom taxa have been described since 1995, many of which were previously regarded to be part of a morphological continuum within a single species. Importantly, it remains to be shown whether the application of a fine-grained taxonomy will result in range-splitting or ecological differentiation rather than merely increasing global species diversity.

The geographic distribution of diatoms: from cosmopolitan genotypes to endemic phenotypes

Under-sampling of suitable habitats and regions, and under-reporting of rare taxa are a real and severe problem for biogeographical studies of microscopic organisms and can lead to underestimation of the geographic ranges of species (Lee and Patterson 2000; Finlay et al. 2002), even if diatom taxa have been well revised or when multimethod approaches are used. However, both phenomena will have a negligible effect if (1) biogeographic studies focus only on habitats that fulfill the environmental requirements of the taxa under investigation, (2) sample sizes are sufficiently large, and (3) the study taxa are locally abundant, making it less likely that they are overlooked elsewhere (assuming there is a close correlation between local and global abundance, Finlay et al. 2002). If these conditions are fulfilled and the species under consideration has consequently turned up in numerous samples, while its distribution is still found to be restricted to a small area, then its endemism is fairly certain (or at least its absence in the other investigated regions).

Phenotype-based approach: flagship taxa and fine-grained morphology

Morphology-based approaches to diatom biogeography can be used provided that published data are reviewed critically and intercalibrated. Given the fact that many descriptions and illustrations of 'difficult' species do not allow unambiguous evaluation of literature data, it is imperative that only conspicuous, easily recognizable taxa are dealt with, the so-called 'flag-ship' taxa (Tyler 1996; Foissner 2006). An alternative is to make a careful new morphological analysis of species in samples from different geographical regions in a single large-scale study, thereby by-passing the need for comparisons with previously published data (Chao et al. 2006). Below, we provide examples of morphologically well-studied, often conspicuous taxa for which there is strong evidence that they have restricted distributions.

While most of the >900 diatom genera are cosmopolitan (Fourtanier and Kociolek 2003), there are some notable exceptions. *Eunophora* (Eunotiophycidae; Fig. 1), a highly distinctive diatom genus comprising at least four species, was described by Vyverman et al. (1998) from Tasmania and New Zealand where it is a widespread and often abundant component of benthic diatom communities in oligo- to dystrophic lakes (Vyverman et al. 1995; Vanhoutte et al. 2006). There are no other reports of similar diatoms from the Northern Hemisphere, where a large number of studies on the benthic diatom flora of similar lakes have been conducted within the framework of coordinated international projects on the effects of acidification and eutrophication (PIRLA, Camburn et al. 1984–1986; SWAP, Stevenson et al. 1991; and EDDI, available on http://craticula.ncl.ac.uk/Eddi/jsp/, databases). Likewise, a query among over 700 diatomists via the diatom listserver (http://www.indiana.edu/~diatom/diatom.html) so far failed to reveal the presence of the genus outside Tasmania and New Zealand, implying that undersampling is of minor importance here. Another conspicuous diatom taxon which appears to display a restricted biogeographical range is the genus *Veigaludwigia* which was introduced by Rumrich et al. (2000) to accommodate two naviculoid species from southern Chile, viz. *Navicula urbana* and *N. willeri* (Krasske 1939), which are uniquely characterized by distinct internal small knobs on the valve face margin (Fig. 2). The genus has now been found in southern South America, Australia, and New Zealand (Vyverman et al. 1995 and unpublished; Villanueva and Maidana 1999; Flower 2005; Vanhoutte et al. 2006). Importantly, most of these studies were the first surveys for their regions, and inevitably involved only a limited number of samples, yet none failed to find *Veigaludwigia*, showing that undersampling and underreporting is not an issue. Notwithstanding the fact that northern temperate regions have been much more thoroughly investigated, *Veigaludwigia* has not yet been found there.

Recent revisions of selected diatom genera at the (morpho)species level corroborate these findings. A recent overview of the genus *Stauroneis* in the Antarctic and sub-Antarctic region (867 samples) and a careful comparison with earlier studies from the Arctic (500

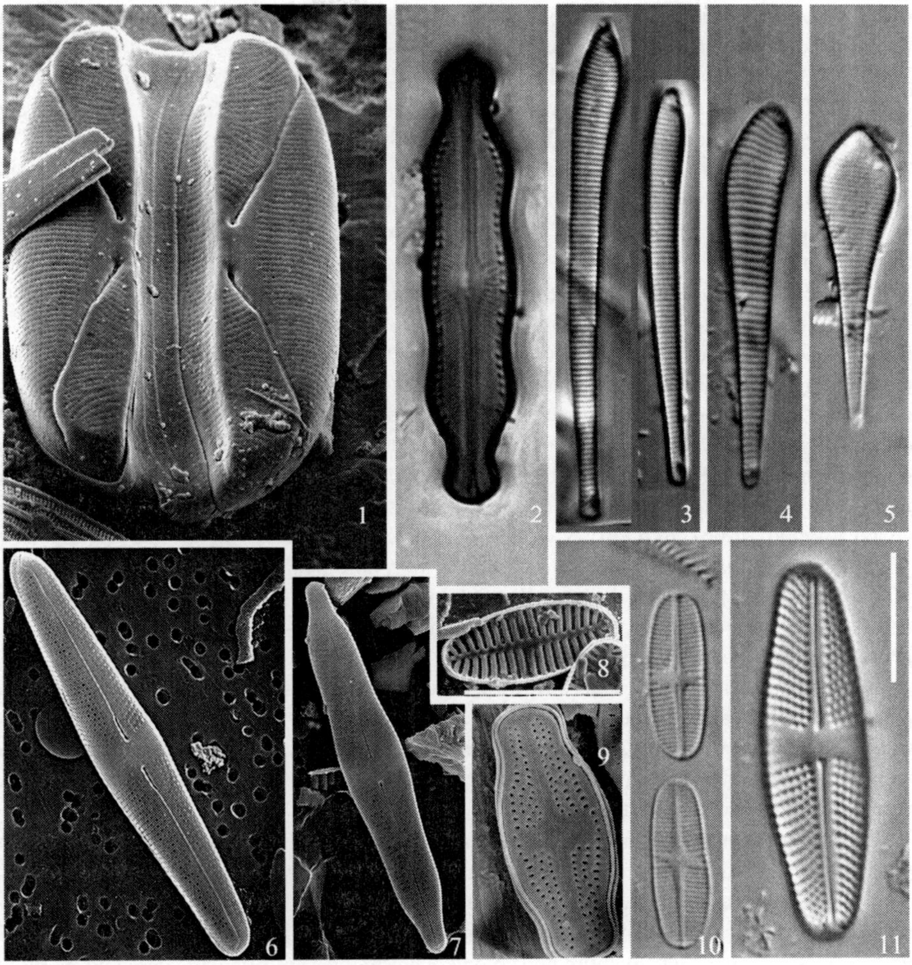

Fig. 1–11 Selected freshwater diatoms endemic to (parts of) the Southern hemisphere. (1) *Eunophora*, a diatom genus with a geographical distribution restricted to freshwater bodies in Tasmania and New Zealand. (2) *Veigaludwigia*, a diatom genus restricted in distribution to temperate regions of the Southern hemisphere. (3) *Actinella pulchella*, endemic to Australia (Tasmania) and New Zealand. (4) *Actinella aotearoaia*, endemic to Australia and New Zealand. (5) *Actinella comperei*, thus far only found in one lake in Tasmania. (6, 7) *Stauroneis* species endemic to the sub-Antarctic islands in the Indian Ocean, viz., *S. pseudomuriella* and *S. pseudosmithii*. (8–11) Diatom species characteristic of the Antarctic region, viz., *Planothidium quadripunctatum*, *Luticola muticopsis*, *Psammothidium abundans*, and *Achnanthes taylorensis*. Figures 3–4, 5, 6–7 and 8–11 were reprinted with permission from Sabbe et al. (2001), Sabbe et al. (2000), Van de Vijver et al. (2004), and Sabbe et al. (2003) respectively. All figures are on the same scale except Fig. 8. Scale bars, 10 μm

samples) revealed the presence of different *Stauroneis* floras in both regions (Van de Vijver et al. 2005; Figs. 6–7). In fact, out of a total of 59 species, only five were common to both regions. This agrees with other recent findings that the Arctic and Antarctic freshwater diatom floras have hardly any species in common (Spaulding and McKnight 1999). Not surprisingly, the five common *Stauroneis* species are recorded from all over the world. Moreover, the distribution of *Stauroneis* species is also restricted within the sub-Antarctic region: the *Stauroneis* flora from the Indian Ocean islands was quite different from that of other sub-Antarctic islands. Although based on a more limited number of samples, similar restricted distribution patterns appear to be present in the genera *Actinella* (Sabbe et al. 2000, 2001) and *Muelleria* (Spaulding et al. 1999), and possibly also *Biremis* (Vyverman et al. 1997), *Fragilariforma* (Kilroy et al. 2003), *Gomphonema* (Kociolek et al. 2004), and *Kobayasiella* (Vanhoutte et al. 2004). In *Actinella*, most of the 43 species appear to be restricted to the Southern Hemisphere, viz. tropical South America (mainly Amazon basin, 18 spp.), sub-Saharan Africa (11 spp.), or the Australasian region (9 spp.), some of which are extremely abundant in acidic freshwater habitats (Sabbe et al. 2001; Figs. 3–5). Only three species occur in Europe, and only two *Actinella* species have as yet been recovered from different continents (namely *Actinella punctata* and *A. brasiliensis*). The genus *Muelleria* is restricted to the high latitudes of the Northern and Southern hemisphere and of the 15 species recorded, all but two are restricted to one of both hemispheres (Spaulding et al. 1999). A floristic and taxonomic survey of the diatom communities in 66 freshwater and saline lakes in three regions in continental eastern Antarctica revealed that at least 40% of the species recovered from one of the regions, the Larsemann Hills, were unknown outside Antarctica (Sabbe et al. 2003; Figs. 8–11). In meltwater streams in the McMurdo Dry Valleys, 60% of the diatoms encountered were taxa only known from Antarctica (Esposito et al. 2006). In a survey of 897 diatom samples across the United States, many gomphonemoid diatoms which were carefully morphologically examined are restricted to the eastern or western United States (Kociolek and Kingston 1999). Also, many examples can be given of species from other genera which are restricted to small geographic areas (Potapova and Charles 2002).

In marine environments, a careful morphological analysis of the Thalassonematiaceae (86 samples from all over the world) revealed that of the few species occurring in temperate to cold waters, one has a cosmopolitan distribution, one is restricted to the northern hemisphere and two others to the Southern Ocean (Hasle 2001). The remaining species are restricted to warmer waters where they may be widespread.

In conclusion, an increasing number of studies based on critical, fine-grained morphological analyses are in favor of the prevalence of restricted distribution patterns among diatom species. At a higher taxonomic level little evidence is found for endemism as, apart from a few convincing exceptions, diatom genera are cosmopolitan. At first glance, this contrasts sharply with macro-organisms where entire genera and even families are often endemic and might be viewed as proof for a higher chance of long-distance dispersal in diatoms. While this might very well be true, higher-order classifications are entirely subjective, and a careful comparison of the average age of genera of diatoms and macro-organism groups is needed before a general statement on the subject can be made.

The importance of species recognition: biological and phylogenetic species

The above-mentioned studies, based on morphology alone, show that different geographic areas can possess distinct diatom floras, even when one might expect that ecological conditions would be similar. On the other hand, many diatom morphospecies are said to be cosmopolitan (Krammer and Lange-Bertalot 1986–1991) and some of these do indeed seem to be very widely distributed. However, many are undoubtedly composed of several (semi)cryptic species. Therefore, it will often be necessary to assess species limits in diatoms based on reproductive compatibility or molecular data, as these are more objective criteria of genetic relatedness than morphological similarity (Mann 1999; Lundholm et al. 2006), and hence also the geographic range of these species. This might especially be true in genera where only a few morphological characters are readily available (e.g., *Nitzschia*, Trobajo et al. 2006; *Eunotia*, Vanormelingen et al. 2007).

Many studies of (semi)cryptic variation patterns in diatoms have thus far focused on sympatric populations (Beszteri et al. 2005; Amato et al. 2007) and only a few studies have incorporated populations from remote locations. Clones assigned to the marine centric diatom *Skeletonema marinoi* (forming a clade with authenticated clones in analyses of LSU sequence data) were found in both the Pacific and Atlantic Oceans and in the Mediterranean Sea (Sarno et al. 2005; Godhe et al. 2006). However, the different populations could be differentiated on the basis of the ITS, D1–D3 regions of the LSU and RAPDs, indicative of restricted dispersal between the different locations (Godhe et al. 2006). In *Pseudonitzschia pungens*, three closely related but distinct ITS clades exist which have different distributions (Casteleyn et al. 2007). One clade is restricted to subtropical latitudes, the two other to temperate latitudes. Of the latter two, one appears to be restricted to the northeast Pacific, the second has a cosmopolitan distribution and was recovered from the northwest and northeast Atlantic, the northwest and northeast Pacific, and coastal waters of New Zealand. These results suggest that at least one ITS clade of this marine diatom is truly cosmopolitan. Importantly, strains from the two clades used for crossing experiments can cross and produce initial cells, independent of ITS differences, although viability and fertility of the F1 progeny was not further investigated. In northeast Atlantic *Pseudonitzschia pungens*, little population differentiation was found in the North Sea over a distance of 100 km and a period of 2 years, indicating high levels of dispersal on such small spatio-temporal scales (Evans et al. 2005). This contrasts strongly with *Ditylum brightwellii*, of which in a single fjord system, different straits, seasons, and years harbored highly differentiated populations, indicative of barriers to gene flow such as partial physical isolation and local adaptation to different environmental conditions (Rynearson and Armbrust 2004; Rynearson et al. 2006). However, there is also a possibility that it concerns closely related species, as there was also a clear divergence in morphology (valve diameter) and ITS rDNA (1.1%).

In freshwater habitats, which are typically more isolated, *Eunotia bilunaris* clones isolated from two locations in New Zealand and one in Tasmania, 450–2000 km apart, were reproductively compatible despite some differentiation in the ITS rDNA region (Vanormelingen et al. 2007). Clones of the *Sellaphora pupula* complex from sites thousands of kilometers apart were similarly capable of interbreeding (i.e., Scotland and Ukraine, Mann 1999; Scotland and the USA, Behnke et al. 2004) and, in one case, their common species identity was also confirmed by ITS sequence analysis (Behnke et al. 2004).

From the examples above it is clear that, thus far, it can only be concluded that at least a few marine diatom species are indeed cosmopolitan while the range of some freshwater species may span several thousands of kilometers (although population divergence apparently takes place over distances of 100s to 1000s of kilometres). Major efforts should therefore be directed towards assessing the distribution and evolutionary history of other morphospecies that have previously been presumed to be cosmopolitan generalists (e.g., *Navicula cryptocephala*) using crossing and molecular data. Until this is done, biogeographical analyses of these taxa are almost meaningless (Poulíčková and Mann 2006).

Human-mediated introductions

For micro-organisms such as diatoms, dispersal between localities is necessarily passive. Potential vectors for passive dispersal fall apart into four main categories; dispersal by water currents, by animals, by wind, and by humans (Kristiansen 1996). While water currents are an obvious dispersal mechanism between directly interconnected habitats, there are also studies indicating that wind and animals can carry smaller amounts of viable cells of at least the more resistant taxa over 10s to 100s of kilometers (Proctor 1959; Maynard 1968; Roy-Ocotla and Carrera 1993), although to our knowledge there is as yet little conclusive evidence for this. Human-mediated introductions of exotic species provide a strong indication that geographic dispersal by these natural vectors was limiting in the past, challenging the ubiquity hypothesis (Foissner 2006). The subsequent spread of these species provides opportunities to study the scale dependence of dispersal (Havel and Shurin 2004). In the absence of a documentary record showing the species' absence in a certain region, the paleoecological records in sediments provide a unique opportunity to assess the history of species occurrences (Willis and Birks 2006).

The best example of an introductory event for diatoms concerns the sudden appearance of *Asterionella formosa* in the sediment records of numerous lakes in New Zealand after 1880 (Harper 1994). *Asterionella formosa* is a widespread planktonic morphospecies and is frequently considered to be cosmopolitan; it is often seasonally dominant in eutrophic lakes (Krammer and Lange-Bertalot 1991). Detailed analysis of fossil material from 21 sediment cores (14 lakes) from New Zealand showed no trace of *A. formosa* in pre-European sediments, although it is now widespread, occurring in 45% of lakes for which phytoplankton records are available. The most likely vector for the introduction of *A. formosa* is the introduction of salmon ova into New Zealand lakes in the second half of the 19th century (Harper 1994). It is highly unlikely that the species was extremely (i.e., not detectably) rare before European settlement as it is a species that can occur across a wide range of environmental conditions, from oligo- to eutrophic (Harper 1994; Van Dam et al. 1994). Interestingly, *A. formosa* is also absent from other lake cores in Australasia, which might rule out environmental change due to the introduction of mammalian grazers as a cause for its sudden appearance. The recent spread in New Zealand of another exotic diatom, *Didymosphenia geminata* (Kilroy et al. 2007), is occurring long after the main human-induced environmental changes. Other convincing evidence for human mediated introduction of species (and hence previous dispersal limitation) among diatoms include the appearance of *Thalassiosira baltica* in the Laurentian Great Lakes (Edlund et al. 2000), and the North-American species *Gomphoneis minuta* and *Encyonema triangulum* in France (Coste and Ector 2000). Well-documented marine introductions include *Odontella sinensis* (Greville) Grunow (Ostenfeld 1908), and *Coscinodiscus wailesii* Gran & Angst (Edwards et al. 2001), which were introduced either via ship's ballast water or by aquacultural practices. *Coscinodiscus wailesii*, a large (175–500 μm diameter) species was previously only known from the Pacific Ocean, and was first recorded (on two independent occasions) in English coastal waters in 1977. It had never before been observed in the monthly plankton samples (Continuous Plankton Recorder, CPR) which were taken in the area since the early 1950s. It subsequently spread into European shelf areas and has now become a significant member of the phytoplankton community (Edwards et al. 2001). It is highly unlikely that *C. wailesii* was already present in very low numbers only to become one of the major phytoplankton species in the NE-Atlantic.

Partitioning the effect of history and environment

One major limitation associated with the above-mentioned studies on diatom biogeography showing that diatoms can have restricted distributions is that limited dispersal is not the only thing that could create a restricted geographic range. Spatial distance is often correlated with environmental distance. Moreover, next to human-mediated introductions, also environmental change through time may allow for shifts in species ranges. A big challenge is therefore to separate the effects of environmental heterogeneity—which everyone accepts will cause biological differences—from community divergences caused by dispersal limitation (Martiny et al. 2006). While it is often implicitly assumed that habitats in different geographic areas (or different points in time) are environmentally similar, e.g., in the Arctic and Antarctic, and that differences in species composition between regions are as a consequence due to restricted dispersal, this is not usually tested. However, there are statistical methods that control for the effects of environment while quantifying the spatial (or temporal) influence on community structure and diversity. These include partial mantel tests and variation partitioning (Borcard et al. 1992; Bonnet and Van de Peer 2002).

The influence of regional processes relative to local environmental factors on stream diatom community structure has been investigated by variation partitioning in two studies. In benthic stream diatom communities in the United States, while 9–26% of the total explained variation could be attributed to spatially correlated environmental parameters, 15–31% was explained by purely geographic variation (Potapova and Charles 2002). This was mainly due to species with restricted distributions not readily explainable by environmental factors. In a study of benthic diatom communities in boreal streams across Finland, a similar 24% of the total explained variation was accounted for by environmentally independent spatial variation, with 40% attributable to spatially structured environmental variables (Soininen et al. 2004).

In a global scale study of lacustrine diatom diversity it was shown that variation in both regional and local genus richness was best explained by geographical variables such as connectivity between lakes and the amount of isolation of lake districts (Vyverman et al. 2007). Local environmental variables were of minor importance, implying that dispersal limitation is important in determining diatom genus diversity. Similarly, Telford et al. (2006) found that local species richness optima in lake regions in Europe and North America are strongly dependent upon regional habitat availability. Lakes with the modal lake pH always had the highest local species richness, regardless of the value of the modal pH (which varied by more than 2.5 units). This suggests that, even at the high lake densities in that part of the northern hemisphere, local species richness can be constrained by dispersal, at least in the less prevalent habitats in a region. While these studies convincingly show a reduced local (and even regional) diversity due to dispersal limitation, they could be complemented by the experimental local introduction of species from the regional species pool in mesocosms installed in the field, as was done for zooplankton (Shurin 2000). This might provide definitive evidence for local diatom diversity to be reduced by extinction–colonization dynamics.

Prospects

As reviewed above, different lines of research have recently challenged the ubiquity hypothesis for diatoms and show that diatom communities are controlled by the same processes affecting macro-organisms, although possibly not to the same degree. These studies therefore underscore the need for conservation and the protection of isolated areas such as Tasmania, New Zealand, and Antarctica against further introductions of exotic species. However, it is also obvious that remarkably little is as yet known about the factors driving diatom species diversity and geographic distributions. Studies focusing on the mechanisms generating species diversity in this very diverse group of organisms are urgently needed. This includes molecular phylogenies to unravel the evolutionary history of species complexes or genera and their correlation with geography. This way, the occurrence and timing of such phenomena as allopatric divergence, long-distance dispersal, and adaptive radiations can be determined. Our efforts should also be further directed towards assessing the distribution of diatom species, using both crosses and molecular methods to confirm conspecificity in apparently cosmopolitan morphospecies. Within species, estimates of population divergence will be useful to assess dispersal distances and frequencies and to identify human-mediated introductions. It is also crucial to assess taxonomic turnover rates over different geographical scales (local to global) based upon taxonomically consistent datasets and to reveal what factors determine whether a diatom is endemic or widely distributed, such as habitat availability, abundance and stress resistance and survival during dispersal. Taken together, these data will allow to determine the relative importance of restricted dispersal in determining the diversity and geographic distribution of diatoms as compared to macro-organism groups, and thus an evaluation of, as not strictly true, how close to the truth the ubiquity hypothesis really was for diatoms.

Acknowledgments Part of the research presented in this review was financed by the Belgian Science Policy Office, the Fund for Scientific Research-Flanders and the Institute for the promotion of Innovation by Science and Technology in Flanders. Elie Verleyen is postdoctoral scientist with the Fund for Scientific Research. We thank Koen Sabbe for many valuable discussions, Wilhelm Foissner and an anonymous referee for their comments on an earlier version of the manuscript, and Bart Van de Vijver for kindly providing *Stauroneis* photographs.

References

Amato A, Kooistra WHCF, Levialdi Ghiron JH et al (2007) Reproductive isolation among sympatric cryptic species in marine diatoms. Protist 158:193–207

Baas-Becking LGM (1934) Geobiologie of inleiding tot de milieukunde. Van Stockum and Zoon, The Hague

Behnke A, Friedl T, Chepurnov VA et al (2004) Reproductive compatibility and rDNA sequence analyses in the *Sellaphora pupula* species complex (Bacillariophyta). J Phycol 40:193–208

Bell T, Ager D, Song J-I et al (2005) Larger islands house more bacterial taxa. Science 308:1884

Beszteri B, Ács É, Medlin LK (2005) Ribosomal DNA sequence variation among sympatric strains of the *Cyclotella meneghiniana* complex (Bacillariophyceae) reveals cryptic diversity. Protist 156:317–333

Bonnet E, Van de Peer Y (2002) zt: a software tool for simple and partial mantle tests. J Stat Softw 7(10):1–12. Available at http://www.jstatsoft.org

Borcard D, Legendre P, Drapeau P (1992) Partialling out the spatial component of ecological variation. Ecology 73:1045–1055

Camburn KEJ, Kingston JC, Charles DE (1984–1986) PIRLA diatom iconograph. PIRLA Unpublished Report Series No 3. Department of Biology, Indiana University, Bloomington

Casteleyn G, Chepurnov VA, Leliaert F et al (2007) Pseudo-Nitzschia pungens (Bacillariophyceae): a cosmopolitan diatom species? Harmful Algae. doi:10.1016/j.hal.2007.08.004

Chao A, Li PC, Agatha S et al (2006) A statistical approach to estimate soil ciliate diversity and distribution based on data from five continents. Oikos 114:479–493

Chepurnov VA, Mann DG, Sabbe K et al (2004) Experimental studies on sexual reproduction in diatoms. Int Rev Cytol 237:91–154

Coste M, Ector L (2000) Diatomées invasives exotiques ou rares en France: principales observations effectuées au cours des dernières décennies. Syst Geogr Plants 70:373–400

Cox EJ (1995) Morphological variation in widely distributed diatom taxa: taxonomic and ecological implications. In: Donato M, Montresor M (eds) Proceedings of the 13th international diatom symposium. Biopress, Bristol, pp 335–345

Cox CB, Moore PD (2005) Biogeography—an ecological and evolutionary approach. 7th edn. Blackwell, Oxford

Créach V, Ernst A, Sabbe K et al (2006) Using quantitative PCR to determine the distribution of a semicryptic benthic diatom, *Navicula phyllepta* (Bacillariophyta). J Phycol 42:1142–1154

Darling KF, Kucera M, Pudsey CJ et al (2004) Molecular evidence links cryptic diversification in polar planktonic protists to Quaternary climate dynamics. PNAS 101:7657–7662

Darling KF, Kucera M, Wade CM (2007) Global molecular phylogeography reveals persistent Arctic circumpolar isolation in a marine planktonic protist. PNAS 104:5002–5007

Edlund MB, Taylor CM, Schelske CL et al (2000) *Thalassiosira baltica* (Grunow) Ostenfeld (Bacillariophyta), a new exotic species in the Great Lakes. Can J Fish Aquat Sci 57:610–615

Edwards M, John AWG, Johns DG et al (2001) Case history and persistence of the non-indigenous diatom *Coscinodiscus wailesii* in the north-east Atlantic. J Mar Biol Ass UK 81:207–211

Esposito RMM, Horn SL, McKnight DM et al (2006) Antarctic climate cooling and response of diatoms in glacial meltwater streams. Geophys Res Lett 33, Art. No. LO7406

Evans KM, Kuhn SF, Hayes PK (2005) High levels of genetic diversity and low levels of genetic differentiation in North Sea *Pseudo-Nitzschia pungens* (Bacillariophyceae) populations. J Phycol 41:506–514

Fenchel T, Finlay BJ (2004) The ubiquity of small species: patterns of local and global diversity. Bioscience 54:777–784

Finlay BJ (2002) Global dispersal of free-living microbial eukaryote species. Science 296:1061–1063

Finlay BJ, Clarke KJ (1999) Ubiquitous dispersal of microbial species. Nature 400:828

Finlay BJ, Fenchel T (2004) Cosmopolitan metapopulations of free-living microbial eukaryotes. Protist 155:237–244

Finlay BJ, Monaghan EB, Maberly SC (2002) Hypothesis: the rate and scale of dispersal of freshwater diatom species is a function of their global abundance. Protist 153:261–273

Flower RJ (2005) A taxonomic and ecological study of diatoms from freshwater habitats in the Falkland Islands, South Atlantic. Diatom Res 20:23–96

Foissner W (2006) Biogeography and dispersal of micro-organisms: a review emphasizing protists. Acta Protozool 45:111–136

Fourtanier E, Kociolek JP (2003) Catalogue of the diatom genera (vol 14, pg 190, 1999). Diatom Res 18:245–258

Godhe A, McQuoid MR, Karunasagar I et al (2006) Comparison of three common molecular tools for distinguishing among geographically separated clones of the diatom *Skeletonema marinoi* Sarno et Zingone (Bacillariophyceae). J Phycol 42:280–291

Green J, Bohannan JM (2006) Spatial scaling of microbial diversity. TREE 21:501–507

Green JL, Holmes AJ, Westoby M et al (2004) Spatial scaling of microbial eukaryote diversity. Nature 432:747–750

Harper MA (1994) Did Europeans introduce *Asterionella formosa* Hassall to New Zealand? In: Kociolek JP (ed) Proceedings of the 11th International Diatom Symposium 1990. California Academy of Sciences, San Francisco, pp 479–484

Hasle GR (1976) The biogeography of some marine planktonic diatoms. Deep-Sea Res 23:319–338

Hasle GR (2001) The marine, planktonic diatom family Thalassionemataceae: morphology, taxonomy and distribution. Diatom Res 16:1–82

Havel JE, Shurin JB (2004) Mechanisms, effects, and scales of dispersal in freshwater zooplankton. Limnol Oceanogr 49:1229–1238

Hustedt F (1927–1966) Die Kieselalgen Deutschlands, Österreichs und der Schweiz. In: Rabenhorst L (ed) Kryptogamen-Flora von Deutschland, Österreich und der Schweiz, Bd. 7(2). Akad Verlagsges, Leipzig

Kilroy C, Sabbe K, Bergey EA et al (2003) New species of *Fragilariforma* (Bacillariophyceae) from New Zealand and Australia. New Zeal J Bot 41:535–554

Kilroy C, Biggs BJ, Vieglais CC (2007) *Didymosphenia geminata* in New Zealand: a science response to help manage an unwanted, invasive freshwater diatom. In: Abstract book, ASLO Aquatic Sciences meeting. Santa Fe (New Mexico), 4–9/02/2007

Kociolek JP, Kingston JC (1999) Taxonomy, ultrastructure, and distribution of some gomphonemoid diatoms (Bacillariophyceae: Gomphonemataceae) from rivers in the United States. Can J Bot 77:686–705

Kociolek JP, Spaulding SA (2000) Freshwater diatom biogeography. Nova Hedwigia 71:223–241

Kociolek JP, Spaulding SA, Sabbe K et al (2004) New *Gomphonema* (Bacillariophyta) species from Tasmania. Phycologia 43:427–444

Krammer K, Lange-Bertalot H (1986–1991) Bacillariophyceae. In: Ettl H, Gerloff, J, Heynig H, Mollenhauer D (eds) Süßwasserflora von Mitteleuropa, vol. 2(1), 876 pp (1986), vol. 2(2), 596 pp (1988), vol. 2(3), 576 pp (1991), vol. 2(4), 437 pp (1991). Fischer, Stuttgart/Jena

Krasske G (1939) Zur Kieselalgenflora Südchiles. Arch Hydrobiol 35:349–468

Kristiansen J (1996) Dispersal of freshwater algae—a review. Hydrobiologia 336:151–157

Lee WJ, Patterson DJ (2000) Heterotrophic flagellates (Protista) from marine sediments of Botany Bay, Australia. J Nat Hist 34:483–562

Leibold MA, Holyoak M, Mouquet N et al (2004) The metacommunity concept: a framework for multiple-scale community ecology. Ecol Lett 7:601–613

Lundholm N, Moestrup Ø, Kotaki Y et al (2006) Inter- and intraspecific variation of the *Pseudo-Nitzschia delicatissima* complex (Bacillariophyceae), illustrated by rRNA probes, morphological data and phylogenetic analysis. J Phycol 42:464–481

Mann DG (1999) The species concept in diatoms. Phycologia 38:437–495

Mann DG, Droop SJM (1996) Biodiversity, biogeography and conservation of diatoms. Hydrobiologia 336:19–32

Martiny JBH, Bohannan BJM, Brown JH et al (2006) Microbial biogeography: putting microorganisms on the map. Nat Rev Biogeogr 4:102–112

Maynard NG (1968) Significance of air-borne algae. Z allg Microbiol 8:225–226

Ostenfeld CH (1908) On the immigration of *Biddulphia sinensis* Grev. and its occurrence in the North Sea during 1903–1907. Meddelelser fra Kommissionen for Havundersogelser, Plankton 1(6):1–25

Papke RT, Ramsing NB, Bateson MM et al (2003) Geographical isolation in hot spring cyanobacteria. Environ Microbiol 5:650–659

Potapova MG, Charles DFC (2002) Benthic diatoms in USA rivers: distributions along spatial and environmental gradients. J Biogeogr 29:167–187

Poulíčková A, Mann DG (2006) Sexual reproduction in *Navicula cryptocephala* (Bacillariophyceae). J Phycol 42:872–886

Proctor VW (1959) Dispersal of freshwater algae by migratory waterbirds. Science 130:623–624

Reche I, Pulido-Villena E, Morales-Baquero R et al (2005) Does ecosystem size determine aquatic bacterial richness? Ecology 86:1715–1722

Ricklefs RE (1987) Community diversity: relative roles of local and regional processes. Science 235:167–171

Round FE, Crawford RM, Mann DG (1990) Diatoms: biology and morphology of the genera. Cambridge Univ. Press, Cambridge

Roy-Ocotla G, Carrera J (1993) Aeroalgae: responses to some aerobiological questions. Grana 32:48–56

Rumrich U, Lange-Bertalot H, Rumrich M (2000) Diatoms of the Andes, from Venezuela to Patagonia/Tierra del Fuego, and two additional contributions. ARG Gartner Verlag KG, Königstein

Rynearson TA, Armbrust EV (2004) Genetic differentiation among populations of the planktonic marine diatom *Ditylum brightwellii*. J Phycol 40:34–43

Rynearson TA, Newton JA, Armbrust EV (2006) Spring bloom development, genetic variation, and population succession in the planktonic diatom *Ditylum birghtwellii*. Limnol Oceanogr 51:1249–1261

Sabbe K, Vanhoutte K, Vyverman W (2000) *Actinella comperei* sp. nov. (Eunotiaceae, Bacillariophyta): a new endemic freshwater diatom from Tasmania (Australia). Syst Geogr Plants 70:237–243

Sabbe K, Vanhoutte K, Lowe RL et al (2001) Six new *Actinella* (Bacillariophyta) species from Papua New Guinea, Australia and New Zealand: further evidence for widespread diatom endemism in the Australasian region. Eur J Phycol 36:321–340

Sabbe K, Verleyen E, Hodgson DA et al (2003) Benthic diatom flora of freshwater and saline lakes in the Larsemann Hills and Rauer Islands, East Antarctica. Antarct Sci 15:227–248

Sarno D, Kooistra WHCF, Medlin LK et al (2005) Diversity in the genus *Skeletonema* (Bacillariophyceae). II. An assessment of the taxonomy of *S. costatum*-like species with the description of four new species. J Phycol 41:151–176

Shurin JB (2000) Dispersal limitation, invasion resistance, and the structure of pond zooplankton communities. Ecology 81:3074–3086

Soininen J, Paavola R, Muotka T (2004) Benthic diatom communities in boreal streams: community structure in relation to environmental and spatial gradients. Ecography 27:330–342

Spaulding SA, McKnight DM (1999) Diatoms as indicators of environmental change in Antarctic freshwaters. In: Stoermer EF, Smol JP (eds) The diatoms: applications for the environmental and earth sciences. Cambridge Univ. Press, Cambridge, pp 245–263

Spaulding SA, Kociolek JP, Wong D (1999) A taxonomic and systematic revision of the genus *Muelleria*. Phycologia 38:314–341

Stevenson AC, Juggins S, Birks HJB et al (1991) The surface waters acidification project paleolimnology programme: modern diatom/lake-water chemistry data set. Ensis pub., London

Syvertsen EE (1977) *Thalassiosira rotula* and *T.* gravida: ecology and morphology. Nova Hedwigia, Beih 54:99–112

Taylor JW, Turner E, Townsend JP et al (2006) Eukaryotic microbes, species recognition and the geographic limits of species: examples from the kingdom Fungi. Philos T R Soc B 361:1947–1963

Telford RJ, Vandvik V, Birks HJB (2006) Dispersal limitation matters for microbial morphospecies. Science 312:1015

Trobajo R, Mann DG, Chepurnov VA et al (2006) Taxonomy, life cycle and auxosporulation *of Nitzschia fonticola* (Bacillariophyta). J Phycol 42:1353–1372

Tyler PA (1996) Endemism in freshwater algae, with special reference to the Australian region. Hydrobiologia 336:127–135

Van Dam H, Mertens A, Sinkeldam J (1994) A coded checklist and ecological indicator values of freshwater diatoms from The Netherlands. Neth J Aquat Ecol 28:117–133

Van de Vijver B, Beyens L, Lange-Bertalot H (2004) The genus *Stauroneis* in the Arctic and (Sub-)Antarctic regions. Biblthca diatomol 51:1–317

Van de Vijver B, Gremmen NJM, Beyens L (2005) The genus *Stauroneis* (Bacillariophyceae) in the Antarctic region. J Biogeogr 32:1791–1798

Vanhoutte K, Verleyen E, Vyverman W et al (2004) The freshwater diatom genus *Kobayasiella* (Bacillariophyta) in Tasmania, Australia. Aust Syst Bot 17:483–496

Vanhoutte K, Verleyen E, Sabbe K et al (2006) Congruence and disparity in benthic diatom community structure of small lakes in New Zealand and Tasmania. Mar Freshw Res 57:789–801

Vanormelingen P, Chepurnov VA, Mann DG et al (2007) Congruence of morphological, reproductive and ITS rDNA sequence data in some Australasian *Eunotia bilunaris*. Eur J Phycol 42:61–79

Villanueva VD, Maidana NI (1999) Diatoms (Bacillariophyta) from Pulmari lake (Nuequen, Argentina). Biologia 54:1–10

Vyverman W, Vyverman R, Hodgson D et al (1995) Diatoms from Tasmanian mountain lakes: a reference data-set (TASDIAT) for environmental reconstruction and a systematic and autecological study. Bibl diatomol 33:1–193

Vyverman W, Sabbe K, Vyverman R (1997) Five new freshwater species of *Biremis* (Bacillariophyta) from Tasmania. Phycologia 36:91–102

Vyverman W, Sabbe K, Mann DG et al (1998) *Eunophora* gen. nov. (Bacillariophyta) from Tasmania and New Zealand: description and comparison with *Eunotia* and amphoroid diatoms. J Phycol 33:95–111

Vyverman W, Verleyen E, Sabbe K et al (2007) Historical processes constrain patterns in global diatom diversity. Ecology 88:1924–1931

Whitaker RJ, Grogan DW, Taylor JW (2003) Geographic barriers isolate endemic populations of hyperthermophilic Archaea. Science 301:976–978

Whitfield J (2005) Biogeography: is everything everywhere? Science 310:960–961

Willis KJ, Birks HJB (2006) What is natural? The need for a long-term perspective in biodiversity conservation. Science 314:1261–1265

Dinoflagellate diversity and distribution

**F. J. R. Taylor · Mona Hoppenrath ·
Juan F. Saldarriaga**

Originally published in the journal Biodiversity and Conservation, Volume 17, No 2, 407–418.
DOI: 10.1007/s10531-007-9258-3

Abstract Dinoflagellates are common to abundant in both marine and freshwater environments. They are particularly diverse in the marine plankton where some cause "red tides" and other harmful blooms. More than 2,000 extant species have been described, only half of which are photosynthetic. They include autotrophs, mixotrophs and grazers. They are biochemically diverse, varying in photosynthetic pigments and toxin production ability. Some are important sources of bioluminescence in the ocean. They can host intracellular symbionts or be endosymbionts themselves. Most of the photosynthetic "zooxanthellae" of invertebrate hosts are mutualistic dinoflagellate symbionts, including all those essential to reef-building corals. Roughly 5% are parasitic on aquatic organisms. The fossil record, consisting of more than 2,500 species, shows a rapid radiation of cysts, starting in the Triassic, peaking in the Cretaceous, and declining throughout the Cenozoic. Marine species with a benthic, dormant cyst stage are confined to the continental shelf and fossil cysts can be used as markers of ancient coastlines. Northern and southern hemispheres contain virtually identical communities within similar latitudes, separated by a belt of circumtropical species. A few endemics are present in tropical and polar waters. Some benthic dinoflagellates are exclusively tropical, including a distinct phycophilic community, some of which are responsible for ciguatera fish poisoning. In lakes chemical and grazing effects can be important. Predatory dinoflagellates co-occur with their prey, often diatoms.

Keywords Alveolate · Dinoflagellates · Dinokaryon · Dinophyceae · Distribution · Fossil · Morphospecies · Resting cyst

---

Special Issue: Protist diversity and geographic distribution. Guest editor: W. Foissner.

F. J. R. Taylor (✉) · M. Hoppenrath · J. F. Saldarriaga
Department of Botany, University of British Columbia, 6270 University Boulevard, Vancouver, BC, Canada V6T1Z4
e-mail: maxt@unixg.ubc.ca

F. J. R. Taylor
Department of Earth and Ocean Sciences, University of British Columbia, 6270 University Boulevard, Vancouver, BC, Canada V6T1Z4

W. Foissner et al. (eds.), *Protist Diversity and Geographical Distribution*
DOI: 10.1007/978-90-481-2801-3_13

Introduction

Dinoflagellates are a highly diverse group of flagellates, consisting of both photosynthetic and non-photosynthetic taxa in equal proportions (Taylor 1987b). Many of the photosynthetic members are mixotrophs, and the heterotrophs feed by a wide variety of mechanisms (Gaines and Elbrächter 1987; Schnepf and Elbrächter 1992). While most are free-living in marine and freshwater environments, others, such as the "zooxanthellae" of reef-building corals, are beneficial endosymbionts and still others are parasites of many protist, invertebrate and vertebrate hosts. Some are luminescent. Some are toxin producers. Their greatest diversity is in the marine plankton where they can produce "red tides" and other monospecific blooms.

Three books have been devoted to the biology of the group: a small volume focused mainly on fossils (Sarjeant 1974), a collection of special topics edited by Spector (1984) and a comprehensive volume edited by Taylor (1987a). The classification of fossil and extant taxa was unified for the first time by Fensome et al. (1993). Several more recent papers concerned primarily with the evolution of the group include Fensome et al. (1999), Saldarriaga et al. (2004) and Taylor (2004).

Most dinoflagellates contain a unique nucleus, the dinokaryon. The DNA is not organized around nucleosomal histones, instead forming fibrillar chromosomes that are always condensed and that divide via a closed mitosis with an external spindle. The nucleus often contains large amounts of DNA (up to 200 pg, 215,000 Mbp). Dinomitosis is closed and consists of an external spindle, the microtubules traversing the nucleus through tunnels or furrows. Some dinoflagellates, such as the noctilucoids with their greatly inflated forms, and the parasitic blastodineans, alternate in their life cycles between a dinokaryon and a more conventional, histone-containing nucleus.

In most dinoflagellates a pair of flagella is laterally inserted (ventral by convention Taylor 1987). This is known as the dinokont arrangement. In desmokonts, mostly *Prorocentrum* species, they are functionally anterior. The transverse flagellum, which winds around the cell in both dino- and desmokonts, is ribbon-like, with the inner edge contracted by a contractile fibre and the axoneme following a spiral path along the outer edge, an arrangement unique to dinoflagellates.

Approximately 13–16% of living dinoflagellates produce a dormant "resting cyst" (Head 1996), the hypnozygote, the only diploid stage in their haplophasic life cycle. The cyst wall contains a tough material, dinosporin, which resembles the sporopollenin of pollen grains.

Dinoflagellate diversity

The number of living species is usually estimated to be approximately 2,000, with 2,500 named fossil species. In a recent revision Gómez (2005) recognized 1,555 free-living marine species. We estimate that there are approximately 160 marine benthic species (psammophilic and phycophilic). Not many freshwater planktonic species have been described since the 220 monographed by Bourrelly (1970). It is probable that there are at least 50 more. Only one explicitly freshwater benthic species has been described. Several dinoflagellates have been recorded from snow and sea ice. One new family, 12 new genera, and 27 new species of living dinoflagellates were described from 2004 until the end of 2006.

For many years the dinoflagellates living as "zooxanthellae" (autotrophic endosymbionts of animals and other protists) were limited to *Zooxanthella nutricola* of colonial radiolarians, referred to by a more recent name *Symbiodinium microadriaticum*, thought to be the sole intracellular inhabitant of all reef-building corals, and a few *Amphidinium* species in turbellarians (Trench 1987). Since then some of the "zooxanthellae" in reef-building corals and other cnidarians have been shown to be stages of common peridinioid genera and *Symbiodinium* has proved to be remarkably heterogeneous. Dinoflagellates are parasites on numerous protist, invertebrate and a few vertebrate (fish) hosts. It is probable that the number of parasitic species is greatly underestimated. Cachon and Cachon (1987) listed 66 parasitic species, mostly blastodineans and syndinians, plus a few un-named taxa.

The species numbers above are "morphospecies", defined in the classical sense. Species concepts as they apply to dinoflagellates or other harmful species have been discussed at length by Taylor (1985) and, most recently, by Lundholm and Moestrup (2006). Dinoflagellate morphospecies can contain considerable genetic diversity, quantifiable with molecular techniques, perhaps including cryptospecies. On the other hand there are cases in which the named species appear to be variants of a single larger gene pool, e.g., the *Alexandrium tamarense* species complex (see Distribution, below).

Fossil diversity

The dinoflagellate fossil record is apparently limited to dinosporin-containing resting cysts. The internal siliceous skeletons and calcareous structures of some vegetative cells can also fossilize. More than 2,500 fossil species, assigned to 400 genera, have been described. Although molecular evidence (Moldowan and Terazina 1998) suggests that the lineage leading to the Alveolata diverged in the Neoproterozoic as part of the eukaryotic "crown" radiation (Baldauf et al. 2000), the first undoubted dinoflagellate fossils have been found in 240 million-year-old Middle Triassic sediments of the Mesozoic (Fensome et al. 1993, 1996, 1999).

From the late Triassic to the Mid-Jurassic the record indicates a rapid radiation of major cyst lineages, with further ones added later in the Jurassic. The earliest appearing groups are suessioids and gonyaulacoids, with peridinioids expanding rapidly in the Cretaceous (Fig. 1). No new major cyst types appear after the Mesozoic. Cyst species richness increases in the Cretaceous to close to 280 (Bujak and Williams 1981) and then declines throughout the Cenozoic to 44 in the Pleistocene. Assuming a ratio of fossilizable to non-fossilizable species similar to present (Quaternary) time, peak species richness could have approached 3,000 in the Cretaceous.

Trends in speciation, extinction, and evolution of tabulation have been discussed at length by Goodman (1987), and Fensome et al. (1993, 1996, 1999).

Diversity of extant taxa

Dinoflagellate morphology is astonishingly diverse (Fig. 2, 3). Typically they are unicellular, but some are colonial, chain formation being common. A few are pseudocolonies (coenocytic cells) or even multicellular. Horns are formed by many marine planktonic dinoflagellates, e.g. *Ceratium* (Fig. 2i, m). They can be athecate or thecate (with multiple cellulosic plates), or pelliculate. The plate pattern is referred to as tabulation. Six basic types of tabulation were recognized by Taylor (1987b, c, 2004) and Fensome et al. (1993): gymnodinoid, suessioid, peridinioid, gonyaulacoid, dinophysoid, and prorocentroid (Fig. 3), on which major orders are based. Phylogenetic studies suggest that thecal

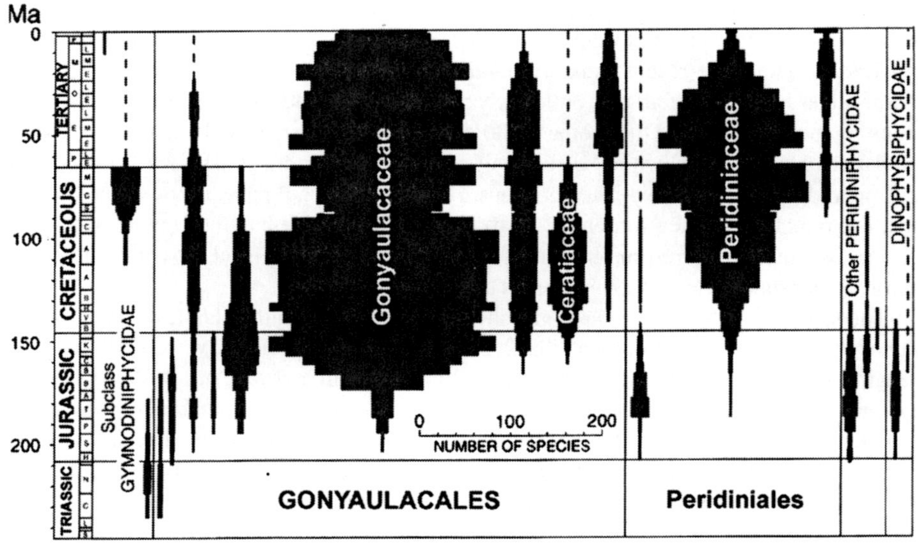

Fig. 1 Fossil dinoflagellate diversity. The number of species per family per time interval is shown as spindle plots (modified from Fensome et al. 1999; extinct groups not labeled)

plates have been lost repeatedly during the evolution of dinoflagellates (Saldarriaga et al. 2004).

Suessiales have nine latitudinal vesicle series, with the vesicles containing delicate plates. Peridiniales and Gonyaulacales, which dominate the living thecate taxa, both have six latitudinal series of clearly recognizable plate-containing vesicles. Members of the Dinophysiales (Fig. 2c–e, g, k) are bilateral, divided by a sagittal suture. In some tropical dinophysoids, such as *Ornithocercus* (Fig. 2d) or *Histioneis* (Fig. 2e), the cingular (girdle) and left sulcal lists project out extensively from the body and are reinforced by ribs. Coccoid cyanobacteria ("phaeosomes") are symbiotic with the more elaborate, non-photosynthetic dinophysoids (Fig. 2d). Prorocentralean species lack a cingulum and sulcus and have the flagella inserted anteriorly (Fig. 2f).

The Noctilucales are unusual, not only because of their nuclei, which alternate between a dinokaryon and a more conventional nucleus, but also because of their strange forms. They achieve very large sizes by inflating their cells with liquid vacuoles and assume unusual, often elaborate forms. *Noctiluca* has a relatively simple spherical shape although it also has a tentacle used in food capture. Others have leaf-like, butterfly-like and medusa-like shapes (Fig. 3).

Fig. 2 Light micrographs of selected extant dinoflagellate taxa showing some of their morphological diversity.(**a**) *Akashiwo sanguinea*, athecate. (**b**) *Erythropsidinium* sp., athecate with anterior ocelloid. (**c**) *Dinophysis miles* var. *schroeteri*, thecate with dinophysoid tabulation. (**d**) *Ornithocercus quadratus*, thecate with dinophysoid tabulation, note the phaeosomes (arrows). (**e**) *Histioneis panda*, thecate with dinophysoid tabulation, with phaeosomes. (**f**) *Prorocentrum micans*, thecate with prorocentroid tabulation and desmokont flagellation (not visible). (**g**) *Oxyphysis oxytoxoides*, thecate with dinophysoid tabulation. (**h**) *Alexandrium catenella*, note the chain-formation and the stained nuclei. (**i**) *Ceratium breve* (middle), *Gonyaulax polygramma* (left) and *Podolampas palmipes* (right), thecate with gonyaulacoid/peridinoid tabulation. (**j**) *Oxytoxum scolopax*. (**k**) *Amphisolenia bidentata*. (**l**) *Pyrocystis fusiformis* showing bioluminescence. (**m**) *Ceratium vultur* with long horns. Scale bars = 10 μm, except (**k, m**) with a scale bar of 50 μm and (**l**) of 100 μm ▶

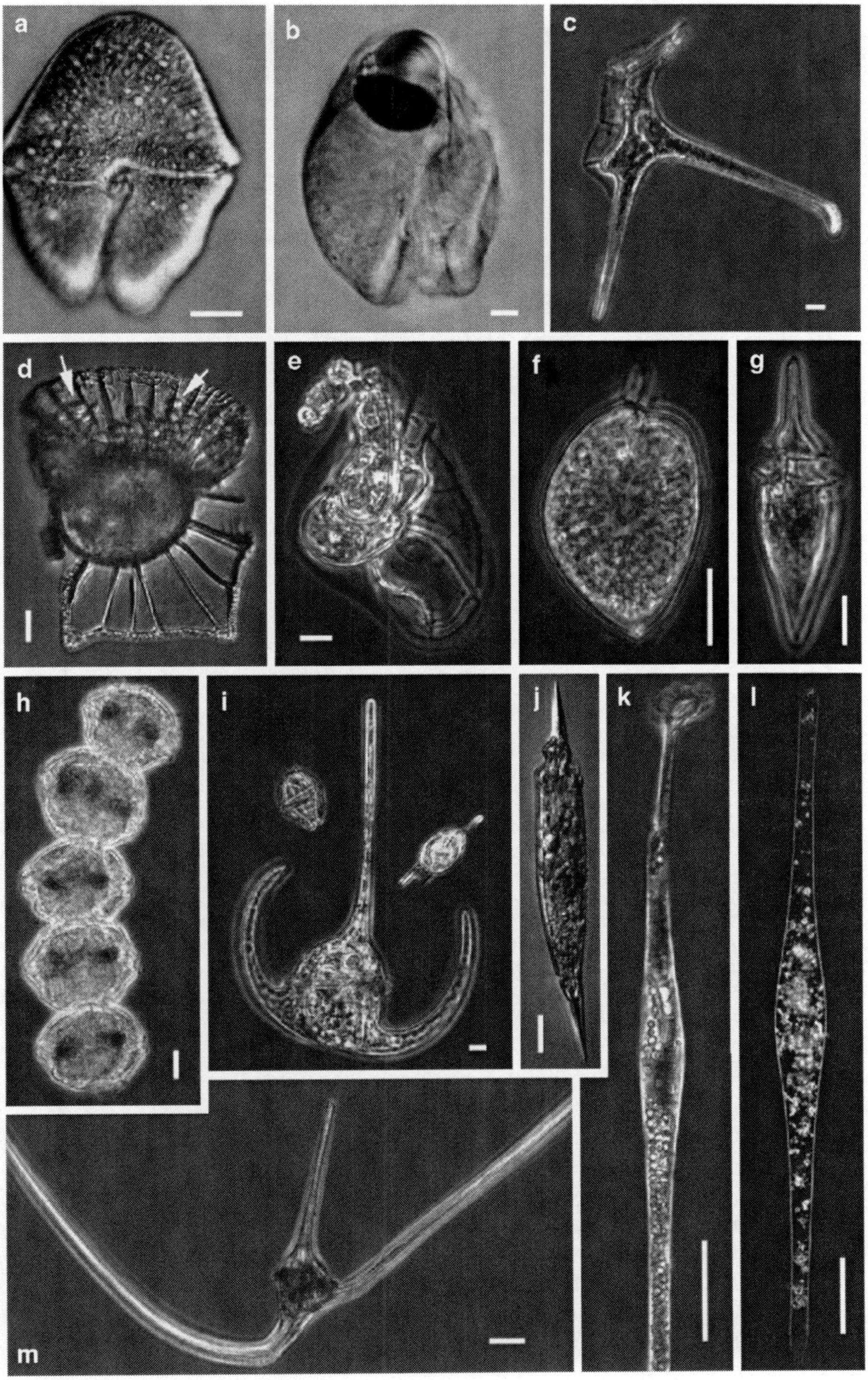
a
b
c
d
e
f
g
h
i
j
k
l
m

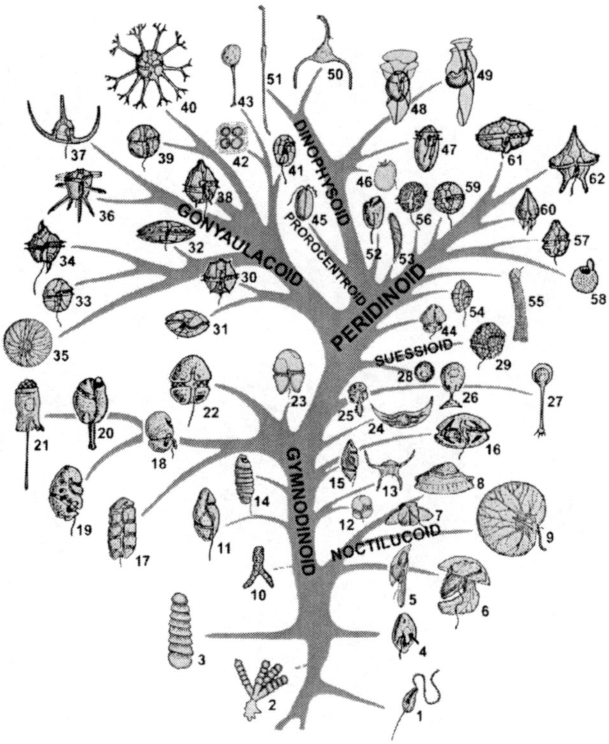

Fig. 3 Extant dinoflagellate diversity. Only about a quarter of the living genera are shown in this tree. (**1**) *Perkinsus*, (**2**) *Thalassomyces*, (**3**) *Amoebophrya*, (**4**) *Oxyrrhis*, (**5**) *Kofoidinium*, (**6**) *Pomatodinium*, (**7**) *Cymbodinium*, (**8**) *Craspedotella*, (**9**) *Noctiluca*, (**10**) *Dinothrix*, *Dinoclonium*, (**11**) *Gyrodinium*, (**12**) *Karenia*, (**13**) *Brachydinium*, (**14**) *Cochlodinium*, (**15**) *Plectodinium*, (**16**) *Actiniscus*, (**17**) *Polykrikos*, (**18**) *Proterythropsis*, *Warnowia*, (**19**) *Nematodinium*, (**20**) *Erythropsidinium*, (**21**) *Greuetodinium* (=*Leucopsis*), (**22**) *Gymnodinium*, (**23**) *Akashiwo*, (**24**) *Dissodinium*, (**25**) *Protoodinium*, (**26**) *Oodinium*, (**27**) *Chytriodinium*, (**28**) *Symbiodinium*, (**29**) *Woloszynskia*, *Polarella*, (**30**) *Triadinium* (=*Heteraulacus*), (**31**) *Gambierdiscus*, also *Coolia*, *Ostreopsis*), (**32**) *Pyrophacus*, (**33**) *Alexandrium*, *Fragilidium*, (**34**) *Pyrodinium*, (**35**) *Pyrocystis*, (**36**) *Ceratocorys*, (**37**) *Ceratium*, (**38**) *Gonyaulax*, (**39**) *Paleophalacroma*, (**40**) *Cladopyxis*, (**41**) *Hemidinium*, (**42**) *Gloeodinium*, (**43**) *Stylodinium*, (**44**) *Katodinium*, (**45**) *Prorocentrum*, *Mesoporus*, (**46**) *Sinophysis*, (**47**) *Dinophysis*, (**48**) *Ornithocercus*, (**49**) *Histioneis*, *Parahistioneis*, (**50**) *Triposolenia*, (**51**) *Amphisolenia*, (**52**) *Amphidinium*, (**53**) *Blastodinium*, (**54**) *Heterocapsa*, (**55**) *Haplozoon*, (**56**) *Peridinium*, (**57**) *Scrippsiella*, (**58**) *Thoracosphaera*, (**59**) *Glenodinium*, (**60**) *Podolampas*, *Blepharocysta*, (**61**) *Diplopsalis*-group (e.g. *Diplopsalopsis*, *Diplopsalis*, *Oblea*, *Preperidinium*), (**62**) *Protoperidinium*. Modified and expanded from Taylor (1987a)

Photosynthetic dinoflagellates show a diversity of chloroplast types acquired by secondary or tertiary symbiosis (Schnepf and Elbrächter 1999; Cavalier-Smith 1999). Molecular phylogenetic analysis suggests multiple losses and replacements of plastids (Saldarriaga et al. 2001). Heterotrophic dinoflagellates have a variety of complex feeding mechanisms (Gaines and Elbrächter 1987; Schnepf and Elbrächter 1992), from simple phagotrophy to pallium- (feeding veil) and peduncle-feeding. Highly complex organelles such as the nematocyst, ocelloid (Fig. 2b), and piston are found in a few related taxa. Their functions and origins are not fully understood (Greuet 1987; Hoppenrath and Leander 2007a, b).

Non-dinokaryotic dinoflagellates

Phylogenies based on molecular data usually put several groups of non-dinokaryotic organisms with dinoflagellate flagellar characteristics (or similar) as sister groups to the dinokaryotes. The *Oxyrrhis* nucleus contains chromosomes that remain condensed in interphase but the chromosomes are not fibrillar. The mitotic spindle in *Oxyrrhis* is internal, unlike dinomitosis.

Syndinians have two distinct life phases: a parasitic, multinucleate stage that lives intra- or extracellularly in crustaceans, tintinnids or other dinoflagellates, and a flagellated motile stage superficially similar to typical dinoflagellates. The nucleus lacks condensed chromosomes in interphase. Alkali-staining suggests that histones could be present. Small-subunit RNA genes place syndinians as a sister group to the dinokaryotic dinoflagellates.

Environmental molecular sequencing revealed two "new" clades within the alveolates (López-García et al. 2001; Moon van-der-Staay et al. 2001, Massana et al. 2002). One of them, the group II alveolates, has since been shown to represent the syndinians (Saldarriaga et al. 2004; Skovgaard et al. 2005), while the identity of the other, the so-called group I alveolates, is still unclear. In SSU-based phylogenetic trees, they tend to branch at the very base of the dinoflagellates.

Biogeographic distribution

The spatial distribution patterns of dinoflagellates are similar to other aquatic protists of similar size (Zeitzschel 1990; Dolan 2005). Dinoflagellates are found in aquatic environments, 90% marine and 10% in freshwater. The general distribution of living dinoflagellates in time and space was last reviewed by Taylor (1987d, marine) and Pollingher (1987, freshwater). The biogeography of harmful dinoflagellate species was included in a general review on harmful algal biogeography by Lundholm and Moestrup (2006; see also comments by Taylor 2004). Dinoflagellate paleobiogeography was extensively reviewed by Goodman (1987) and its applications and limitations recently discussed by Pross et al. (2004). It is obvious that differing species concepts have a strong bearing on the consideration of biogeography. These have been discussed by Taylor (1992) and Lundholm and Moestrup (2006). The distributions described below deal primarily with morphospecies.

Dinoflagellate distribution has been termed "modified latitudinal cosmopolitanism" (Taylor 1987d, 2004), meaning that the same morphospecies occur within similar climatic zones in both northern and southern hemispheres. It is little appreciated that the entire microplankton community of northern and southern temperate ocean waters are virtually identical (Taylor 2001, 2004) despite being separated by the circumtropical community, at least since the Miocene (20Ma). Superimposed on this geographic pattern are clear differences between neritic (coastal) versus oceanic plankton. Many neritic dinoflagellate and diatom species include a benthic resting stage (resting cysts or resting spores respectively) in their life cycle and are therefore confined to the shallower waters of the continental shelf. Neriticism in some species may also be related to a requirement for land-derived nutrients or other products, such as humic acids.

It seems that the harsher the environment, the more cosmopolitan the protist species (Taylor 2004). The primary basis for the major biogeographic zones (Fig. 4) seems to be temperature, current systems extending the temperature boundaries to the north, e.g., the Gulf Stream, or south, e.g., the Agulhas Current. A broad tropical belt in the Atlantic and Indo-Pacific Oceans contains the same species throughout, such as those recorded and

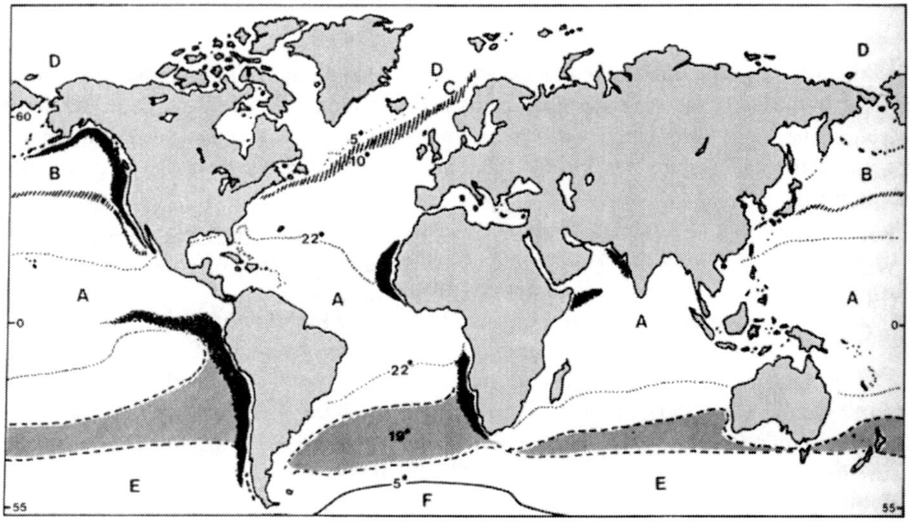

Fig. 4 Major dinoflagellate biogeographic zones. A = tropical-temperate macrozone, B = subarctic North Pacific, C = subarctic North Atlantic, D = arctic zone, E = subantarctic zone, F = Antarctic zone, black areas = substantial seasonal upwelling, hatched areas = regions of mixed character. From Taylor (1987d)

illustrated for the Indian Ocean by Taylor (1976). Taylor identified 286 species in formalin-preserved net samples in which most athecate species, other than noctilucoids and *Pyrocystis*, were not preserved. Consequently the total number of species in such waters is probably at least double this. Genera such as morphologically complex *Ornithocercus* (Fig. 2d) and *Histioneis* (Fig. 2e) are confined to the tropics, the latter showing a preference for deeper water. Some mangrove-lined bays, such as Bahia Fosforescente in Puerto Rico, host persistent luminescent blooms of *Pyrodinium bahamense*. Members of the luminescent genus *Pyrocystis* (Fig. 2l) are common in open waters. In tropical oceanic regions dinoflagellates are the dominant protist group and show their greatest diversity, particularly long-horned members of *Ceratium* (Fig. 2m). The Indo-Pacific region is richer in dinoflagellate species than the tropical Atlantic. For example, *Dinophysis miles* is unknown from the Atlantic Ocean and the var. *schroeteri* (Fig. 2c) is more restricted, occurring only in S.E. Asian waters. The var. *indica* is more widely distributed in the Indo-Pacific and its rare appearance in the eastern Mediterranean is perhaps an indication of transit through the Suez Canal.

True endemicism (species exclusively present in only one region) is rare. *Ceratium dens* appears to be endemic to northern Indian Ocean and S.E. Asian waters (Taylor 1987d). Reports of this species off southern California appear to be due to confusion with *C. divaricatum*. Judging by species lists alone one might think that there are many endemics, but single records have to be excluded from consideration for this purpose because of the possibility of error or inadequate description or misidentification. For example, in a recent study of Mediterranean dinoflagellate biogeography Gómez (2006) found that, although 44 species were recorded only from the Mediterranean and thus were potential endemics, when species recorded only once by their describer were excluded, there were none exclusive to the Mediterranean.

Endemicism (northern or southern only) is more common in polar waters, although some common temperate species also extend into polar waters. Heterotrophic polar dinoflagellates feed on the summer diatom bloom. These are mainly members of *Protoperidinium* and *Gymnodinium* (see McMinn and Scott 2005 for Antarctic species). Some photosynthetic dinoflagellates can form blooms in polar waters. For example, the planktonic *Peridiniella catenata* can form blooms in sea ice in Siberian Arctic waters (Okolodkov 1999) and also in the Mackenzie River delta (Taylor, unpubl.). There are some well established bipolar species, such as *Polarella glacialis* (Montresor et al. 2003).

Some species are much more common in one ocean than another. For example, *Oxyphysis oxytoxoides* (Fig. 2g) is a distinctive, common North Pacific species, but it is found very rarely in the northern North Atlantic.

The distribution of those species responsible for "harmful algal blooms" (HABs such as fish kills, human poisoning), have received the most detailed study. The relatively recently described fish-killer *Pfiesteria piscicida* (Steidinger et al. 1996) is known only from brackish waters of the east coast of the United States from Delaware to North Carolina to Texas but is probably much more widespread.

Species of *Alexandrium* are found in coastal waters around the world where many are responsible for paralytic shellfish poisoning. Some, such as the closely related toxic species *A. catenella* and *A. tamarense*, are very widespread in temperate waters of both northern and southern hemispheres, the former found particularly in frontal zones and the latter in more estuarine waters. In the tropics *A. tamiyavanichi* is a source of poison. The two former species are members of the so-called "*tamarense* species complex", along with *A. acatenella, A. excavatum* and *A. fundyense*. Although these are recognisable morphospecies in the wild (the latter known only from the Gulf of Maine and further north), in culture under the same conditions they can be difficult if not impossible to distinguish. An early attempt to use biochemical characters, isozyme electrophoresis, was made by Cembella and Taylor (1985, 1986). This found multiple isozymes within, and considerable variation between strains, which did not correlate with the morphospecies. Similarly, toxin spectra or lack of toxin production do not correlate with morphology (Cembella et al. 1987) or biogeography (Taylor 1984). A similar cluster of species form the "*minutum* species complex" (*A. minutum, A. lusitancum, A. angustitabulatum*).

More recently small and large subunit ribosomal DNA gene sequences have shown a similar lack of correlation with some of these named species. However, they do usually show a strong biogeographic pattern: isolates from similar geographic regions are generally more similar to each other than those from more remote locations, e.g., Scholin (1998) for the *tamarense* complex. John et al. (2003) used rDNA signatures and a molecular clock, together with the fossil record, to infer the evolution of the present distribution pattern in *Alexandrium*.

Regionally anomalous signatures have been used to support the "global spreading hypothesis" (Wyatt 1995). This asserts that the increased frequency and more widespread records of HABs in recent years is a reflection of a real increase in occurrence related to eutrophication and artificial spread. Ballast water was proposed as the primary mechanism for spread (Hallegraeff and Bolch 1991, 1992). Dinoflagellate cysts were found in 50% of the sediments in the ballast tanks of ships in Australian ports and a third of them could be excysted, including toxic *Alexandrium* species. On the basis of DNA and toxin spectrum comparisons Lilly et al. (2002) suggested that the presence of *A. catenella* on the coast of France was due to an introduction from the west Pacific. Similar suggestions have been made for introductions into other areas, particularly from the northern to the southern hemisphere. There are several difficulties with these claims, not the least of which is that presence in both hemispheres conforms to the expected norm, not the exception. Furthermore, these types of data are rare or completely absent for many coastal countries. The ribotype differences have also raised the question of "cryptic species" being present within single morphospecies. This question can be better addressed when "species" itself is clearly defined in protists (Taylor 1992).

Sand (psammophilic) dinoflagellates have received relatively less study than planktonic species. Many appear to be very widely distributed, the same species occurring on even remote, disjunct islands. *Prorocentrum lima* is widespread and common in both tropical and temperate benthic environments. However, tropical benthic communities contain species of *Gambierdiscus* and *Ostreopsis* associated with ciguatera fish poisoning, that do not extend into non-tropical waters. The former is strongly associated with seaweeds, living attached to their surfaces. Taylor (1987d) termed such an association phycophilic.

Fresh water species also appear to be widespread in geographic occurrence (Pollingher 1987). For example, *Peridinium gatunense*, described from a tropical lake in Panama, is common in some British Columbia lakes (Taylor, unpubl) and is the major bloom-former in Lake Kinneret in Israel (at first erroneously under the name *P. cinctum* var. *westii*). In contrast to temperate coastal and freshwaters, the blooms in Lake Kinneret occur in the late winter to spring. Although described from Africa, *Peridinium africanum* forms dense blooms in Lake Biwa in Japan. Differences in lake chemistry, such as acidic or alkaline, or nutrient status, can strongly influence which species are present. In general dinoflagellates are alkalinophile (Pollingher 1987). There appears to be dinoflagellate endemicism in some large lakes, such as Lake Baikal. Interestingly, this lake also has endemic ciliate species (Foissner 1999).

Dinoflagellates are highly diverse, not only in species richness but in morphological and biochemical ways as well. There is still much to be learned about this fascinating and ecologically important group.

References

Baldauf SL, Roger AJ, Wenk-Siefert I et al (2000) A kingdom-level phylogeny of eukaryotes based on combined protein data. Science 290:972–977

Bourrelly P (1970) Les algues d'eau douce. Initiation à la systématique. Tome III: Les algues bleues et rouges. Les Eugléniens, Peridiniens et Cryptomonadines. Éditions N. Boubée and Cie, Paris

Bujak JP, Williams GL (1981) The evolution of dinoflagellates. Can J Bot 59:2077–2087

Cachon J, Cachon M (1987) Parasitic dinoflagellates. In: Taylor FJR (ed) The biology of dinoflagellates. Bot Monogr 21:571–610

Cavalier-Smith T (1999) Principles of protein and lipid targeting in secondary symbiogenesis: euglenoid, dinoflagellate and sporozoan plastid origins and the eukaryote family tree. J Eukaryot Microbiol 46:347–366

Cembella AD, Taylor FJR (1985) Biochemical variability within the *Protogonyaulax tamarensis/catenella* species complex. In: Anderson DM, White AW, Baden DG (eds) Toxic dinoflagellates. Elsevier, Holland, pp 55–60

Cembella AD, Taylor FJR (1986) Electrophoretic variability within the *Protogonyaulax tamarensis/catenella* species complex: pyridine linked dehydrogenases. Biochem System Ecol 14:311–323

Cembella AD, Sullivan JJ, Boyer GG, Taylor FJR, Andersen RJ (1987) Variation in paralytic shellfish toxin within the *Protogonyaulax tamarensis/catenella* species complex; red tide dinoflagellates. Biochem System Ecol 15:171–186

Dolan J (2005) An introduction to the biogeography of aquatic microbes. Aquat Microb Ecol 41:30–48

Fensome RA, Taylor FJR, Norris G, Sarjeant WAS, Wharton DI, Williams DL (1993) A classification of living and fossil dinoflagellates. Am Mus Nat Hist, Micropal Spec Publ 7

Fensome RA, MacRae RA, Moldowan JM, Taylor FJR, Williams GL (1996) The early Mesozoic radiation of dinoflagellates. Paleobiol 22:329–338

Fensome RA, Saldarriaga JF, Taylor FJR (1999) Dinoflagellate phylogeny revisited: reconciling morphological and molecular-based phylogenies. Grana 38:66–80

Foissner W (1999) Protist diversity: estimates of the near-imponderable. Protist 150:363–368

Gaines G, Elbrächter M (1987) Heterotrophic nutrition. In: Taylor FJR (ed) The biology of dinoflagellates. Bot Monogr 21:224–268

Gómez F (2005) A list of dinoflagellates in the world oceans. Acta Bot Croatica 84:129–212

Gómez F (2006) Endemic and Indo-Pacific plankton in the Mediterranean Sea: a study based on dinoflagellate records. J Biogr 33:261–270

Goodman DK (1987) Dinoflagellate cysts in ancient and modern sediments. In: Taylor FJR (ed) The biology of dinoflagellates. Bot Monogr 21:649–722

Greuet C (1987) Complex organelles. In: Taylor FJR (ed) The biology of dinoflagellates. Bot Monogr 21:119–142

Hallegraeff GM, Bolch CJ (1991) Transport of diatom and dinoflagellate resting cysts via ship's ballast water. Mar Pollut Bull 22:27–30

Hallegraeff GM, Bolch CJ (1992) Transport of diatom and dinoflagellate resting spores in ships ballast water: implications for plankton biogeography and aquaculture. J Plankton Res 14:1067–1084

Head M (1996) Modern dinoflagellate cysts and their biological affinities. In: Jansonius J, McGregor DC (eds) Palynology: principles and applications. Dallas, USA, pp 1197–1248

Hoppenrath M, Leander BS (2007a) Character evolution in polykrikoid dinoflagellates. J Phycol 43:366–377

Hoppenrath M, Leander BS (2007b) Morphology and phylogeny of the pseudocolonial dinoflagellates *Polykrikos lebourae* and *Polykrikos herdmanae* n. sp. Protist 158:209–227

John U, Fensome RA, Medlin LK (2003) The application of a molecular clock based on molecular sequences and the fossil record to explain biogeographic distribution within the *Alexandrium tamarense* "species complex" (Dinophyceae). Mol Biol Evol 20:1015–1027

Lilly E, Kulis D, Gentien P et al (2002) Paralytic shellfish poisoning in France linked to a human-introduced strain of *Alexandrium catenella* from the western Pacific: evidence from DNA and toxin analysis. J Plankt Res 24:443–452

López-García P, Rodríguez-Valera F, Pedrós-Alió C et al (2001) Unexpected diversity of small eukaryotes in deep-sea Antarctic phytoplankton. Nature 409:603–607

Lundholm N, Moestrop Ø (2006) The biogeography of harmful algae. In: Granelli E, Turner JT (eds) Ecology of harmful algae. Ecol Stud 189:23–35

Massana R, Guillou L, Díez B et al (2002) Unveiling the organisms behind novel eukaryotic ribosomal DNA sequences from the Ocean. Appl Environ Microbiol 68:4554–4558

McMinn A, Scot FJ (2005) 3. Dinoflagellates. In: Scott FJ, Marchant HW (eds) Antarctic marine protists. Austral Biol Res Study, Canberra, pp 202–250

Moldowan JM, Terazina NM (1998) Biogeochemical evidence for dinoflagellate ancestors in the Early Cambrian. Science 281:1168–1170

Montresor M, Lovejoy C, Orsini L et al (2003) Bipolar distribution of the cyst-forming dinoflagellate *Polarella glacialis*. Polar Biol 26:186–194

Moon van-der-Staay SY, De Wachter R, Vaulot D (2001) Oceanic 18S rDNA sequences from picoplankton reveal unsuspected eukaryotic diversity. Nature 409:607–610

Okolodkov YB (1999) An ice-bound planktonic dinoflagellate *Peridiniella catenata* (Levander) Balech: morphology, ecology and distribution. Bot Mar 42:333–341

Pollingher U (1987) Freshwater ecosystems. In: Taylor FJR (ed) The biology of dinoflagellates. Bot Monogr 21:502–529

Pross J, Kotthof U, Zonnefeld K (2004) Organic-walled dinoflagellate cysts as paleoenvironmental indicators in the Meso- and Cenozoic: potential and limits. Palaentol Zeitschr 78:5–39

Rossignol M (1962) Analyse pollenique de sediments Quaternaires en Israël. II. Sédiments Pleistocénes. Pollen et Spores 4:121–148

Saldarriaga JF, Taylor FJR, Keeling P et al (2001) Dinoflagellate nuclear SSU rRNA phylogeny suggests multiple plastid losses and replacements. J Mol Evol 53:204–213

Saldarriaga JF, Taylor FJR, Cavalier-Smith T et al (2004) Molecular data and the evolutionary history of dinoflagellates. Eur J Protistol 40:85–111

Sarjeant WAS (1974) Fossil and living dinoflagellates. Academic Press, London

Schnepf E, Elbrächter M (1992) Nutritional strategies in dinoflagellates. A review with emphasis on cell biological aspects. Eur J Protistol 28:3–24

Schnepf E, Elbrächter M (1999) Dinophyte chloroplasts and phylogeny —a review. Grana 38:81–97

Scholin CA (1998) Morphological, genetic, and biogeographic relationships of the toxic dinoflagellates *Alexandrium tamarense*, *A. catenella*, and *A. fundyense*. In: Anderson DM, Cembella AD, Hallegraeff GM (eds) Physiological ecology of harmful algal blooms. NATO ASI Ser G: Ecol Sci 41:13–27

Skovgaard A, Massana R, Balagué V, Saiz E (2005) Phylogenetic position of the copepod-infesting parasite *Syndinium turbo* (Dinoflagellata, Syndinea). Protist 156:413–423

Spector D (1984) Dinoflagellates. Academic Press, Orlando

Steidinger KA, Burkholder JM, Glasgow HB et al (1996) *Pfiesteria piscicida* gen. et sp. nov. (Pfiesteriaceae), a new toxic dinoflagellate with a complex life cycle and behavior. J Phycol 32:157–164

Taylor FJR (1976) Dinoflagellates from the International Indian Ocean Expedition. Biblthca Bot 132: 1–222 + 46 plates

Taylor FJR (1984) Toxic dinoflagellates: taxonomic and biogeographic aspects with emphasis on *Protogonyaulax*. In: Ragelis E (ed) Seafood toxins. Am Chem Soc Symp Ser 262:77–97

Taylor FJR (1985) The taxonomy and relationships of red tide flagellates. In: Anderson DM, White AW, Baden DG (eds) Toxic dinoflagellates. Elsevier, Holland, pp 11–26

Taylor FJR (ed) (1987a) The Biology of Dinoflagellates. Bot Monogr 21:i–xii, 1–785

Taylor FJR (1987b) General group characteristics, special features of interest; and a short history of dinoflagellate study. In: Taylor FJR (ed) The biology of dinoflagellates. Bot Monogr 21:1–23

Taylor FJR (1987c) Dinoflagellate morphology. In: Taylor FJR (ed) The biology of dinoflagellates. Bot Monogr 21:24–91

Taylor FJR (1987d) Dinoflagellate ecology: general and marine ecosystems. In: Taylor FJR (ed) The biology of dinoflagellates. Bot Monogr 21:398–502

Taylor FJR (1992) The species problem and its impact on harmful phytoplankton studies, with emphasis on dinoflagellate morphology. In: Smayda TJ, Shimizu Y (eds) Toxic phytoplankton blooms in the sea. Elsevier, New York, pp 81–86

Taylor FJR (2001) Conference overview: harmful algal bloom studies enter the new millenium. In: Hallegraeff GM, Blackburn SI, Bolch CJ et al (eds) Harmful algal blooms 2000. Proc 9th Int Conf Harmful Algal Blooms, IOC, UNESCO, Paris, pp 3–5

Taylor FJR (2004) Harmful dinoflagellate species in space and time and the value of morphospecies. In: Steidinger KA, Landsberg JH, Tomas CR et al (eds) Harmful algae 2002. Fla Fish Wildl. Cons Comm, Fla Inst Ocean, IOC UNESCO, St. Petersburg, Fla, pp 555–559

Trench RK (1987) Dinoflagellates in non-parasitic symbioses. In: Taylor FJR (ed) The biology of dinoflagellates. Bot Monogr 21:530–570

Wyatt T (1995) Global spreading, time series, models and monitoring. In: Lassus P, Arzul G, Erard-LeDenn F et al (eds) Harmful marine algal blooms. Lavoisier, Paris, pp 755–764

Zeitzschel B (1990) Zoogeography of marine protozoa: an overview emphasizing distribution of planktonic forms. In: Capriulo GM (ed) Ecology of marine protozoa. Oxford University Press, Oxford, pp 139–185

Dispersal and biogeography of silica-scaled chrysophytes

Jørgen Kristiansen

Originally published in the journal Biodiversity and Conservation, Volume 17, No 2, 419–426.
DOI: 10.1007/s10531-007-9259-2

Abstract The silica-scaled chrysophytes—here mainly represented by the freshwater genera *Mallomonas* and *Synura*—have special problems in dispersal from one habitat to another because they cannot tolerate desiccation. Their dispersal is limited by the fragile construction and aquatic habit. Dispersal from one water body to another involves dangerous changes of the environment, and the ability to avoid desiccation during transport is crucial. So, air-borne and ectozoic dispersal by birds or mammals can only work at short distances. This danger may be avoided by endozoic dispersal of thick-walled cysts; as far as they can tolerate the digestion fluids in the intestine. In spite of these difficulties, Chrysophytes have been dispersed worldwide, but they display various distinct distribution patterns, e.g., cosmopolitan, arctic-northern temperate, bipolar, and tropical. Quite a large proportion may be considered endemic, occurring only within a restricted area. Even if the exact dispersal methods are elusive, the distribution of chrysophytes around the world proves their ability for dispersal. On the other hand, the different degree of distribution shows the varying success of the individual species. The distribution of a species at a given time depends on several factors: dispersal capacity—available vectors—suitable available habitats—and most important: sufficient time for dispersal. It is remarkable that the chrysophytes—in spite of their fragile cell construction and apparently low dispersal capacity—show distribution types comparable to those found in, e.g., blue–greens and desmids, whose cell construction appears much better adapted for dispersal.

Keywords *Mallomonas* · *Synura* · Environmental factors · Dispersal vectors · Long distance transport · Distribution types · Endemic species

---

Special Issue: Protist diversity and geographic distribution. Guest editor: W. Foissner

---

J. Kristiansen (✉)
Biological Institute, Section of Phycology, University of Copenhagen, Øster Farimagsgade 2 D, Copenhagen 1353, Denmark
e-mail: joergenk@bi.ku.dk

Introduction

"Chrysophytes" is a joint name for the two classes Chrysophyceae and Synurophyceae, within the algal phylum Chromophyta. A general account of the chrysophytes has been published by Kristiansen (2005), and an identification survey of the Synurophyceae has recently appeared as a volume of the "Süsswasserflora Mitteleuropas" (Kristiansen and Preisig 2007).

The chrysophytes are characterized by three main sets of characters: (1) heterokont flagellation: one flagellum with tripartite tubular hairs, the other smooth and often reduced; (2) pigments are chlorophylls a and c and the yellow fucoxanthin; and (3) resting stages are stomatocysts, i.e., endogenously produced survival stages surrounded by a thick, silicified wall.

The differences between the two classes are in the pigment composition (chlorophylls c_1 and c_2, versus c_2 only) and in the construction and biogenesis of the silica scales (Kristiansen 1996a). In the Chrysophyceae, silica scales—when present—are homopolar and formed in Golgi vesicles contacting the ER. In the Synurophyceae, the scales are heteropolar, bilaterally symmetric, and formed in vesicles contacting the chloroplast ER.

This review is based on the silica scaled forms because the scales of these organisms—when viewed in the electron microscope—yield distinct and stable characters which make exact identifications possible. Accordingly, world-wide investigations have given reliable material for distribution studies (Fig. 1).

In the Chrysophyceae, scaled forms are only present in the family Paraphysomonadaceae (mainly the genera *Chrysosphaerella* and *Paraphysomonas*), whereas all Synurophyceae bear scales; the most important genera are *Mallomonas* and *Synura* (Figs. 2–4).

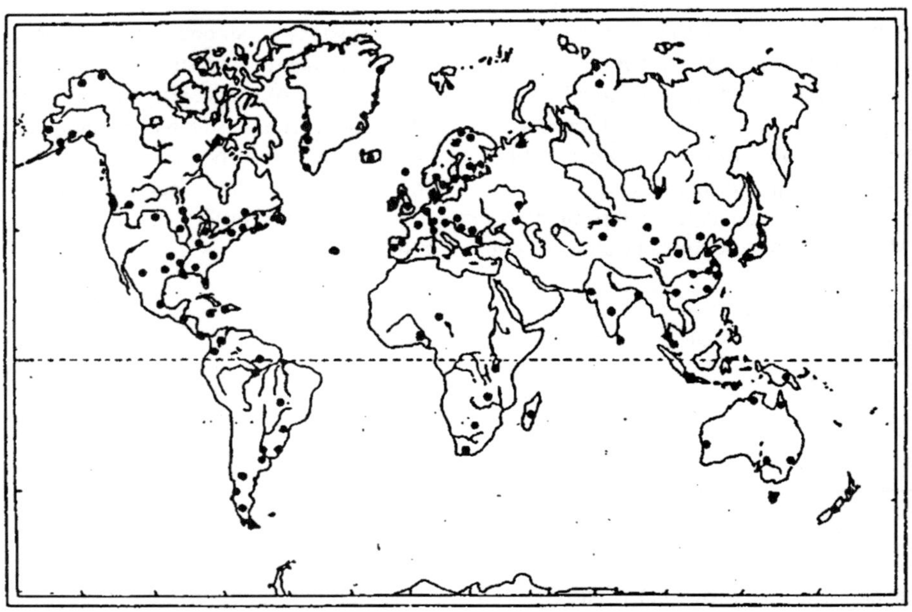

Fig. 1 World map of investigation sites for silica-scaled chrysophytes

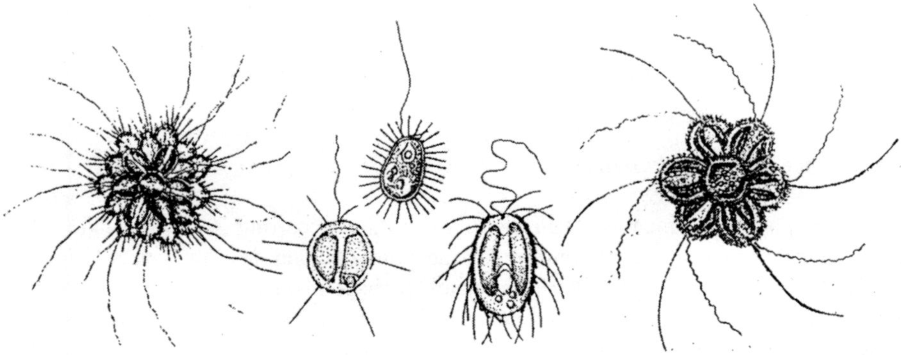

Fig. 2 Genera of silica-scaled chrysophytes: *Chrysosphaerella*, *Spiniferomonas*, *Paraphysomonas*, *Mallomonas*, and *Synura*

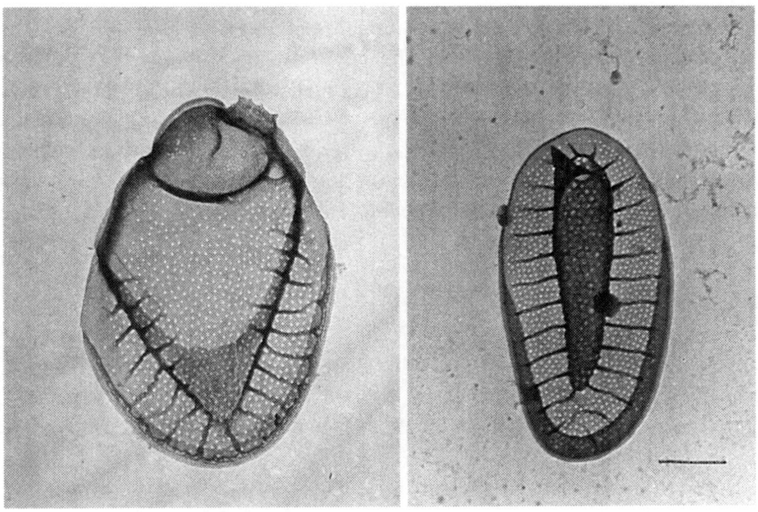

Figs. 3–4 Examples of silica scales of common species of Synurophyceae: *Mallomonas acaroides* and *Synura petersenii*. Bar 1 μm

Life histories

The silica-scaled chrysophytes are unicellular flagellates, solitary or in colonies, in the phytoplankton. Vegetative reproduction takes place by means of longitudinal cell division. Under certain circumstances, immotile cells remain together forming gelatinous palmella stages (Wee et al. 2005). During or at the end of the vegetative period, formation of stomatocysts takes place (Sandgren 1988). These more or less globular resting stages are formed endogenously and are surrounded by a silicified wall with an apical porus closed by a pectic stopper. The cysts sink to the bottom and germinate there when conditions become favourable, often next spring. The protoplast then escapes through the porus. The thick wall makes them resistant, and germination can be postponed for several years.

Sexuality is known in relatively few cases, both in species of *Mallomonas* and *Synura* (Kristiansen 1961; Wawrik 1972; Sandgren and Flanagin 1986). Specimens of the appearance of vegetative cells fuse and form a zygote, which then encysts for later germination.

Occurrence in different environments

The chrysophytes are aquatic organisms, most of them occurring in phytoplankton. Most are freshwater organisms, some few are marine mainly belonging to the genus *Paraphysomonas*. Chrysophytes occur under all climates, perhaps mainly in the temperate regions, but also in arctic and tropical environments. Some are found the whole year, others mainly in spring and summer, depending on their temperature preferences, e.g., species of the genus *Synura* (Kristiansen 1975). In temperate regions, the main occurrence is in the cold spring waters, often just after the disappearance of the ice. In tropical regions, other factors may be important, such as the shift between dry and rainy seasons (Kristiansen and Menezes 1998).

Environmental factors determine the regional occurrence, and many species have very distinct occurrence spectra (Siver 1991). This is most evident as regards their occurrence in relation to pH, for instance, in species of *Synura* (Kristiansen 1975). Some are distinctly acidobiontic, for instance, *Synura sphagnicola.* A great proportion of the species is circumneutral to alkaliphilic, e.g., *Synura spinosa*; some are almost alkalibiontic, such as *Synura uvella. Synura petersenii* is almost indifferent and has a very wide occurrence range. A similar pattern is seen in the genus *Mallomonas* (Smol et al. 1984).

Dispersal

A general account on dispersal of freshwater algae has been given by Kristiansen (1996b). However, chrysophytes most often meet difficult problems for their dispersal from one place to another, not only due to their aquatic habit but also because of the fragile construction of the vegetative stages.

Only dispersal within the same water body offers no particular problems—whether by water currents or by other organisms. This is also the case when chrysophytes are transported in water from one water body to another. Marine organisms have the advantage that their habitats more or less constitute a continuum.

Dispersal out of the water from one water body to another normally involves changes of the environment—from water to air and back to water again, with danger of desiccation. If the transport, e.g., on feet or feathers of water birds, is very short, desiccation can be avoided. If the transport takes place in the intestine of an animal, there is no danger of desiccation, but cells are exposed to digestion juices.

This means that the ability to tolerate dispersal is just as important as available vectors and as the possibility of being deposited in a suitable environment. Together, these factors constitute the dispersal capacity of the species (Kristiansen 2001b). For many algae, especially blue–greens, desmids and chlorococcalean greens, there is ample evidence for dispersal—based on cultures from dust and fur, feathers and feet, as well as from droppings. For chrysophytes, however, evidence is scant (Schlichting 1960).

Dispersal can be effected by abiotic and biotic vectors. Abiotic vectors include those by water and wind. Water currents and other water movements can carry propagules from one place to another in a water system. Wind can carry droplets of water with included organisms from one place to another. Dry sediments can be carried for long distances, but propagules must be able to tolerate desiccation.

Biotic factors include ectozoic dispersal mainly by mammals in wet fur, and by waterfowl on their feathers and feet (Schlichting 1960). Vegetative naked cells have difficulties in surviving, so long distances require tolerance to desiccation. Palmella stages with their gelatinous envelopes, such as shown in *Synura*, should be able to keep moist on feathers during transport and thus be able to survive (Wee et al. 2005). However, there is only one reported case of *Synura* cultivated from feathers of waterfowl (Schlichting 1960). Endozoic dispersal is effected by birds engulfing phytoplankton when feeding or drinking. Survival in the intestine is crucial, and best adapted are thick-walled cells such as cysts. However, it has not been possible to germinate chrysophyte cysts from bird droppings or gut contents. One problem is that many birds prefer to empty their intestine before long flights, rather when they have reached their destination.

Dispersal by man is an important factor, but only known from indirect proofs. It has been postulated that *Mallomonas vannigera* has been introduced from the Baltic area to the Great Lakes area in Canada by ship ballast water (Nicholls 2001). Rinsing and filling drinking water barrels in remote islands may be a factor, such as in Easter Island, where such events supposedly changed the stomatocyst flora in lake sediments, occurring at the times of visits of early European explorers (Zeeb and coworkers 1998). Not unimportant is contamination with plankton organisms from one locality to another during scientific investigations by poorly rinsed nets and bottles. Dispersal by man and birds visiting or homesteading is reflected in changes of the species contents in sediments (Munch 1980), and the occurrence of certain species along the migratory routes (Wee et al. 1993; Péterfi and Momeu 1996).

It is perhaps also significant that remote islands have very scant floras of silica scaled chrysophytes in the plankton, whereas there is a remarkably rich flora of stomatocysts in moss cushions in Antarctic islands (van de Vijver and Beyens 1997) and in lake sediments, such as found in the Azores islands (Hansen 2001).

This documents the difficulties in long distance dispersal, and it may show that the successful species are those spending most of their life in the encysted stage; such species have been positively selected. Only under special, satisfactory conditions they germinate to ephemeral flagellates which readily encyst again. No viable cysts have been found in dust spread by wind: obviously desiccation has been fatal. Thus transport conditions and time is crucial.

Distribution of chrysophytes

The distribution patterns are best known for the silica-scaled forms because the scales are robust and distinct taxonomic markers. A controversial and much disputed issue is that all microorganisms are ubiquitous—that is to say that they occur everywhere the environment is suitable (Finlay and Clark 1999). Similarly, Řezáčová and Neustupa (2007) have tried to evaluate statistically the concept of ubiquitous dispersal, based on studies on *Mallomonas* in ponds of the Czech Republic. Ubiquity is certainly the case in most of the minute species of the genus *Paraphysomonas*, of which some are marine (Lee and Patterson 1998). This is also true for some of the larger forms, such as species of *Synura* and *Mallomonas* where about 20 are cosmopolitan, and further 35–40 are widely distributed, found on almost all continents (Kristiansen 2001b).

Figs. 5–7 Examples of scales from endemic chrysophytes: *Mallomonas marsupialis* from Australia; *M. palaestrica* from Denmark–Netherlands; and *Synura punctulosa* from Finland–Russia

However, the ubiquity concept is not valid for the silica scaled chrysophytes in general. The majority of the species are not ubiquitous, but have distinct, restricted distribution patterns (Kristiansen 2001a). The different degrees of distribution show that the various species have had varying success in dispersal.

The northern temperate-subarctic-arctic group is very large, found in North America–Europe–North Asia. Several of these species have a circumpolar distribution. Other species found in this area have a bipolar distribution, also found in the southern temperate zone (southernmost South America). But no circumpolar southern temperate species have been found (Kristiansen and Vigna 1996). Chrysophytes have previously been thought to be restricted to temperate waters, but investigations during the last about 25 years have documented a fairly rich tropical flora of about 20 taxa, several of which are pantropical and some others endemic (Cronberg 1989).

About one third of the taxa are endemic, i.e., have only been found in a restricted area, for instance, 69 of the 172 *Mallomonas* species. In fact, almost all taxa originally started as endemics, but had sooner or later lost this status, either because of dispersal, but nowadays most often due to more intense research. A list of all endemic taxa, and of those previously considered endemic, has been given by Kristiansen and Lind (2005, compare also Tyler 1996). The endemic taxa (examples are shown by Figs. 5–7) have not been found in so special environments as to account for their endemicity. They are not confined to isolated habitats, such as distant oceanic islands, but most of them have been found in very well investigated areas, such as eastern North America (Nicholls 1989, 2001) and Denmark (Hansen et al. 1993). Their restricted occurrence must be due to poor dispersal capacity, or perhaps rather due to lack of time. Oceanic islands are poor in planctic chrysophytes, certainly because of difficulties in dispersal. On the other hand, as mentioned, rich stomatocyst floras have been found at such places, perhaps indicating that species spending most of their time in the encysted stage are better adapted for dispersal.

It is remarkable that the chrysophytes—in spite of their fragile cell construction and apparently low dispersal capacity—show distribution types comparable to those found in, e g., blue–greens and desmids, whose cell construction appears much better adapted for dispersal (Hoffman 1996; Coesel 1996).

References

Coesel PFM (1996) Biogeography of desmids. Hydrobiologia 336:41–53

Dumont HJ, Cocquit C, Fontugne M, Arnold M, Reyss J-L, Bloemendal J, Oldfield F, Steenbergen CLM, Korthals HJ, Zeeb BA (1998) The end of moai quarrying and its effect on Lake Rano Raraku, Easter Island. J Paleolimnol 20:409–422

Finlay B, Clark KJ (1999) Apparent global ubiquity of species in the protist genus *Paraphysomonas*. Protist 150:419–430

Hansen P (2001) Chrysophyte stomatocysts in the Azores—biogeographical implications and 110 new morphotypes. Opera Bot 138:1–96

Hansen P, Johansen JE, Skovgård A et al (1993) *Mallomonas palaestrica* sp. nov., a new member of sect. Torquatae from Denmark. Nord J Bot 13:107–110

Hoffman L (1996) Geographic distribution of freshwater blue-green algae. Hydrobiologia 336:33–40

Kristiansen J (1961) Sexual reproduction in *Mallomonas caudata*. Bot Tidsskr 57:306–309

Kristiansen J (1975) On the occurrence of the species of *Synura*. Verhandl Internat Verein Limnol 19:2709–2715

Kristiansen J (1996a) Silica structures in the taxonomy and identification of scaled chrysophytes. Nova Hedwigia, Beih 112:355–365

Kristiansen J (1996b) Dispersal of freshwater algae—a review. Hydrobiologia 336:121–126

Kristiansen J (2001a) Biogeography of silica-scaled chrysophytes. Proceedings of the fifth international chrysophyte symposium. Nova Hedwigia, Beih 122:23–39

Kristiansen J (2001b) Cosmopolitan chrysophytes. Syst Geogr Plants 70:291–300

Kristiansen J (2005) Golden algae—a biology of chrysophytes. Gantner Verlag, Koenigstein

Kristiansen J, Lind JF (2005) Endemicity in silica-scaled chrysophytes. Proceedings of the sixth international chrysophyte symposium. Nova Hedwigia, Beih 128:65–83

Kristiansen J, Menezes M (1998) Silica-scaled chrysophytes from an Amazonian flood-plain lake, Mussurá lake, northern Brazil. Algol Stud 90:97–118

Kristiansen J, Preisig HR (2007) Süsswasserflora von Mitteleuropa, vol 1, part 2, Synurophyceae. Spektrum Akad Verlag, Heidelberg

Kristiansen J, Vigna MS (1996) Bipolarity in the distribution of silica-scaled chrysophytes. Hydrobiologia 336:151–157

Lee W, Patterson DJ (1998) Diversity and geographic distribution of free-living heterotrophic flagellates—analysis by PRIMER. Protist 149:229–244

Munch S (1980) Fossil diatoms and scales of Chrysophyceae in the recent history of Hall Lake, Washington. Freshw Biol 10:61–66

Nicholls KH (1989) Description of four new *Mallomonas* taxa (Mallomonadaceae, Chrysophyceae). J Phycol 25:292–300

Nicholls KH (2001) New and little known *Mallomonas* (Synurophyceae) taxa from Ontario, Canada. Nord J Bot 21:551–560

Péterfi LS, Momeu L (1996) Three Rumanian *Mallomonas* species (Synurophycae) with special distribution patterns. Hydrobiologia 336:143–150

Řezáčová M, Neustupa J (2007) Distribution of the genus *Mallomonas* (Synurophyceae)—ubiquitous dispersal in microorganisms evaluated. Protist 158:29–37

Sandgren CD (1988) The ecology of chrysophyte flagellates: their growth and perennation strategies as freshwater phytoplankton. In: Sandgren CD (ed) Growth and reproduction strategies of freshwater phytoplankton. Cambridge University Press, Cambridge, pp 9–104

Sandgren CD, Flanagin J (1986) Heterothallic sexuality and density dependent encystment in the chrysophyte alga *Synura petersenii*. J Phycol 22:675–692

Schlichting HE (1960) The role of waterfowl in the dispersal of algae. Trans Am microsc Soc 79:160–166

Siver PA (1991) The biology of *Mallomonas*. Kluwer, Dordrecht

Smol JP, Charles DF, Whitehead DR (1984) Mallomonadacean (chrysophycean) assemblages and their relationships with limnological characteristics in 38 Adirondack (New York) lakes. Can J Bot 62:911–923

Tyler PA (1996) Endemism in freshwater algae. Hydrobiologia 336:127–135

Van de Vijver B, Beyens L (1997) The chrysophyte stomatocyst flora of the moss vegetation from Strømness Bay area, South Georgia. Arch Protistenk 148:505–520

Wawrik F (1972) Isogame Hologamie in der Gattung *Mallomonas*. Nova Hedwigia 23:353–362

Wee JL, Booth DJ, Bossier MN (1993) Synurophyceae from the southern atlantic coastal plain of North America: a preliminary survey in Louisiana, USA. Nord J Bot 13:95–106

Wee JL, James AM, Hood CS (2005) Exploratory investigations of palmella-stage formation in the Synurophyceae. Nova Hedwigia, Beih 128:257–266

Conservation of protists: is it needed at all?

F. P. D. Cotterill · Khaled Al-Rasheid ·
Wilhelm Foissner

Originally published in the journal Biodiversity and Conservation, Volume 17, No 2, 427–443.
DOI: 10.1007/s10531-007-9261-8 © Springer Science+Business Media B.V. 2007

Abstract Protists have scarcely been considered in traditional perspectives and strategies in environmental management and biodiversity conservation. This is a remarkable omission given that these tiny organisms are highly diverse, and have performed as key ecological players in evolutionary theatres for over a billion years of Earth history. Protists hold key roles in nearly all ecosystems, notably as participants in fluxes of energy and matter through foodwebs that centre on their predation on microbes. In spite of this, they have been largely ignored in conservation issues due to a widespread, naive belief that protists are ubiquitous and cosmopolitanously distributed. Nevertheless, recent research shows that many protists have markedly restricted distributions. These range from palaeoendemics (Gondwanan-Laurasian distribution) to local endemics. Our ignorance about the ultimate and proximate causes of such acute disparities in scale-dependent distributions of protists can be flagged as a singular reason to preserve these more cryptic participants in ecological and evolutionary dynamics. This argument is disturbing when one considers anthropogenic modifications of landscapes and the very poorly understood roles of protists in ecological processes in soils, not least in agroecolandscapes and hydrological systems. Major concerns include host specific symbiotic, symphoric and parasitic species which become

---

F. P. D. Cotterill: Formerly Principal Curator of Vertebrates, Natural History Museum of Zimbabwe, P.O. Box 240, Bulawayo, Zimbabwe

---

Special Issue: Protist diversity and geographic distribution. Guest editor: W. Foissner

---

F. P. D. Cotterill
AEON – African Earth Observatory Network, Departments of Geological Sciences, and Molecular and Cell Biology, University of Cape Town, Rondebosch 7701, South Africa

K. Al-Rasheid
Zoology Department, King Saud University, P.O. Box 2455, Riyadh 11451, Saudi Arabia

W. Foissner (✉)
FB Organismische Biologie, Universität Salzburg, Hellbrunnerstrasse 34, 5020 Salzburg, Austria
e-mail: Wilhelm.Foissner@sbg.ac.at

W. Foissner et al. (eds.), *Protist Diversity and Geographical Distribution*
DOI: 10.1007/978-90-481-2801-3_15

extinct, unseen and largely unknown, alongside their metazoan hosts; change or loss of habitats; massive change or loss of type localities; and losses of unique genetic resources and evolutionary potential. These concerns are illustrated by examples to argue that conservation of protists should be integral to any strategy that traditionally targets vascular plants and animals. The ongoing decline in research capacity to inventory and classify protist diversity exemplifies a most acute symptom of the failures, at local, national and international levels, to support scientific responses to the biodiversity crisis. Responsible responses to these severe problems need to centre on the revival of natural history as the core discipline in biology.

Keywords Biodiversity · Genetic resources · Habitat loss · Idiographic and nomothetic science · Protist protection and conservation · Type locality · Taxonomic inventories

Protist conservation and conservation values in the climate of deficient knowledge

Contrasting conservation values

The conservation of protists has not been emphasized in their own right. This paper seeks to remedy this deficiency, within the constraints of space and insufficient knowledge. Coherent arguments to conserve protists invoke aesthetic, epistemic and utilitarian themes. We describe four value systems, biased to one of these three themes. We conclude by acknowledging that each system–aesthetic, biophilic, ecosystem goods and services, evolutionary potential–focuses a node interlinking an encompassing argument in support of protist conservation. The profile of a conservation strategy for protists awaits very overdue reception by human society. Several actions await attention which needs to be considered by conservation biologists and environmental managers, without further neglect.

Questions that centre on conservation of selected biota and/or landforms invoke arguments whose compatibility is not always recognized. One such distinct school of thought is underpinned by aesthetic values we hold about the natural world, and what is now called biodiversity. It is epitomized by what Aldo Leopold called the "wilderness ethic" (Meine and Knight 1999). Such aesthetic values extend to practical actions and activities in which environments and biodiversity are managed toward aesthetically based goals (Meffe et al. 1997). Aesthetically based conservation values also equate closely with biophilia (Wilson 1984).

The second argument for conservation is considered more practicable, and is invariably seen to be at odds with any aesthetic-based school. This second school argues that conservation be moulded to maintain more utilitarian values of the environment that benefit humankind primarily (Western and Pearl 1989; Heywood and Watson 1995). It is epitomized by the concepts and arguments developed in ecological economics (Smith 1996) and centres on the values of 'ecosystem goods and services' (Westman 1977; Arrow et al. 1995; Baskin 1997; Costanza et al. 1997; Tilman et al. 1997). Its protagonists argue that demands by socio-economies on the ecological goods and services necessary to sustain humans and our activities (as interacting participants in the biosphere) overrule any argument to preserve a species, habitat or ecological landscape solely on any perceived aesthetic values. The ecosystem goods and services argument shares many similarities with the 'use it or lose it' arguments for sustainable utilization of biology, and especially biodiversity prospecting (Janzen 1993, 1998).

A third school of values motivating biodiversity conservation is also underpinned by an aesthetic argument. We term this the "biophilia" school. Its protagonists value particular groups of organisms, rather then biodiversity and environment per se. Although, this biophilia school is founded on aesthetic rationales, it extends to more academic interests in biodiversity and biology. At its more superficial, biophilia manifests in people's identification with 'flagship species' (Figs. 1–8). Likewise, its aficionados spend millions on what is popularly called the petshop trade. It also manifests in the private and professional lifetimes of scientists who devote considerable resources to research and know their favourite organisms intimately. Biophilia expresses in the profound satisfaction, and values, founded in the detailed knowledge gained by lifetimes of studies of organisms and the natural world. These manifest as focused studies of selected taxa, a deeply professed fascination with, for example, ants (Holldobler and Wilson 1994) and protists (Kreutz and Foissner 2006). Such infatuations with the natural history of organisms, which we identify with this biophilic school, most certainly embrace conservation values placed on protists (Kreutz and Foissner 2006). This might indeed have been barely developed, considering the multifarious attractiveness of protists. The aesthetics resplendent in the astounding morphological variety of these minute organisms (especially in all their superb details revealed by modern technology) matches those of larger, more charismatic organisms.

For consider the aesthetic enjoyment aroused by paging through superb colour or scanning electron micrographs of protists, resplendent in the diversity of their translucent adaptations to microcosmic environments (Figs. 1–8). We venture to suggest that only a truly sad human being would not share in such biophilic experiences?! Biophilia is indeed characterized by marked individualism frequently based in passionate affinities for particular organisms, epitomized by the orchid cognoscenti (Hansen 2000). Nevertheless, it is a mistake to overlook the social, political and economic profundity represented in biophilia. It motivates lifetimes of research commitments, and is equally expressed when its aficionados are roused to conservation action.

Even so, economic and anthropocentric practicalities in conservation biology can sideline aesthetic and biophilic arguments, and confine any argument for protist conservation firmly in the ecosystem goods and services school. This introduces questions pertaining to 'What do protists do in ecosystems?' Such questions extend further to flesh out an evolutionary dimension to the ecosystem goods and services school. This deeper perspective on conservation values characterises the fourth school—termed 'evolutionary potential'. This is structured by robust knowledge of the evolutionary history of biodiversity. In this context, questions and decisions in conservation biology employ knowledge which elucidates where and when species and clades evolved, and their relative abilities to persist in ecosystems altered by disturbances. The rationale of the evolutionary potential singles out clades and regions that exhibit phylogenetic uniqueness (Vane-Wright et al. 1991), and evolutionary vibrancy (Erwin 1991). A major argument for biodiversity conservation is founded on asymmetries of phylogenetic diversity (Faith 1994; Faith and Walker 1996; Faith and Trueman 2001). Equally important attention is demanded of what Erwin (1991) termed 'evolutionarily vibrancy'. Taxa that form the rapidly evolving tips of phylogenetic trees, invariably represent the species rich, bushy clades that formed through recent radiations. It is these evolutionarily vibrant clades that can be expected to maintain the evolutionary potential to ride out environmental perturbations. These are better disposed to maintaining the performance of ecosystems than older, more phylogenetically distinctive, but less speciose clades (Erwin 1991).

Clades and species that score highly against both these criteria constitute prime candidates for conservation. More practicably, conservation actions should maintain the ecological

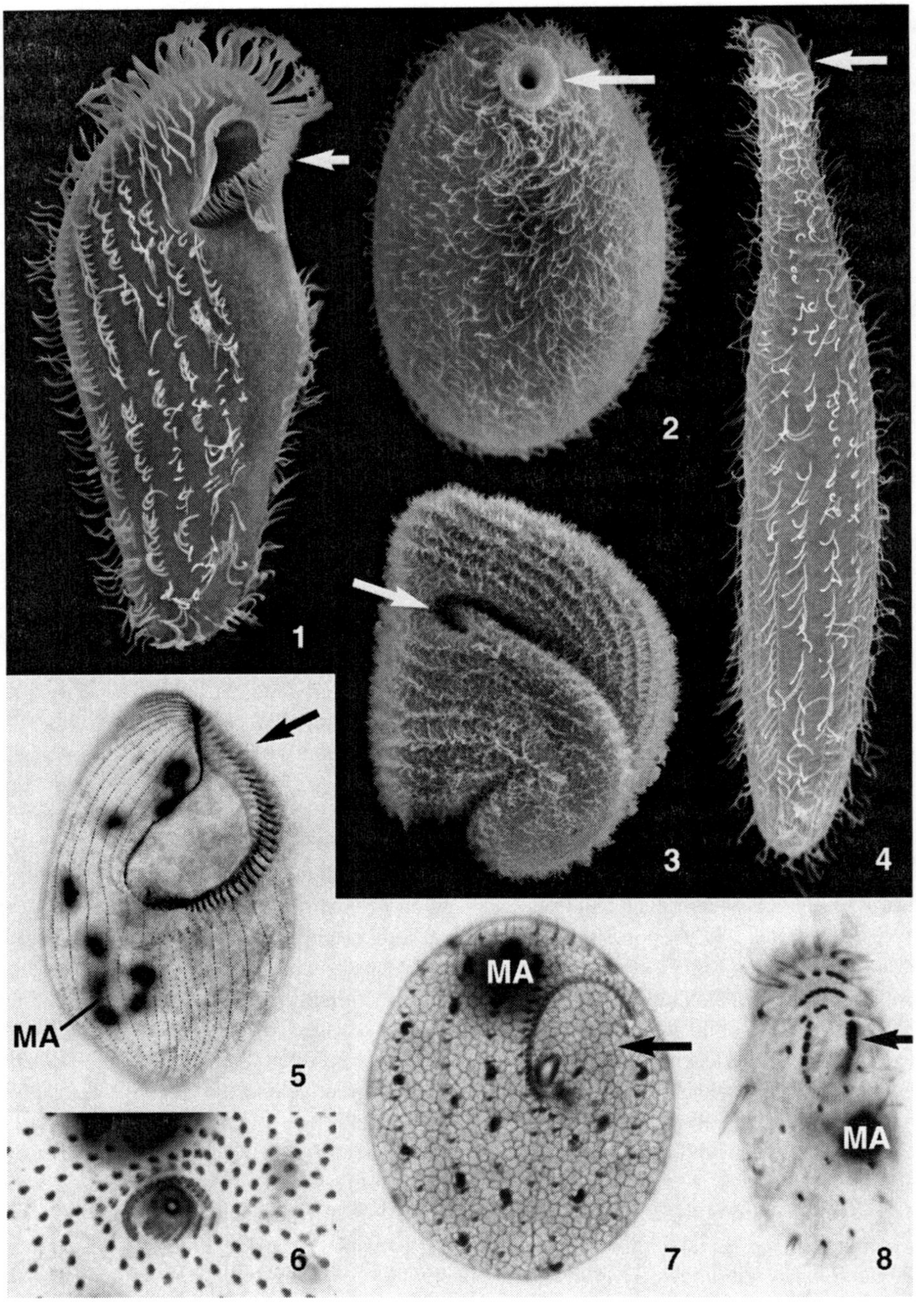
1
2
3
4
MA
5
6
MA
7
MA
8

◀ **Figs. 1–8** Examples of soil ciliate flagship species with, likely, restricted geographic distributions. Both, scanning electron microscopy (**1–4**) and silver impregnation (**5–8**) were used for the identification of the ciliates. These methods reveal finest features and are thus indispensable in modern ciliate taxonomy. Arrows mark mouth area. MA – macronucleus **1:** So far, *Saudithrix terricola*, an about 270 μm long, highly characteristic stichotrich ciliate, has been found only in field soil from Saudi Arabia and China (from Berger et al. 2006). **2:** *Enchelydium blattereri* was discovered in floodplain soil from Australia. This conspicuous species, which belongs to the haptorid gymnostomes, has a length of about 240 μm and a highly characteristic oral bulge (from Foissner et al. 2002). **3:** This is a not yet described colpodid flagship from a green river bed (an ephemeral part of the Chobe River which becomes a savannah during the dry season) in Botswana, Africa. It has a length of about 300 μm and is distinctly spiralized. **4:** A not yet described, about 200 μm long *Spathidium* (haptorid gymnostome) from soil of the Galapagos Islands. **5:** A not yet described, about 250 μm long heterotrich ciliate from soil of a mangrove forest in Venezuela. This species, which belongs to the genus *Condylostomides*, is a flagship because it is large and green due to countless cortical granules. **6:** *Fungiphrya strobli* is a functional flagship that belongs to the obligate mycophagous colpodids. So far, this species has been found only in soil from the Table Mountain in Cape Town, Republic of South Africa. The unique oral apparatus is recognizable in the centre of the micrograph. It comprises a semicircular undulating membrane on the upper margin of the oral area and seven short adoral ciliary rows on the lower margin of the oral area. Between undulating membrane and adoral ciliary rows, there is a black circle with a bright centre, which is the ~2 μm long feeding tube, used to penetrate fungal hyphae and to transport their contents into the ciliate (from Foissner 1999a). **7:** *Apocolpodidium (Phagoon) macrostoma* is only 50 μm long, but conspicuous due to the huge oral apparatus with a semicircular undulating membrane. As yet, this species has been found only in soil from the Everglades of Florida, USA (from Foissner et al. 2002). **8:** *Pseudokreyella etoschensis* was discovered in the Etosha Pan, Namibia. Although it is only 20 μm long, it is a morphological flagship due to the complex somatic and oral ciliary pattern (from Foissner et al. 2002)

integrity of the biomes and habitats in which these organisms have evolved. Any measure of evolutionary potential requires a robust phylogeny of any biodiversity indicators that serves to identify landscapes with high phylogenetic uniqueness (Brooks and McLennan 2002). Thus, phylogenies of protists are critical to apply scientific criteria to distinguish clades and habitats that possess reservoirs of evolutionary potential most likely to accommodate environmental disturbances.

So far we have outlined four arguments to justify conservation of protists. These invoke reasons and values grounded in four schools: the aesthetic, ecosystem goods and services, biophilia, and evolutionary potential. We argue that the answers to conservation problems reside largely in the details of the biodiversity we seek to manage, and we conclude that an ecological, utilitarian imperative, based in maintaining ecosystem services, does not constitute the singular reason to justify conservation of protists. In fact, our argument for protist conservation is built on profound matters of aesthetics and knowledge. If any such argument purports to be scientifically informed, then its conservation actions and plans cannot escape strictures imposed by all four value systems. Before discussing conservation values of protists any further, we need to acknowledge our state of knowledge of biodiversity and protists.

Biodiversity science, knowledge of protists and the taxonomic impediment

It cannot be overemphasized that many microbial assemblages have escaped discovery, let alone description (Margulis et al. 1986; Wall 1999). In all respects, our species' habitual overlooking of microorganismal biodiversity can be termed the "problem writ large" afflicting appreciation and cognisance of protists. It especially impacts on perceptions of their conservation importance. The plight of microorganisms, especially protists, is an apt example of taxonomic chauvinism and methodological challenges that bias the biodiversity sciences (Pawar 2003). These deficiencies magnify the importance of arguments that endorse investments in systematic biology founded on natural history (Cracraft 1996, 2002), to rectify the Taxonomic Impediment (Hoagland 1996; Foissner 1999b; Cresswell and Bridgewater 2000; Cotterill 2002; Wheeler 2004). Any such discussion encounters acute uncertainties, with answers to such debate grounded (and indeed ubiquitously weakened) by deficiencies in our knowledge. What we continue to discover about protists highlights the conclusions of Donoghue and Alverson (2000) that biology is challenged by a new age of discovery.

Conservation decision-making pivots on understanding how the biota will respond to stresses and disturbances of ecosystems. In this respect, our knowledge of any biota (a species or an assemblage) ranges from the poor to non existent, no matter our scale of enquiry. Our ignorance of biodiversity dynamics reflects on glaring gaps in our knowledge, which extend from genes to ecosystems. In this respect, two aspects especially weaken conservation decisions. One, we have a very incomplete knowledge of where species occur in the biosphere. Two, the idiographic intricacies (unique facts) pertaining to species' natural history are known only for very few species, mostly angiosperms and vertebrates, and some parasites of humans and our domesticates. So we can only guess at how even the most common organisms influence ecological processes. As summarized by Wall (1999) for the ecology and biota of soil, what we think we might know about roles of protists in the biosphere exemplifies acute inadequacies of this knowledge, in all its multifariousness. Our ignorance is highlighted by recent discoveries of the phylogenetic and functional diversity of marine microbes (Moreira and López-García 2002; Gross 2007), and notably of protists along hydrothermal vents (Moreira and López-García 2003). These extreme magnitudes of undiscovered biodiversity constitute an especially pertinent example, given this essay's focus on conservation of the underappreciated microscopic realms of life! They raise many questions, not least pertaining to roles of marine protists in these hyperdiverse foodwebs. We repeat that considered in the context of how little we know about protists, these microscopic examples emphasize how biology is challenged by a new age of discovery (Donoghue and Alverson 2000).

Similarly to bacterial biodiversity, too few researchers have commented on conservation issues couched in anything approaching a protist-centric perspective! This contrasts against the mushrooms (macrofungi) for which a vast local and global literature exists (Hawksworth 1991). Thus, with few exceptions (Wells et al. 1983; Foissner 1994; Mann and Droop 1996; Staley 1997; Rondon et al. 1999), conservation of protists awaits attention in Red Data books, let alone focused scientific discussion. We suggest that a significant reason for this neglect is the widespread belief that most, or even all, protists exhibit a cosmopolitan distribution. Despite very poor biogeographic data, this belief has fostered sweeping statements that no protist can possibly be endangered (Finlay et al. 1996). Yet, evaluation of actual data (reviewed in Foissner 2006, 2007, and this issue) reveals that up to a third of protists have restricted distributions. These range from palaeoendemics (Laurasian and Pangaea endemics) to species exhibiting a highly local endemism. These spatially-nested patterns of endemism constitute a singular reason, in their own right, to place protists as a central concern in conservation biology, debate and policy.

Recently, the biotechnological potential of protists has been recognized. The genetic and metabolic diversity of protists represents a rich source of valuable compounds, such as omega-3 fatty acids, pigments, polymers and enzymes (Kiy 1997; Beck 2002; Hausmann et al. 2003). When considered against the framework of the hugely under explored diversity of protists (see below), we conclude that biodiversity prospecting of these microorganisms has barely been inaugurated.

According to Corliss (2000), a grand total of at least 213,000 protist species have been described, of which about 113,000 are fossil forms. However, Corliss (2000) included in his estimation only few of the 130,000 described fungi (Hawksworth 2001). Corliss (2000) emphasized that his figures are likely gross underestimates: "On the basis of personal communication with many protistologists, I am obliged to draw the conclusion that, for numerous groups, vast numbers of unique protists do await description. Perhaps we have only scratched the surface regarding the biodiversity of these organisms".

We agree! Furthermore, preliminary surveys of genetic diversity of protists point to significant levels of cryptic species that have been overlooked (Von der Heyden and Cavalier-Smith 2005; Scheckenbach et al. 2006).While there are likely 1.5 million fungal species, with ~130,000 (~9%) described (Hawksworth 2001) and up to 40,000 free-living ciliate species, of which only 4,500 (~11%) have been described (Foissner et al. this issue). The global magnitude of eukaryotic diversity likely encompasses 20 million arthropods alone; this is a conservative estimate, combining global estimates of insects (Stork 1997) with arachnids (Andre et al. 1994; Behan-Pelletier and Newton 1999).

Now, what do these statistics imply for the global diversity of protists? In contrast to the widespread belief that free-living protists are uninteresting and of little ecological significance, research through the past 20 years has shown that most eukaryotic biodiversity consists of protists (Hausmann et al. 2003). Protists are of great importance in all ecosystems, for instance, as producers of oxygen and, in microbial foodwebs, they are critical to fluxes of energy and matter through ecosystems (Fenchel 1987; Foissner 1987; Sorokin 1999). Furthermore, beyond important model organisms, most protists of epidemiological concern remain poorly known (Palmer et al. 1998; Beck 2002; Hausmann et al. 2003). It follows that the global diversity of protists approximates tens of millions of species, but this inference of global protist diversity needs further qualification. We argue that any such assessment of global protist diversity is challenged to incorporate two potentially significant determinants. Both reflect on the ecological opportunities open to adaptive radiations of protists since the Proterozoic (Cavalier-Smith 2006). One relates to the geomorphological extent and complexity of soil habitats available to protists. The second involves coevolutionary opportunities, and pertains to all the multicellular hosts potentially available to single celled eukaryotes. The significance of both determinants is enhanced by the fractal structuring of ecological resources available to protists at microscopic and mesoscopic spatial scales.

So firstly, we point out (as argued by Richter and Markewitz 1995) that the depths and complexities of soils and their biodiversity are still being bottomed (Andre et al. 1994). This especially applies to where ever deep regolith (Clarke 2003) supports microbes to depths of tens and hundreds of metres below the surface (as in tropical landscapes). These considerations obviously cannot exclude groundwater, especially in granitic and karstic landscapes. We ask what are the implications of such spatial complexity, inherent compartmentalized, in these subterranean landscapes, for protist diversity? How does such geomorphological complexity—at these fine-grained fractal dimensions—influence protist habitats, especially given the magnitudes of microbial biodiversity already discovered at these depths?

Secondly, we ask the same questions of the fine-grained habitats represented in the guts and tissues of multicellular organisms, which are even more compartmentalized than those in soil and regolith. Consider, furthermore, that these microbial and protist habitats, distributed within and amongst the Eucarya, are principally allopatric; because the diversity of these lineages has been parsed into tens of millions of extant species, each of which exhibits a unique evolutionary trajectory. Clearly, the extent and complexity of these subterranean and commensal resources, available to protists, present fertile research fields centred in protistology. Any such study of these significant regions of the biosphere is challenged to elucidate the ecological and evolutionary dynamics entailed in fluxes of matter and turnovers of organisms at these fine-grained micro- and meso-scales. Complementary questions and research challenges pertain to how ecological resources are divided up in underground landscapes and amongst multicellular organisms, respectively. We anticipate that this complexity of geomorphological and commensal niches has exercised profound controls on the evolution of protist diversity. It cannot be overemphasized the research challenges pertaining to these frontiers of ignorance are critical to understand and manage the biosphere, and especially geomorphological, edaphic and hydrological processes.

So far we have sketched out four significant arguments why protists cannot continue to be excluded, but need to become focal concerns in conservation plans and actions. In summary, these encompass: the biotechnology argument with respect to biodiversity prospecting (the importance of properties of protists in basic and applied science); their ecological roles in foodwebs; the restricted distributions—and thus overlooked endemism—of many species; and above all, the pervasive inadequacy of knowledge, especially considered in the light of inferred magnitudes of global biodiversity, and the microscopic complexity of protist habitats across the biosphere. This framework of our argument sets the stage to move on, and consider additional, major concerns bearing on the conservation of protists.

Major issues in protist conservation

At least one protist species becomes extinct with each metazoan lost to extinction

Our argument for greatly underestimated protist diversity highlights the significance of rates of speciation and extinction among multicellular organisms. Most higher animals (Metazoa) are associated with at least one host specific symbiotic, symphoric or parasitic protist species; often they host several or even many. Accordingly, the number of endangered or extinct species is at least twice as high as generally assumed; mycologists assume a rate of 1:6, that is, on average, six fungal species are found on each species of plant (Hawksworth 1991). For instance, more than 50 specific ciliate species were discovered on and in water beetles in the surroundings of the town of Erlangen, Germany (Matthes and Guhl 1975). In Austria, as elsewhere, such investigations are rare, but 60 out of the 127 species of water beetles known from Austria are endangered (Wewalka 1984), together with their specific ciliates. This situation parallels that in crustaceans and fish, which host a rich diversity of protists, many of which are still undescribed species. Specialized habitats in structurally complex landscapes are not only confined to soil and underlying regolith; for recently new species of ciliates were described in microhabitats supported within bromeliads. These specialized protists represent previously overlooked clades of protists endemic to Neotropical forest (Foissner 2002; Foissner et al. 2003). Wholesale losses of such specialized plant habitats would simultaneously extinguish the less obvious biodiversity they support.

Considered at a global scale, these examples point to an alarming trend, because most potential hosts never have been systematically investigated for protists; whilst those that were associated with already extinct host species are lost for ever. Unfortunately, this situation will persist in the near future because alpha-taxonomists are themselves threatened by extinction (Ziegler et al. 1997; Cotterill 2002). The losses of natural science collections and taxonomic expertise is caused by their neglect, and failures to appreciate the values of capacity building in these principal foundations of the biodiversity sciences. The continued neglect of taxonomic resources equates to the losses of irreplaceable information in the Alexandrian libraries, but is perhaps more insidious (Cotterill 1997, 1999, 2002).

Change or loss of habitats

Disturbance and loss of habitats threaten not only plants and animals, but also protists which are sensitive indicators of environmental changes (Foissner 1987; Foissner et al. 1995; Lange-Bertalot 1997). Of special concern is the devastation and loss of rain forests, where many groups of protists are heavily under-researched, both in limnetic and terrestrial habitats. This deficiency becomes obvious when one looks at the earth's biodiversity hotspots, where 44% of all species of vascular plants and 35% of species in four vertebrate groups (birds, reptiles, amphibians, mammals) are found (Fig. 9): most hotspots are rain forests in the tropics and subtropics. This figure further highlights our drastic ignorance of protist diversity and biogeography. With the partial exception of the Mediterranean Basin, no or very sparse data are available on ciliates and many other protist groups from these hotspots.

Moreover, rain forests are not the only concern with respect to protist conservation; for consider important ecosystems in moderate climates, notably peatlands and bog ponds which contain highly specific and diverse protist assemblages. These have been investigated for ciliates only in Europe, but remain poorly known, pertinently across Asia. Kreutz and Foissner (2006) found about 800 protist species in a few *Sphagnum* ponds in the surroundings of the town of Constance, Germany. At least of these 100 species were undescribed, mainly amoebae and ciliates, and some of the undescribed species might be local endemics. Thus, Kreutz and Foissner (2006) concluded that this unique area be protected by law. A similar situation is known from European diatoms: most of the 417 oligo- or slightly mesotraphentic taxa are members of the "Red list" because oligotrophic and dystrophic habitats are especially prone to disturbance (Lange-Bertalot 1997).

Other poorly explored habitats are agroecolandscapes and groundwaters which are threatened by pollution and pesticide application (Foissner 1987; Novarino et al. 1994; DeLeo and Baveye 1997). We doubt whether anyone has extended the concept of biological

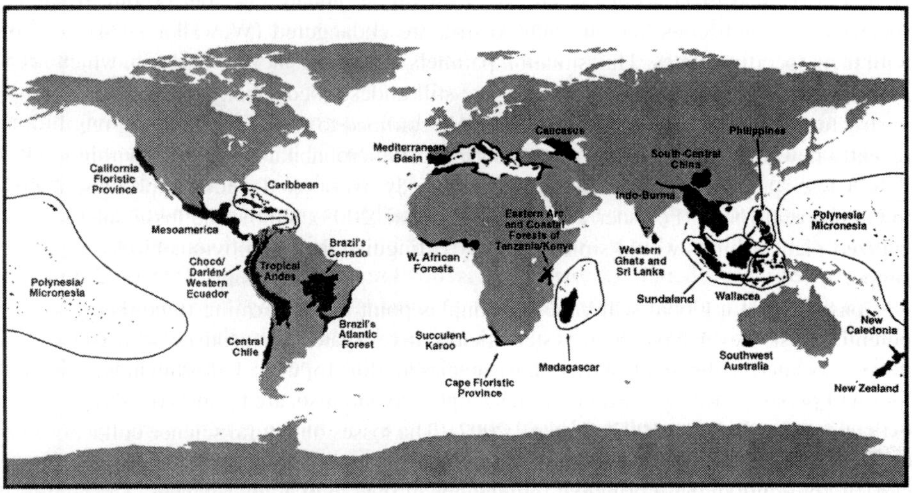

Fig. 9 The 25 biodiversity hotspots (after Myers et al. 2000)

magnification to consider possibilities of the accumulation of inorganic pollutants or residual pesticides in protists!

Loss of type localities

The type locality is the site where a previously unknown organism was first discovered and described for the first time; it is critical to taxonomy as the geographical place of capture or collection of the name-bearing type of a nominal species or subspecies. The extinction or massive habitat change of the type locality of a species by anthropogenic activities is always a great scientific loss for any described taxa it represents. Such losses of biodiversity are magnified in scale dependent manner in the case of protistology, because most protists are difficult to preserve, and precise resampling from the type locality is often required to solve taxonomic problems. For example, Nilsson (1986) was only able to recognize a new species of African *Stentor* after comparing it with a similar species resampled in Austria.

Moreover, protist type localities are not only threatened by such local destruction of habitats, but also by massive environmental changes. A pertinent example is the testate amoeba *Difflugia biwae*, a supposed local endemic of Lake Biwa in Japan. It was discovered in 1918 but disappeared in the 1980s, likely due to the heavy eutrophication of the lake (Ichise et al. 2004). Although, fortunately, it has been found recently in some lakes in China (Yang and Shen 2005), the genetic status of these geographically isolated populations awaits elucidation.

Loss of unique genetic resources

Protists are the ancestors of plants and animals (Cavalier-Smith 2006). It is unknown how well these tiny organisms faired in the global extinction crises that decimated biodiversity, notably the events that characterize the Permian-Triassic and Cretaceous-Tertiary boundaries, when up to 96% of biodiversity disappeared (Wilson 1989; Hallam and Wignall 1997). Persistence of many protist organization types and species through such long periods of geological time is likely the main reason for their great diversity. Thus, protists are not only important for understanding evolution of life, but also represent a unique reservoir of genetic peculiarities. Among others, three examples include: the dual genome of ciliates and some foraminiferans; the absolute strand polarity in kinetoplastid flagellates; and the unique nuclear apparatus of the Dinoflagellates (Puytorac et al. 1987; Hausmann et al. 2003; McGrath and Katz 2004).

The threats to any of the major suites of unique adaptations exhibited among protists (some mentioned above) appear insignificant, assuming it is unlikely that a whole clade will be driven to extinction by human activities. Nevertheless, local endemics may constitute exceptions to this generalization, and especially for monotypic and/or taxa that are very locally confined. This is exemplified by the plight of the curious soil fungus *Geosiphon pyriforme*. Although *G. pyriforme* is of almost macroscopic size (1–2 mm) and known since 1915, a region in the Spessart Mountains (Germany) remains its only known natural habitat (Schuessler and Kluge 2000). Usually, however, the genetic variation of individual species and, especially, of populations within species is threatened. Indeed, in micro-organisms (emphasizing the significance attached to losses of type localities) the destruction of a single small pool may cause the loss of unique genetic variation for ever, because such a pool might represent the entirety of the biodiversity represented by a specific protist genotype or morphotype (Mann and Droop 1996; Lange-Bertalot 1997; Shayler and Siver 2004).

The conservation relevance of protist biodiversity and evolutionary history

Deep and shallow evolutionary history of protists

Knowledge of the evolutionary relationships among protists is still hampered severely by lack of primary inventory data; for we can only guess at how many deeply unique species and clades of protists await formal discovery by taxonomists. Nevertheless, according to the evolutionary and ecology theory on which the biodiversity sciences are structured (Erwin 1991; Heywood and Watson 1995; Ormond et al. 1997; Reaka-Kudla et al. 1997; McKinney and Drake 1998), what we do know about the evolutionary history of protists reveals that they exhibit key credentials, which elevates their pertinence in considerations of evolutionary potential. Considered in light of the incidence of local endemics, protists appear to be comprised of at least some evolutionarily vibrant clades, with these speciose bushes rooted deep in geological time. Humanity would be wise to maintain all such evolutionary potential biodiversity in all its representativeness within protected area systems, and especially within landscapes managed according to the principles of ecosystem management. This especially applies to agroecolandscapes, drainage systems, and reservoirs of groundwater (Brooks and McLennan 2002).

Deep earth history, protists, and the evolution of predation

The deepest origins of protists extend beyond the Precambrian well into the Proterozoic (Cavalier-Smith 2006). Tantalizing questions arise as to when the first protists appeared on Earth's early continents, and began to prey on microbes. For this event constituted an evolutionary threshold in its own right, for it marked the origin of predation with all its associated complexities of coevolution (Bengtson 2002). This profound importance of protists in Earth history, and thus our study of it, gains a most intimate immediacy when we consider the origins of *Homo* in the context of parasitism and predation. Brain (1981) argued that exposure to predation was a significant agent that acted on the diversification of hominids in Neogene Africa. Moreover, our ancestors not only evolved in environments rich in vertebrate predators of primates, because *Homo* has interacted persistently with particular protozoa, namely *Plasmodium* and *Trypanosoma*. These two genera present as dominant parasites of *H. sapiens* today. Massive investments in research, toward their control, can only benefit from an evolutionary perspective (Lambrecht 1985). Yet, the literature too rarely reflects an appreciation of coevolutionary history we share with our parasites, nor the benefits of centering such study on a scaffolding of hominid evolution.

Nevertheless, a caveat must be emphasized here in how we conceptualize such evolutionary patterns. As with any taxa that we like to conceptualize as basal (Krell and Cranston 2004; Jenner 2006; Jenner and Wills 2007), and in absence of superbly preserved fossils, the attributes of ancestral protists will always remain subjects of intrigue. This situation is reinforced when we acknowledge the highly derived richness of adaptations exhibited amongst the morphologies and life histories of those relatively few protists known to science.

As inaugurated by Gause (1934), free-living protists and their prey present a rich resource to study coevolutionary dynamics. As importantly, it is postulated that ancestral protozoans invaded the plethora of habitats now exploited by parasites and mutualists, which include the organs and tissues of primates and all other vertebrates. This immense richness of ecological interactions and coevolutionary associations that are centred on free-living protists constitute the benchmarks against which to judge our idiographic discoveries, and construct nomothetic generalisations. This cornucopia of biodiversity that has evolved in fine-grained patches across landscapes, at microscopic scales through deep geological time, presents 21st biologists with rich opportunities in comparative biology and parasitology. Biologists have barely begun to conceptualize these arenas of opportunity, yet alone explore them.

The origins of predation constitutes yet another sound argument for why we are behoven to keep rich assemblages of protists around for a little while longer in a biosphere that we humans increasingly exploit and change. We reiterate that protist-microbe assemblages constitute one of the deepest and richest legacies of coevolution; presumably they originated on the biosphere's oldest continents. It follows that vestiges of their ancient diversity persist across all continents descended from Laurasia and Gondwana, and indeed the latter's progenitors such as Rodinia. The scientific tools available to study these complexities of evolutionary history continue to improve rapidly. The rapid developments demonstrated in microbial genomics (Venter et al. 2003, 2004; Eisen 2007; Gross 2007) can greatly augment studies of protist taxonomy and ecology challenged to tease out autecologies, morphologies and life histories of the actual organisms.

We predict that surveys of the genomic and functional diversity of protists and their prey will especially benefit from shotgun-sequencing strategies (Venter et al. 2003, 2004; Eisen 2007). Nevertheless, any such genetic inventory methods are challenged to survey the complexity of eukaryote genome. Recent discoveries in protist taxonomy have highlighted organisms with respect to their unusual morphological adaptations, so these stand out as 'flagships' (Foissner 2006). It is fascinating to consider how much we stand to learn from more comprehensive and thorough inventories that collate knowledge of protist natural history in idiographic detail, and integrate these characterizations of their properties into generalizations detailing genomic and organismal trends in protist biology. It is important to emphasize that these discoveries and their formal description (cited above by Foissner) represent the efforts of a sole researcher challenged to inventory the immense diversity of soil ciliates, distributed across the fine-scaled patches that constitute the world's protist habitats. It cannot be overemphasized that these discoveries of the organismal diversity resplendent in protists have barely even begun, considering the richness of unexplored habitats. It is especially important to acknowledge how very little is known about any adaptations, ancient and derived, of protists in all their genomic, biochemical and physiological intricacies.

Protists and the taxonomic impediment: natural history as the fundamental life science

It is most unfortunate that too few scientists distinguish between the idiographic and nomothetic properties of a science (Ghiselin 1997; Jenner and Wills 2007); and that all sciences are built on idiographic knowledge (unique particulars) which, when assembled to reveal sufficient patterns, allow us to derive nomethetic (law like) generalizations. For this dichotomy is mutualistic, and underpins the growth and integrity of knowledge, especially in the life sciences. Unfortunately a widespread failing among biologists is to ignore and denigrate these so called descriptive sciences that provide theorists and experimenters with primary knowledge in the form of idiographic details. Inadequate idiographic coverage in a discipline hamstrings nomothetic synthesis. We reiterate that the acute lack of knowledge (constrained primarily by the Taxonomic Impediment) afflicts all aspects of protist biology. This constraint on generation of idiographic knowledge is an especially acute example of a pervasive hindrance to scientific progress. Stated bluntly, it is impossible to derive nomothetic generalizations when one has too few idiographic facts. Clearly, major deficiencies need to be rectified in how science is taught and supported to rectify a most serious crisis in biology, and especially the biodiversity sciences.

These deficiencies in scientific policy and capacity especially hamstring any policy and/or actions tasked to apply scientific knowledge of protists to their conservation. Our knowledge of the biology of protists will only advance from intensive inventories and field studies of these microorganisms across the world's habitats. Mandatory research actions in a global conservation strategy for protists need to place a premium on the development of these inventories. This can be optimally structured in a nested sampling design that targets representative habitats across continents. This research can borrow existing tactics and techniques from established initiatives and biodiversity research programmes, notably the All-Taxa-Biota Inventories (ATBIs), especially where they target commensals and parasites of multicellular organisms. In this respect, Wheeler (1995, 2004) advocated a combination of ATBIs and ABTIs (All-Biota-Taxonomic Inventories), which we suggest can be adapted to the challenges of protist inventories. Above all, the arguments of Herman (2002) and Schmidly (2005) especially apply to the plight of protist biology; for there is no better example than protistology for why biology has to return to teaching and support of natural history as the *sine qua none* of the biodiversity sciences. Until this situation is improved radically, it will remain very difficult, if not impossible, to study and monitor protist biodiversity.

Protist taxonomy and conservation: the future

Several interlinked threads of argument deserve reiteration. This essay has rallied evidence to scope out a conservation strategy that targets protists throughout the biosphere. It is grounded in an infallible utilitarian-based argument for protist conservation, where benefits of biodiversity centre "on potable water, clean air and fertile soils" (Gee 1992, p. 639). The persistence, resilience and resistance of biodiversity ultimately hinges on whether the integrity of ecosystems is maintained: across forests, savannahs, wetlands and agroecolandscapes, and especially the underlying regolith in which critical edaphic and hydrological processes occur. Only at their peril, can biology and society continue to ignore the diversity of protists in these landscapes, in all their intimate ecological dynamics as predators and mutualists. The integrity of soil ecosystems is founded on their processing of nutrients and matter across microscopic and mesoscopic scales. These and more encompassing ecological processes confer critical ecosystem services (Carson 1962; Daily 1997). These benefits of protist-microbial ecology and evolution are by no means confined only to terrestrial soils, but are equally important in hydrological processes. The very existence of this coevolutionary legacy remains very poorly appreciated, let alone researched, even though it has persisted in oceans and on continents since the Proterozoic. We can only guess at what we might learn about coevolution (with spin offs to understanding disease) from research that focuses on foodwebs involving protist predators and parasites in aquatic, marine and terrestrial habitats.

An immediate step toward remedying our global ignorance of protists, given their significance, is to embrace the research objectives of bold initiatives that emphasize the inventory and description of biodiversity as focal activities in 21st century biology. These were inaugurated in Systematics Agenda 2000 (1994) and subsequent initiatives (Wheeler 1995, 2004; Hoagland 1996; Foissner 1999b; Cresswell and Bridgewater 2000; Cotterill 2002). We single out contributions of the All-Species Foundation toward biotic inventory and taxonomy, which pivot on providing an on line compendium of biodiversity information.

Knowledge of each species is consolidated in a webpage to optimize the efforts of a very inadequate taxonomic workforce, and highlights its pivotal importance to all society (Wilson 2003; Kirkland-Berger 2005). An achievable goal is to collate all such information about the world's described protists: with salient aspects of the natural history of each species graphically depicted on web pages. Such an on line library of biodiversity will serve several integrated purposes. One is to highlight the diversity of protists in all their aesthetic attributes. As committed taxonomists, we are convinced that the success and persistence demanded of the scholarly study of biodiversity is best fostered at a young age; and we agree with Wilson (1984) that a deep-seated biophilia inaugurates the growth of such skills. What more apt subject for aspiring biologists than the study of protists? Such a synopsis of knowledge about protists and their biology (preserved in natural science collections, Cotterill 2002) serves as the foundation for 21st century researchers to investigate ecological and evolutionary questions, and solve conservation problems.

It is sad to consider that the infrastructure to research the natural history of protists is declining to extinction, largely driven by forces of ignorance. We can no longer allow such neglect and ignorance to isolate the appreciation and study of protists and their remarkable adaptations from exciting developments in 21st century science. Conservation of protists should obviously be integrated into environmentally sound systems of ecosystem management. Any such activity hinges on scientific monitoring of the biodiversity it seeks to conserve. It will continue to be impossible to monitor the integrity of protist diversity in any such habitat so long as institutions and nations continue to suffer from the near universal lack of expertise and commitments to survey and identify these organisms.

Acknowledgements Supported by the Austrian Science Foundation (FWF, P-15017 and P-19699-B17) and by the King Saud University, Saudi Arabia. Cotterill gratefully thanks essential support for interdisciplinary research from the Claude Leon Foundation, Cape Town, and a Biodiversity Leadership Award from the Bay Foundation, and the Josephine Bay Paul and C. Michael Paul Foundations, New York City. This is AEON Contribution No 0030. The technical assistance of Mag. Gudrun Fuss is gratefully acknowledged.

References

Andre HM, Noti M-I, LeBrun P (1994) The soil fauna: the last biotic frontier. Biodivers Conserv 3:45–56

Arrow K, Bolin B, Costanza R et al (1995) Economic growth, carrying capacity, and the environment. Science 268:520–521

Baskin Y (1997) The work of nature: how the diversity of life sustains. Island Press, Washington DC

Beck E (eds) (2002) Faszination Lebenswissenschaften. Wiley-VCH, Weinheim

Behan-Pelletier V, Newton G (1999) Linking soil biodiversity and ecosystem function – the taxonomic dilemma. Bioscience 49:149–153

Bengtson S (2002) Origins and early evolution of predation. In: Kowalewski M, Kelley PH (eds) The fossil record of predation. The Paleontological Society Papers 8. The Paleontological Society, pp 289–317

Berger H, Al-Rasheid KAS, Foissner W (2006) Morphology and cell division of *Saudithrix terricola* n. gen., n. sp., a large, stichotrich ciliate from Saudi Arabia. J Eukaryot Microbiol 53:260–268

Brain CK (1981) The hunters or the hunted? An introduction to African cave taphonomy. Univ Chicago Press, Chicago

Brooks DR, McLennan DA (2002) The nature of diversity: an evolutionary voyage of discovery. Univ Chicago Press, Chicago

Carson R (1962) Silent spring. Houghton Mifflin Co., Boston

Cavalier-Smith T (2006) Cell evolution and earth history: stasis and revolution. Phil Trans R Soc B 361:969–1006

Clarke JDA (2003) The limits of regolith: a planetary scale perspective. In: Roach IC (eds) Advances in regolith. CRC, LEME, pp 74–77

Corliss JO (2000) Biodiversity, classification, and numbers of species of protists. In: Raven PH, Williams T (eds) Nature and human society. The quest for a sustainable world. National Academy Press, Washington, pp 130–155
Costanza R, d'Arge R, de Groot R et al (1997) The value of the world's ecosystem services and natural capital. Nature 387:253–260
Cotterill FPD (1997) The Second Alexandrian Tragedy, and the fundamental relationship between biological collections and scientific knowledge. In: Nudds JR, Pettitt CW (eds) The value and valuation of natural science collections. Geological Society, London, pp 227–241
Cotterill FPD (1999) Toward exorcism of the ghost of W.T. Thistleton-Dyer: a comment on "overduplication" and the scientific properties, uses and values of natural science specimens. Taxon 48:35–39
Cotterill FPD (2002) The future of natural science collections into the 21st century. Conferencia de clausura. In: Actas del I Simposio sobre el Patrimonio Natural en las Colecciones Públicas en España (Vitoria, 25–27 Septiembre 2001). Departamento de Cultura, Diputación Foral de Alava, Vitoria, pp 237–282
Cracraft J (1996) Systematics, biodiversity science, and the conservation of the Earth's biota. Verh Dt Zool Ges 89:41–47
Cracraft J (2002) The seven great questions of systematic biology: an essential foundation for conservation and sustainable use of biodiversity. Ann Miss Bot Gard 89:127–144
Cresswell ID, Bridgewater P (2000) The global taxonomic initiative – quo vadis? Biol Int 38:12–16
Daily GC (1997) Nature's services: societal dependence on natural ecosystems. Island Press, Washington DC
DeLeo PC, Baveye P (1997) Factors affecting protozoan predation of bacteria clogging laboratory aquifer microcosms. Geomicrobiol J 14:127–149
Donoghue MJ, Alverson WS (2000) A new age of discovery. Ann Miss Bot Gard 87:110–126
Eisen JA (2007) Environmental genome shotgun sequencing: its potential and challenges for studying the hidden world of microbes. PLoS Biol 82:384–388
Erwin TL (1991) An evolutionary basis for conservation strategies. Science 253:750–752
Faith DP (1994) Genetic diversity and taxonomic priorities for conservation. Biol Conserv 68:69–74
Faith DP, Trueman JWH (2001) Towards an inclusive philosophy for phylogenetic inference. Syst Biol 50:331–350
Faith DP, Walker PA (1996) How do indicator groups provide information about the relative biodiversity of different sets of areas? On hotspots, complementarity and pattern-based approaches. Biodivers Lett 3:18–25
Fenchel T (1987) Ecology of protozoa. Springer, Berlin and Tokyo
Finlay BJ, Corliss JO, Esteban G et al (1996) Biodiversity at the microbial level: the number of free-living ciliates in the biosphere. Q Rev Biol 71:221–237
Foissner W (1987) Soil protozoa: fundamental problems, ecological significance, adaptations in ciliates and testaceans, bioindicators, and guide to the literature. Prog Protistol 2:69–212
Foissner W (1994) Kommentar zur Gefährdungssituation der Einzeller (Protozoa). In: Gepp J (ed) Rote Listen gefährdeter Tiere Österreichs, Band 2. Moser, Graz, pp 317–319
Foissner W (1999a) Description of two new, mycophagous soil ciliates (Ciliophora, Colpodea): *Fungiphrya strobli* n. g., n. sp. and *Grossglockneria ovata* n. sp. J Eukaryot Microbiol 46:34–42
Foissner W (1999b) Protist diversity: estimates of the near-imponderable. Protist 150:363–368
Foissner W (2002) Morphology and ontogenesis of *Bromeliophyra brasiliensis* gen. n., sp. n, a new ciliate (Protozoa: Ciliophora) from Brazilian tank bromeliads (Bromeliacae). Acta Protozool 42:55–70
Foissner W (2006) Biogeography and dispersal of micro-organisms: a review emphasizing protists. Acta Protozool 45:111–136
Foissner W (2007) Dispersal and biogeography of protists: recent advances. Jpn J Protozool 40:1–16
Foissner W, Xu K (2007) Monograph of the Spathidiida (Ciliophora, Haptoria). Volume II: Spathidiidae and Pharyngospathidiidae. Monographiae biol (in preparation)
Foissner W, Berger H, Blatterer H et al (1995) Taxonomische und ökologische Revision der Ciliaten des Saprobiensystems - Band IV: Gymnostomatea, *Loxodes*, Suctoria. Informationsberichte des Bayer. Landesamtes für Wasserwirtschaft 1/95:1–540
Foissner W, Agatha S, Berger H (2002) Soil ciliates (Protozoa, Ciliophora) from Namibia (Southwest Africa), with emphasis on two contrasting environments, the Etosha Region and the Namib Desert. Denisia 5:1–1459
Foissner W, Strueder-Kypke M, van der Stay GWM et al (2003) Endemic ciliates (Protozoa, Ciliophora) from tank bromeliads (Bromeliacae): a combined morphological, molecular and ecological study. Eur J Protistol 39:365–372
Gause GF (1934) The struggle for existence. Williams and Wilkins, Baltimore
Gee H (1992) The objective case for conservation. Nature 357:639
Ghiselin MT (1997) Metaphysics and the origin of species. State Univ New York, New York

Gross L (2007) Untapped bounty: sampling the seas to survey microbial biodiversity. PLoS Biol 85:371–376
Hallam A, Wignall P (1997) Mass extinctions and their aftermath. Oxford Univ Press, Oxford
Hansen E (2000) Orchid fever: a horticultural tale of love, lust and lunacy. Pantheon Books, New York
Hausmann K, Hülsmann N, Radek R (2003) Protistology. Schweizerbart, Berlin Stuttgart
Hawksworth DL (2001) The magnitude of fungal diversity: the 1.5 million species estimate revisited. Mycol Res 105:1422–1432
Herman SG (2002) Wildlife biology and natural history: time for a reunion. J Wildl Manage 66:933–946
Heywood VH, Watson RT (1995) Global biodiversity assessment. Cambridge Univ Press, Cambridge
Hoagland KE (1996) The taxonomic impediment and the convention on biodiversity ASC News 24:61–62, 66–67
Holldobler B, Wilson EO (1994) Journey to the ants: a story of scientific exploration. Harvard Univ Press, Cambridge, MA
Ichise S, Morita T, Wakabayashi T et al (2004) Disappearance and its past distribution of the endemic protista *Difflugia biwae* Kawamura, 1918 in lake Biwa, Central Japan. Rep Shiga Pref Inst Pub Health & Environ Sci 39:57–63 (in Japanese)
Janzen DH (1993) Taxonomy: universal and essential infrastructure for development and management of tropical wildland biodiversity. In: Sandlund OT, Schei PJ (eds) Norway/UNEP expert conference on biodiversity. NINA, Trondheim, pp 100–113
Janzen DH (1998) Gardenification of wildland nature and the human footprint. Science 279:1312–1313
Jenner RA (2006) Unburdening evo-devo: ancestral attractions, model organisms, and basal baloney. Dev Genes Evol 216:385–394
Jenner RA, Wills MA (2007) The choice of model organisms in evo-devo. Nat Rev Genet 8:311–319
Kirkland-Berger J (2005) Mission possible: All-Species Foundation and the call for discovery. Proc Calif Acad Sci Suppl 56:114–118
Kiy T (1997) Biotechnology of Protozoa. In: Abstracts of the 10th international congress of protozoology, Sydney
Krell F-T, Cranston PS (2004) Which side of the tree is more basal? Syst Entomol 29:279–281
Kreutz M, Foissner W (2006) The *Sphagnum* ponds of Simmelried in Germany: a biodiversity hot-spot for microscopic organisms. Protozool Monogr 3:1–267
Lambrecht FL (1985) Trypanosomes and hominid evolution. Bioscience 35:640–646
Lange-Bertalot H (1997) A first ecological evaluation of the diatom flora in Central Europe. Species diversity, selective human interactions and the need of habitat protection. Lauterbornia 31:117–123
Mann DG, Droop SJM (1996) Biodiversity, biogeography and conservation of diatoms. Hydrobiologia 336:19–32
Margulis L, Chase D, Guerrero R (1986) Microbial communities: invisible to the scrutiny of naturalists, most microbial communities have escaped description. Bioscience 36:160–170
Matthes D, Guhl W (1975) Systematik, Anpassungen und Raumparasitismus auf Hydrophiliden lebender operculariformer Epistyliden. Arch Protistenk 117:110–186
McGrath CL, Katz LA (2004) Genome diversity in microbial eukaryotes. Trends Ecol Evol 19:32–38
McKinney ML, Drake JA (eds) (1998) Biodiversity dynamics: turnover of populations, taxa and communities. Columbia Univ Press, New York
Meffe G, Contributors (1997) Principles of conservation biology, 2nd edn. Sinauer Press, Sunderland
Meine C, Knight RL (1999) The essential Aldo Leopold: quotations and commentaries. Univ Wisconsin Press, Wisconsin
Moreira D, López-García P (2002) The molecular ecology of microbial eukaryotes unveils a hidden world. Trends Microbiol 10:31–38
Moreira D, López-García P (2003) Are hydrothermal vents oases for parasitic protists? Trends Parasitol 19:556–558
Myers N, Mittermeier RA, Mittermeier CG et al (2000) Biodiversity hotspots for conservation priorities. Nature 403:853–858
Nilsson JR (1986) The African heterotrich ciliate, *Stentor andreseni* sp. nov., and *S. amethystinus* Leidy. Biol Skr Dan Vid Selsk 27:1–43
Novarino G, Warren A, Kinner NE et al (1994) Protists from a sewage-contaminated aquifer on Cape Cod, Massachusetts. Geomicrobiol J 12:23–36
Ormond RFG, Gage JD, Angel MV (1997) Marine biodiversity: patterns and processes. Cambridge Univ Press, Cambridge
Palmer SR, Soulsby L, Simpson DIH (1998) Zoonoses: biology, clinical practice, and public health control. Oxford Univ Press, Oxford
Pawar S (2003) Taxonomic chauvinism and the methodologically challenged. Bioscience 53:861–864
Puytorac P de, Grain J, Mignot J-P (1987) Précis de protistologie. Boubée et Fondation Singer Polignac, Paris

Reaka-Kudla ML, Wilson DE, Wilson EO (1997) Biodiversity II: understanding and protecting our biological resources. Joseph Henry Press, Washington DC

Richter DD, Markewitz D (1995) How deep is soil? Bioscience 45:600–609

Rondon MR, Goodman RM, Handelsman J (1999) The Earth's bounty: assessing and accessing soil microbial diversity. Trends Biotechnol 17:403–409

Scheckenbach F, Wylezich C, Mylnikov AP et al (2006) Molecular comparisons of freshwater and marine isolates of the same morphospecies of heterotrophic flagellates. Appl Environ Microbiol 72:6638–6643

Schmidly DJ (2005) What it means to be a naturalist and the future of natural history at American universities. J Mammal 86:449–456

Schuessler A, Kluge M (2000) *Geosiphon pyriforme*, an endocytobiosis between fungus and cyanobacteria, and its meaning as a model system for arbuscular mycorrhizal research. In: Hock B (ed) The Mycota, vol 9. Springer, Berlin, pp 151–161

Shayler HA, Siver PA (2004) Description of a new species of the diatom genus *Brachysira* (Bacillariophyta), *Brachysira gravida* sp nov from the Ocala National Forest, Florida, USA. Nova Hedwigia 78:399–409

Smith F (1996) Biological diversity, ecosystem stability and economic development. Ecol Econ 16:191–203

Sorokin YI (1999) Aquatic microbial ecology. Backhuys Pub, Leiden

Staley JT (1997) Biodiversity: are microbial species threatened? Comment. Curr Opin Biotechnol 8:340–345

Stork N (1997) Measuring global biodiversity and its decline. In: Reaka-Kudla ML, Wilson DE, Wilson EO (eds) Biodiversity II: understanding and protecting our biological resources. Joseph Henry Press, Washington DC, pp 41–68

Systematics Agenda 2000 (1994) Systematics Agenda 2000. Charting the biosphere. Technical report. Amer Mus Nat Hist, New York

Tilman D, Naeem S, Knops J et al (1997) Biodiversity and ecosystem properties. Science 278:1866–1867

Vane-Wright RI, Humphries CJ, Williams PH (1991) What to protect?–Systematics and the agony of choice. Biol Conserv 55:235–254

Venter JC, Levy S, Remington K et al (2003) Massive parallelism, randomness and genomic advances. Nat Genet 33:219–227

Venter JC, Remington K, Heidelberg JF et al (2004) Environmental genome shotgun sequencing of the Sargasso Sea. Science 304:66–74

Von der Heyden S, Cavalier-Smith T (2005) Culturing and environmental DNA sequencing uncover hidden kinetoplastid biodiversity and a major marine clade within ancestrally freshwater *Neobodo designis*. Int J Syst Evol Microbiol 55:2605–2621

Wall DH (1999) Soil biodiversity – life in soil. In: Cracraft J, Grifo F (eds) The living planet in crisis: biodiversity science and policy. Columbia Univ Press, New York, pp 124–128

Wells SM, Pyle RM, Collins NM (1983) The IUCN invertebrate red data book. Gresham Press, Old Woking, Surrey, UK

Western D, Pearl M (1989) Conservation for the twenty-first century. Oxford Univ Press, Oxford

Westman WE (1977) How much are nature's services worth? Science 197:960–964

Wewalka G (1984) Rote Liste der gefährdeten Schwimmkäfer Österreichs (Dytiscidae, Coleoptera). In: Gepp J (ed) Rote Listen gefährdeter Tiere Österreichs, 2nd edn. BM Gesundheit und Umweltschutz, Wien, pp 123–126

Wheeler QD (1995) Systematics, the scientific basis for inventories of biodiversity. Biodivers Conserv 4:476–489

Wheeler QD (2004) Taxonomic triage and the poverty of phylogeny. Phil Trans Roy Soc Lond 359:571–583

Wilson EO (1984) Biophilia: the human bond with other species. Harvard Univ Press, Cambridge MA

Wilson EO (1989) Bedrohung des Artenreichtums. Spektrum der Wissenschaft 11:88–95

Wilson EO (2003) Encyclopedia of life. Trends Ecol Evol 18:77–80

Yang J, Shen Y (2005) Morphology, biometry and distribution of *Difflugia biwae* Kawamura 1918 (Protozoa: Rhizopoda). Acta Protozool 44:103–111

Ziegler W, Bode H-J, Mollenhauer D et al (1997) Biodiversitätsforschung. Ihre Bedeutung für Wissenschaft, Anwendung und Ausbildung. Fakten, Argumente und Perspektiven. Kleine Senckenberg-Reihe 26:1–68

TOPICS IN BIODIVERSITY AND CONSERVATION

1. David L. Hawksworth and Alan T. Bull (eds): *Arthropod Diversity and Conservation*. 2006 ISBN 978-1-4020-5203-0

2. David L. Hawksworth and Alan T. Bull (eds): *Forest Diversity and Management*. 2006 ISBN 978-1-4020-5207-3

3. David L. Hawksworth and Alan T. Bull (eds): *Human Exploitation and Biodiversity Conservation*. 2006 ISBN 978-1-4020-5282-0

4. David L. Hawksworth and Alan T. Bull (eds): *Marine, Freshwater, and Wetlands Biodiversity Conservation*. 2006 ISBN 978-1-4020-5733-4

5. David L. Hawksworth and Alan T. Bull (eds): *Vertebrate Conservation and Biodiversity*. 2007 ISBN 978-1-4020-6319-0

6. David L. Hawksworth and Alan T. Bull (eds): *Plant Conservation and Biodiversity*. 2007 ISBN 978-1-4020-6443-2

7. David L. Hawksworth and Alan T. Bull (eds): *Biodiversity and Conservation in Europe*. 2007 ISBN 978-1-4020-6864-5

8. W. Foissner and David L. Hawksworth (eds): *Protist Diversity and Geographical Distribution*. 2009 ISBN 978-90-481-2800-6

Lightning Source UK Ltd.
Milton Keynes UK
14 March 2010

151358UK00007B/165/P

9 789048 128006